Towards a Wireless Connected World: Achievements and New Technologies

Al-Sakib Khan Pathan
Editor

Towards a Wireless Connected World: Achievements and New Technologies

Springer

Editor
Al-Sakib Khan Pathan
Department of Computer Science
and Engineering
United International University
Dhaka, Bangladesh

ISBN 978-3-031-04320-8 ISBN 978-3-031-04321-5 (eBook)
https://doi.org/10.1007/978-3-031-04321-5

This Springer imprint is published by the registered company Springer Nature Switzerland AG
The registered company address is: Gewerbestrasse 11, 6330 Cham, Switzerland

Dedicated to …
"My family."

—Al-Sakib Khan Pathan

Preface—The Impact of Wireless Connectivity

While the deep impact of the Internet was realized even in the past decades, during the COVID-19 pandemic period, its utility has been realized even more meaningfully. During the lockdown phases and due to taking their own precautionary measures, millions of Internet users preferred staying at home and doing whatever possible to do via online mode. Perhaps, for similar future cases (if any, again), the same mode of communication would be very crucial. We have experienced academic institutions, regular corporate offices, businesses, political meetings, and even judicial proceedings running online. For many of these activities, people often use a wired connection to get Internet services; however, it has been experienced in many places that when the wired connection fails, a more reliable form of connection could be achieved via wireless technology. Also, it gives more flexibility to stay connected with the Internet when the users are on the move. For instance, the mobile phone's data services can be turned on and a smartphone can be made a hot spot to connect a computer to the Internet—all via wireless technology. During online exams, for better reliability, some students can use this method to get uninterrupted Internet connectivity.

In some parts of the globe, regular electricity supply is a problem and often the Internet connection gets cut when power failure occurs. Wireless Internet via smartphones can solve the problem in this case. Many students could attend online classes, businessmen can attend meetings or listen or watch important briefings while traveling via bus or vehicles and still remain connected to the Internet via wireless communication. The same experience is heard from the journalists who may use mobile data for their job-related tasks for interacting with people when covering some events outside. This is what we call the Wireless Internet, which could be facilitated via Wireless Local Area Networks (WLAN), Cellular network, and Satellite communications technologies. To enhance the coverage and service features of the Wireless Internet (and, general wireless connectivity), many works have been done from various angles on various wireless technologies. The challenges that remain in this field are still manifold.

The objective of this book was to collate the significant contributions in the fields related to Wireless Internet technologies and, in general, wireless connectivity. After the call for chapters, a good number of chapter proposals were received. After the due quality check and review process, finally, we could select a total of 14 chapters for the book, which are put under separate parts (with separate headings) based on their addressed topics. Authors representing at least 13 different countries have written about their impactful solutions, trends, and technologies that would make the Wireless Internet (in general, wireless connectivity for distance communication) more reliable and secure for the coming years. It is our expectation that this book could serve as a premier reference source for various aspects of wireless connectivity and associated cutting-edge technologies.

Dhaka, Bangladesh

Prof. Al-Sakib Khan Pathan, Ph.D., SMIEEE
spathan@ieee.org

Acknowledgments

First of all, I would like to express my sincere gratitude to the Almighty Allah for giving me the stamina and time to complete this work. During the extended COVID-19 pandemic period (more than 2 years as of today officially), working on this has not been that easy. Since dedicated office hours are not available nowadays, the lifestyle and modes of work need to be changed significantly, and at home, multiple types of tasks and responsibilities do complicate things. Also, a real concern was about obtaining a sufficient number of quality contributions. At the end, it is good to see that several submissions came in and some of the top quality chapters could be accepted. I express my sincere gratitude to all the authors who contributed to this book project. Special thanks to the publishing staff and the publishing editor for giving me this opportunity.

Prof. Al-Sakib Khan Pathan, Ph.D., SMIEEE
spathan@ieee.org

Contents

Intelligent Wireless Communication Issues

Intelligent Decision Making Issues

About the Editor

Al-Sakib Khan Pathan is a Professor at Computer Science and Engineering department, United International University (UIU), Bangladesh. He is also serving as a Ph.D. Co-supervisor (external) at Computer Sciences Department, University Ferhat Abbas Setif 1, Algeria. He received a Ph.D. degree in Computer Engineering in 2009 from Kyung Hee University, South Korea, and a B.Sc. degree in Computer Science and Information Technology from Islamic University of Technology (IUT), Bangladesh, in 2003. In his academic career so far, he worked as a faculty member with various capacities in various institutions like the CSE Department of Independent University, Bangladesh (IUB), during 2020–2021, Southeast University, Bangladesh, during 2015–2020, Computer Science Department, International Islamic University Malaysia (IIUM), Malaysia, during 2010–2015; at BRACU, Bangladesh, during 2009–2010, and at NSU, Bangladesh, during 2004–2005. He served as a Guest Professor at the Department of Technical and Vocational Education, Islamic University of Technology, Bangladesh, in 2018. He also worked as a Researcher at Networking Lab, Kyung Hee University, South Korea, from September 2005 to August 2009, where he completed his M.S. leading to his Ph.D. His research interests include wireless sensor networks, network security, cloud computing, and e-services technologies. Currently, he is also working on some multidisciplinary issues. He is a recipient of several awards/best paper awards and has several notable publications in these areas. So far, he has delivered 32 Keynotes and

Invited speeches at various international conferences and events.

He was named on the List of Top 2% Scientists in the World, 2019, and Top 2% Scientists in the World, 2020, by Stanford University, USA, in 2020 and 2021. He has served as a General Chair, Organizing Committee Member, and Technical Program Committee (TPC) member in numerous top-ranked international conferences/workshops like INFOCOM, GLOBECOM, ICC, LCN, GreenCom, AINA, WCNC, HPCS, ICA3PP, IWCMC, VTC, HPCC, SGIoT, etc. He was awarded the IEEE Outstanding Leadership Award for his role in IEEE GreenCom'13 conference and IEEE Outstanding Service Awards twice in recognition and appreciation of the service and outstanding contributions to the IEEE IRI'20 and IRI'21. He is currently serving as the Editor-in-Chief of International Journal of Computers and Applications and Journal of Cyber Security Technology, Taylor & Francis, UK; Editor of Ad Hoc and Sensor Wireless Networks, Old City Publishing, International Journal of Sensor Networks, Inderscience Publishers, and Malaysian Journal of Computer Science, Associate Editor of Connection Science, Taylor & Francis, UK, International Journal of Computational Science and Engineering, Inderscience, Area Editor of International Journal of Communication Networks and Information Security, Guest Editor of many special issues of top-ranked journals, and Editor/Author of 30 books. One of his books has been included twice in Intel Corporation's Recommended Reading List for Developers, 2nd half of 2013 and 1st half of 2014; 3 books were included in IEEE Communications Society's (IEEE ComSoc) Best Readings in Communications and Information Systems Security, 2013, several other books were indexed with all the titles (chapters) in Elsevier's acclaimed abstract and citation database, Scopus and in Web of Science (WoS), Book Citation Index, Clarivate Analytics, at least one has been approved as a textbook at NJCU, USA, in 2020, one is among the Top Used resources on SpringerLink in 2020 for UN's Sustainable Development Goal 7 (SDG7)—Affordable and Clean Energy and one book has been translated to simplified Chinese language from English version. Also, 2 of his journal papers and 1 conference paper were included under different categories in IEEE Communications Society's

(IEEE ComSoc) Best Readings Topics on Communications and Information Systems Security, 2013. He also serves as a referee for many prestigious journals. He received some awards for his reviewing activities like: one of the most active reviewers of IAJIT several times; Elsevier Outstanding Reviewer for Computer Networks, Ad Hoc Networks, FGCS, and JNCA in multiple years. He is a Senior Member of the Institute of Electrical and Electronics Engineers (IEEE), USA.

Security, Trust, and Reliability Issues for Wireless Connectivity

Trust Issues with Wireless Internet Devices

Robert Gordon

Abstract More people are keeping their mobile devices with them at all times than ever before. Mobile devices are becoming the primary means for accessing the Internet. Given this shift in use, one would expect that more people would invest in protecting their data on their wireless devices. However, research has shown that few people have invested in protecting their mobile devices. This situation is because more people trust their mobile devices more than their computers. Since mobile devices are being carried everywhere and people even become more anxious when separated from their mobile devices, it becomes a trusted item. This sense of trust in our mobile devices becomes problematic as more bad actors exploit mobile devices because of less security. It becomes even more critical than ever for people to remain vigilant when utilizing a mobile phone to download Apps and when accessing the Internet.

Keywords Communication · Cyber · Cybercrime · Cybersecurity · Digital · Mobile · Security · Smartphone · Trust · Virtual

1 Introduction

Mobile devices have successfully drawn users away from desktop computers coupled with competitive Internet connectivity becoming standard worldwide [1]. As more people use their mobile devices as their primary means to access the Internet, more bad actors are taking advantage of this shift. Consider that cybercrime is an action where a computer network, such as a mobile device used to access the Internet, is used to execute some criminal act [2]. Mobile cybercrime has been growing steadily, but many cases go unreported. Unreported mobile cybercrime makes the risks seem lower than using other Internet devices. Mobile devices allow people to stay in contact with distant friends, read global news, and even be used for educational purposes.

R. Gordon (✉)
American Public University System, 919 Bethany Court, Annapolis, MD 21403, USA
e-mail: rgordon@apus.edu

A. K. Pathan (ed.), *Towards a Wireless Connected World: Achievements and New Technologies*, https://doi.org/10.1007/978-3-031-04321-5_1

Furthermore, most people always carry their mobile devices with them. Given all the positive feedback people get from mobile devices, it is no surprise that people develop a level of trust with their mobile devices. This feedback loop can create confidence in all things mobile when people should be vigilant. People's psychological bond with their mobile devices requires a greater education and understanding of trust and security in the wireless world.

One might wonder why we connect with our mobile devices so well. Digital natives have always known a world with computers and the Internet, while Digital citizens, who became more active online, have become more virtual than ever before. Laptops have been around for nearly forty years, and tablets and smartphones have been around since the mid-1990s. Given that laptops have been around longer, why do we feel more connected to our smartphones than laptops? The reasons are complex, but the overarching reality is that people have developed greater trust with their smartphones than with any other device.

1.1 What is Trust?

Trust is the feeling of connectedness to something or someone that allows an individual to believe the individual or organization. Trust is often given to those with an essential or respected role; however, trust is earned over time in many cases. Trust is also created with engagement. Engaged people will trust an organization and feel connected to the people, things, and locations. People want to believe that something more significant to belong and connect to. Consider that trust is why people work for a leader and continue to work for a leader. Trust creates the bond between the leader and follower so that it allows the follower to move towards an uncertain reality willingly. Followers trust leaders who can describe a future that has not arrived and then help bring those followers to that same reality. This bond also develops between people and things, particularly things that are kept regularly on our person.

1.2 Trust and the Digital Citizen

The COVID-19 pandemic changed a great many things. COVID-19 forced the creation of more Digital citizens than any other human event in history. The pandemic caused people that were never expected to work virtually to have to work virtually. COVID-19 caused people to be thrust into a virtual workplace, and few understood how to navigate these uncharted waters. Digital natives could adapt quickly, but people who were used to working face-to-face had to learn to become more comfortable in the virtual environment.

Organizations with good communication, reliable means of transmitting information, and flexibility had a good start as those elements are essential for trust in a virtual organization. Those who lacked these elements learned quickly that they needed to define better these areas to be a successful virtual organization. Although creating teams in person is similar, creating effective virtual teams is not as easy as many thought. Suddenly organizations were charged with developing, training, and assisting Digital citizens navigate a new virtual world of work.

One of the issues with Digital citizenship is loneliness. Organizations learned quickly that making roles clear will help alleviate some of the tension with the loneliness when working virtually. If one does not have a clear role, one will make mistakes, duplicate effort, and potentially create more problems that might have otherwise happened. Information, communication, flexibility are the foundation of trust for the Digital citizen.

Digital citizens and digital natives are more prevalent than ever. Working virtually has become mainstream in the US. Some organizations have publicly announced that they will never require people to return to the office and work face-to-face again. Duarte & Snyder were pioneers in working virtually and described trust in the virtual environment and found that competence, integrity, and concern for others were crucial [3]. Duarte & Snyder did not envision that these factors could be connected to an inanimate object like a mobile device [3].

These elements of trust are necessary to make an organization successful because they create a connection. This connection is not limited to connecting to people or an organization; it also includes connection to things. No virtual organization can be successful, let alone effective, without the individuals involved having a high level of trust [3]. This connection has been growing between people and their mobile devices. More people use their mobile devices for more extended periods than ever before. It has become such an issue that mobile carriers allow parental controls to limit screen time for children's mobile devices.

2 Creating Trust in Things

Human communication is complex and vital to the Digital citizen. Traditional organizations worked on the assumption that people would work together in the same building or proximate area. Traditional organizations believed that physical centralization was the only way to run an organization. However, organizations are not the tangible places they once were, as now more organizations are connected virtually by technology. As far back as the 1990s, improvements in technology were pushing people to implement virtual organizations. COVID-19 has made virtual organizations the norm, and the brick-and-mortar organization is becoming a relic of the past.

As the Digital citizen becomes more familiar with the virtual world, some people might reject the need for greater face-to-face communication. In fact, many virtual programs such as Zoom®, and Microsoft Teams® have begun to replace face-to-face communication. As people are less connected to others, more people are connecting to mobile devices they carry with them at all times.

As more people use more digital means to connect, there comes a level of trust with these devices to access the virtual environment. More than ever before, the percentage of website visitors is greater with mobile applications than with desktops [4]. As more people become comfortable and trust the virtual tools, this can create an aura of trust in all things virtual. With Digital citizens becoming more familiar with communication in the virtual world, it creates a bond between themselves and their tools. Digital natives have already developed this connection; however, digital natives have also been exposed to the risks more than Digital citizens.

Furthermore, Digital citizens must learn to communicate better in the virtual environment. This communication skill without being face-to-face becomes more important as some people are compelled to operate virtually. However, this is not without risks, as Digital citizens are less aware of the dangers of online and using mobile devices. Individuals need to consider the hazards of operating virtually and take the necessary steps as bad actors seek to steal and exploit individuals online. Cybercrime is rising as more people use mobile devices at an increasing rate [5]. Cybercriminals are starting to focus on mobile devices as they are often less protected than other networks and can access the same information [5].

3 Cybercrime

Digital citizens have become high-value targets for cybercrime. What happens is that Digital citizens are new to operating virtually and are not as versed as digital natives in the risks that come with being more virtual. In addition, many Digital citizens are lonely, and loneliness is one of the elements that cybercriminals will leverage. Consider that cybercrime continues to grow, and although there are some records about how much crime has cost individuals, there are likely even more crimes that go unreported. After all, if one pays the ransom to a cybercriminal, the individual is likely not to report the crime as they would be admitting that they were naïve or, worse, foolish. Cybercrime has been growing steadily over the years, and despite this fact, many cases go unreported [6]. Unreported cybercrime gives the false sense that it is not as risky and that individuals might be tempted to be less prepared due to a belief that the risks are low. In addition, people have the feeling that it could not happen to them, which is a risky position to talk to. Given this rise, there need to be more people to report these crimes, and agencies need to take steps against these bad actors. Furthermore, few consumers currently pay for security for their mobile devices [6].

The significance is that all devices are a target. Mobile devices have the same vulnerabilities as laptop devices, and mobile devices now account for 55% of the network device market share [6]. There are currently 1.5 million cyber-attacks on US companies per year [7]. 1.5 million cyber-attacks per year translates to roughly three cyber-attacks every minute [7]. It is estimated that the cost of cyber-attacks has cost organizations three trillion dollars in losses in 2015 [8]. It is estimated that by 2021 losses from cyber-attacks will cost companies six trillion dollars [8]. Ransomware has done 11.5 billion dollars in damage, and it is expected to rise to 20 billion dollars by 2021 [8]. Phishing attacks via email have cost companies over two billion dollars per year [8]. Mobile devices are particularly vulnerable, and attackers can gain valuable intellectual property, sensitive data, or credit card information [6].

Furthermore, mobile devices are used for payment and can be attacked. As the US becomes more cashless, mobile devices become a more prominent target as more people pay via phone rather than a physically chipped credit card. The US is the primary target for cyber-attacks, and mobile devices are becoming a larger target than ever before [8]. This information alone should shock every person to take action and invest in security for mobile devices. However, people have been slow to take action. Part of the problem is that people use their mobile devices more and trust and connect with the device.

Given all these opportunities for bad actors, cyber criminals will increase the targeting of mobile devices [5]. Given that few cybercrimes on mobile devices are reported, the statistics will remain low and give people a false sense of security. The World Economic Forum stated that 85% of cybercrime remains unreported [9]. So, reporting these crimes needs to become a priority for individuals to help give greater visibility to this growing issue.

4 Why Do We Trust Our Mobile Devices?

Tablets and smartphones came about around the same time in the mid-1990s, yet more people connect with their mobile devices than their tablets. It is a curious phenomenon but consider a child's connection with a favorite stuffed animal, blanket or pillow, or some other favorite toy. A child would always carry that favorite thing with them. That item represented a level of security, and hence why many parents use the term security blanket that the individual would have with an inanimate object. Furthermore, many people travel with a favorite pillow to ensure a good night's sleep when traveling.

Even adults form strong connections with items, such as wedding rings, watches, favorite jewelry, and other psychological security items. Furthermore, even the older population is using phones more than ever before. Older Americans are trusting mobile devices with private information such as banking information [10].

All demographic areas are expanding their use of mobile devices. A study of mobile device use showed that two-thirds of Americans over 50 own a mobile device and have used it to browse the web [10]. Even younger Americans are expanding their use of mobile devices. 82% of 5 to 7-year-olds go online for average usage of 9.5 h per week, while 99% of 11 to 15-year-old children are online for an average of 20.5 h per week [11]. In 2022, the US will have the most smartphone users than ever before [5].

Although tablets are small, smartphones fit in our pockets and can be carried around at all times. There is documented evidence that some people have anxiety when separated from their phones for too long. A UK study in 2008 found that 53% of people that used mobile phones suffered from anxiety from being separated from their phone too long, and this anxiety level was similar to wedding day jitters

[12]. Consider there are no cases of laptop or tablet anxiety. The reason is that a smartphone connects us to others virtually, and in many cases, we communicate with our smartphone as if it were another person.

Personal virtual assistants have become commonplace, and Alexa and Siri have entered into the collective consciousness. In addition to this physical proximity that can create a bond, people also communicate with their devices. Now, instead of navigating complex Apps or knowing what button to push, Digital citizens and digital natives can just ask. A person can seek out information about any topic imaginable by voice command. People have developed a relationship where they talk to their phone, and it knows things. Smartphones will often know more than we do, such as with navigation. Using Apps for car navigation will know the obvious way to the destination and upcoming traffic conditions to offer alternatives that the driver might not have known. In this way, artificial intelligence and the virtual assistant will understand better that many people trust navigation information, even when it appears to be going in the wrong direction. This two-way communication of a person asking for directions and then getting verbal information to navigate has become quite popular and widely accepted. Many states have already made it illegal to drive while using a mobile device, so verbal commands become the most common means of accessing navigation information. In addition, some interfaces, such as Apple Play®, will only allow verbal commands to prevent the driver from typing on the keyboard while driving. This communication adds to the bond that people already have from carrying around their smartphones.

4.1 The Bond of Good Communication

Maintaining good communication is an important aspect of trust. Communication is a fundamental element of trust and a strong relationship. People communicate with others all the time, and those that people communicate the most are likely people that we trust the most. Individuals connect with those that share information with us, those that make us feel good, and those that give us the facts.

Communication and trust can be transient with people, but our mobile assistants and smartphones are always there to share information. It might mean asking about a local restaurant, tickets to a show, or information about a historical fact; a smartphone is on call to share information. Good communication and information sharing can lead to trust, but those developed close bonds will decay away if that communication fades. This decay often happens when there is no time for being social. Everyone knows that without maintenance, any machine will ultimately fail. When people feel taken for granted for too long, they are likely to move on. Gradually, trust starts to break down, and people begin to look elsewhere for communication. People might get busy, but our reliable smartphones will always be available to share information and entertain.

People are social beings and require a solid feedback loop of communication to remain engaged. Often, at work, people are only focusing on the results (or lack of results) without looking further to find out why the people were not successful, people start to get very defensive. Trust comes over time and is done through positive communication. If that positive communication drops, the trust and relationship can fade. Interestingly, our communication with our mobile devices is always positive. After all, Siri and Alexa are always available and helpful, regardless of the time or place. Even if we have not spoken to our digital assistants or thanked them, they remain available and attentive. This positive two-way communication reinforces our trust in these devices.

4.2 *Mobile Device Interaction Increases Our Trust in Mobile Devices*

Mobile devices are also connected to share facts with anyone at a moment's notice. It no longer requires a web search, but our smartphone can give us information about every conceivable topic with voice command. In addition, the data is highly accurate and factual in most cases. There can be times when a smartphone will purposefully offer different information. Still, generally, that is due to either ambiguity in the request, multiple answers, or possibly to keep a person from harm. A smartphone virtual assistant might offer information based on what they thought was needed and seek greater clarity. Multiple answers are also possible. If one asks for nearby options for coffee, that might lead to a variety of options. Finally, our smartphones will also help protect us at times. Requesting a personal assistant for options to commit suicide will not result in options but will instead offer the person sources to help those with suicidal thoughts.

All of this information from mobile devices is made possible by the artificial intelligence of these virtual personal assistants. This artificial intelligence has been designed to meet many needs and adapt and recognize the owner's voice. With this communication comes the potential that a person may bond with that item. Digital assistants seem like human companions since people communicate with them, and they maintain confidentiality and integrity. Also, as people upgrade their mobile device, their data and digital assistant are retained. This continuity gives people a long time to connect with their mobile device and offers an expanded relationship as a person continues to have their mobile device.

People may lose sight of maintaining good communication with others, but virtual assistants are always available. This situation might make it seem that artificial intelligent virtual assistants might make better associates. However, they are only programmed to do as they are told and lack true human intelligence. Still, this connection between people and their smartphones causes people to be open to more significant risks by cybercriminals. This risk comes when one puts too much trust in a mobile device.

People inherently want to trust others. People want to trust the police, firefighters, and others in authority to feel safe. People also want to trust institutions. An example would be a personal bank. Although a person might have a home branch of their bank, they can walk into any bank branch and access their funds. It might be via an in-person visit or using one of the bank's Automated Teller Machines (ATM). The bank is trusted with our money, and we trust the bank's institution to give us access to our money when needed. This trust transference happens with people as well. A person will trust their friends, and they will likely trust a friend of a friend because of that bond with the friend. People feel that if a friend who is trusted has the trust of another, then that friend of a friend is also trustworthy. Surprisingly, this transference of faith applies to our smartphones as well.

Trusting our smartphone is already risky but trusting all smartphones carries even greater risk. Organizations already understand this growing problem as mobile device attacks can disrupt an organization and is an increasing threat worldwide. Securing mobile devices at the organizational level is challenging and is already a massive issue for the Department of Homeland Security Science and Technology Directorate (DHS S&T). The DHS S&T has a considerable research and development project for just this area. If this is a priority at the national level, imagine the risks to other organizations and individuals [13].

5 Cyber Threats and Cyber-Attacks on Mobile Devices

Cyber threats are out there, and people need to take these risks seriously. Despite the apparent fact that people trust their mobile devices, greater risk mitigation needs to be done. This situation is reminiscent of when personal computers first came out; devices were used without any concern for viruses or other cyber-attacks. The worst offenders were Apple users, who initially felt that viruses could only be developed to infect PCs. Of course, that was wrong, but many Apple users were surprised when the first Apple virus was successful. As viruses became more prevalent and more organizations fell victim to these early attacks, people started to take more significant notice and more vigorous action to protect their assets.

Over time, given the bad publicity of viruses and other virtual attacks, manufacturers and operating system companies realized the need to bundle protection and security along with the purchase of a computer. In addition, the entire Y2K concern about programs shutting down at the start of the new millennium boosted the computer security industry. The shifts in understanding the need changed how people looked at computer security.

Cybercrime continues to threaten users in a variety of manners. Cybercrime has become so pervasive and destructive that nations work together to combat this threat [14]. Mobile devices have become such high-value targets with less protection than networks and computers. With the expanded number of items connected in cyberspace, mobile devices and other technologies are increased the attack surface for cybercrime [2]. Cybercriminals recycle old hacking tools to exploit mobile devices. What might not work on a computer, laptop or network might work on a smartphone with no protection. The risks are real, and many are known, so there is no reason that individuals should not take immediate action to protect their data and their interests.

In the meantime, individuals will need to take security steps to protect their mobile devices. In addition, there needs to be more reporting of cybercrimes against mobile devices. It seems a little paranoid given the amount of reported crime, but if only one in seven crimes is being reported, smartphones are very insecure and a target for bad actors in this space. The greater the visibility of the threat to an industry, the more likely the industry will take widespread security steps. Even insurance companies are starting to require more disclosure regarding the potential of a cyber incursion before determining the cost of insurance to recover from a cyberattack. Like what happened in the personal computer market, security will become a standard with smartphones. Right now, the public sees this type of security as an expense that they do not need. A person will not purchase snow tires if they never drive in the snow. Sadly, security for mobile devices will have to become standard to make people understand the risks involved because the threat to the public is currently misunderstood.

5.1 How to Protect Yourself from Cyber Threats to a Mobile Device?

The first thing that everyone with a mobile device must consider is viewing the request very carefully whenever you are asked to share personal information. Consider if everything appears correct and legitimate. Is it being sent from a recognized domain? To start, never share private and confidential information with anyone, especially with people you do not know. Even if the individuals appear helpful, do not share that information. So, whenever a person unknown to an individual asks for important information, it is best not to share it with them. In the end, secure information is only safe if you keep it hidden from everyone.

Social engineering is the primary way that cybercriminals attempt to exploit their victims. Social engineering is when an individual will try to appear trustworthy to solicit important security information from a person to steal something of value. Social engineering is most successful when they are designed to target specific individuals. For example, a bad actor might exploit our interest in a new job. While posing as a recruiter, they might ask for personal information to be considered for a new position. The request may seem reasonable, but other people should not be asking for private information. One needs to consider the person asking for the information before blindly clicking away. It is always best not to share the information.

Also, consider who is contacting whom. When contacting a bank and performing a bank transfer, one should expect the bank to ask security questions to verify your identity. If someone reaches you and says they are with your bank and want your security questions, this should be a red flag that this is very likely to be a scam. Institutions understand security and would not be calling people these matters. If you are unsure or think it could be genuine, hang up and contact your bank directly. Once being positive about speaking with an authorized representative, then one should be willing to share secure information.

Another area of security that needs to be considered is downloading Apps. First, one should make sure they are from a trusted source. Make sure that you know where the request is coming from. Consider why that person might need the information.

Second, even if they are from a trusted source, if the App appears too good to be true, then one should be very suspicious. Keep in mind that malicious Apps can find their way on safe sites as they might be sleeper-type Apps that deploy their payload much later. In addition, some secure Apps could be hijacked by bad actors to deploy a malicious payload.

Third, do you need this App? Cluttering a phone with unused Apps just increases the chance of an incident. Many of us have many more Apps than we use, and downloading a hot new App has some risks. Consider waiting to download an App to see if early subscribers have any issues.

Malicious Apps are becoming more common and are offered on the same platforms as safe Apps. Over 813,000 Apps have been removed from the Apple Store© and Google Play© in 2021 [15]. Many were reported to be targeting children and lacked a privacy policy; however, Google Play reported removing Apps that could be used to skim users' information to exploit elsewhere [15]. The report does not include details on how many were malicious. However, given the high number of Apps that were removed, it is clear that at least a small percentage of these were found to be dangerous. There is no doubt that bad actors are using smartphone Apps to exploit individuals.

5.2 *How to Remain Vigilant Against Cyber Attacks*

The most important part of remaining safe in the digital world is to stay educated about the risks. Understanding that individuals have a relationship with their smartphone is unlike their connection with no other device means that people feel safer when using their smartphone. The average person checks their mobile device 47 times during the day [16]. Furthermore, people take their mobile phones everywhere, and as such, our phones are filthy. A study by the University of Arizona has found that the average mobile phone is ten times dirtier than the average toilet set [16]. People have a peculiar relationship with their mobile devices as people take them everywhere. Few people would take a laptop or tablet to the toilet, but most people take their smartphone with them to the toilet. Education and understanding are important because it is unlikely that our habits will change soon.

To remain vigilant, since smartphones are here to stay with us, one needs to realize the risks. First, cybercriminals work all day and all night to exploit and steal from individuals. Given how much information can be on the average cell phone, it seems that everyone is a target of value. Second, avoid lowering your guard. There are times when we naturally feel safer or want to believe, such as when a person is on vacation. Cybercriminals will exploit that vulnerability and strike when we least expect it. Third, a common tactic of cybercriminals is haste. They want a person to hurry up and hand over the information they need before logically considering the consequences. If a person is rushed, that is the time to step back and think about the situation objectively and logically. Cybercriminals prey upon our emotions. Fourth, cybercriminals will also try to use high-pressure tactics. They will try to make a person act now without thinking to move on to their next victim. After all, if a person were to stop and think, why is a Nigerian Price contacting me rather than someone else? The person would realize that it is likely a scam.

Remaining vigilant at all times is vital because training and education will often keep us safer than acting impulsively. Consider the risks and the rewards. Consider the situation and make sure it is one that one controls rather than another person controls. Understand how giving this information away might have a negative impact. Vigilant means taking steps to prevent a disaster. Vigilant means not allowing others to take advantage of our nature to want to help and do good.

6 Trust in a Dangerous Digital World

After learning the trust put in our mobile devices, the trust that can sometimes be given to strangers, it makes a person wonder where they can place their trust. What many forget is that a person needs to trust themselves first. Put faith in being educated about the risks when using a smartphone. An individual must be confident in making the right decisions in a world where people are trying to take advantage of those that are naïve to the dangers.

Trust yourself first in the way that you want to trust other people. People need to consider their actions and make sure what they do does not have negative consequences. Just as a person knows the danger of using a bare hand to pull out a pan from the oven, one needs to understand that the digital world is dangerous. Just because the digital world might have our favorite games, it does not mean that it is safe. Understand where to place trust and make sure it is done safely.

Second, there are times to trust but verify. As stated by President Ronald Regan, trust but verify is the mantra of the digital world. The person calling me from my trusted smartphone might not be who they say they are. A cybercriminal could spoof a phone number or email and pretend to be someone else [17]. Just because they claim to be a long-lost aunt looking for money to pay a lawyer to inherit millions that they are willing to share does not make it true. Protect yourself by verifying all claims made by strangers who solicit private information.

Third, a person needs to assume the worst in the digital world and assume that others are trying to steal things. Just as in different games online, there are allies and villains. It is hard to tell the difference at times. One must assume they are all villains until proven otherwise. Remain on patrol and make your passwords strong and make secret questions difficult for others to find out.

Remaining on patrol means that a person must make strong passwords at all times, make strong secret questions, utilize two-factor authentication, and use a virtual private network (VPN) whenever possible. If a person uses a password manager, then use those strong passwords and keep the password to the password manager the safest of all. Although some people will say that passwords are ineffective, they are often the only defense available. Also, make it a habit to change your passwords from time to time. Please put it on the calendar on a mobile device so the remainder will be there to have you take action.

Remaining on patrol also means keeping a secret question secret. Do not put any private questions anywhere that others can see. It may seem innocuous to share favorite books on social media, but it is no longer secret if a person uses that book as a secret question. In addition, with the proliferation of private questions, it is best to use multiple different ones rather than reusing the same ones repeatedly. In addition, consider making your secret question a secret from yourself. Secret questions are often typical personal details like where one went to school or where you met your spouse. A quick check on social media should give cybercriminals all the information they need to hack into personal accounts. Consider using your second favorite book that you do not share with others, or consider making the secret answer harder to guess, such as instead of first pet, maybe use favorite pet and use first pet for the favorite pet. Make it as difficult as one can for strangers to sneak into your accounts.

Use two-factor authentication wherever possible. It may seem like an inconvenience but using two-factor authentication makes it exceedingly difficult for cybercriminals to gain access [17]. It would mean that criminals would need not only your password but they will need your mobile device or other means of authentication. If the two-factor authentication fails, then the cybercrime has been stopped.

Many people feel that a VPN is the best way to keep people from accessing a network, but there are still those that will try to access a VPN. If a VPN is available, then one should use that. Personal VPNs are also available as part of a package of defenses, including virus protection. Virus protection is important but keeping your network traffic secure is also important. One should use as many defenses as possible to keep their information safe from bad actors.

7 Conclusions

In conclusion, trust in the digital world is difficult for Digital citizens and digital natives. There are bad actors out there trying to take or extort people. Digital crime is rising because the majority remains unreported, and few criminals are brought to justice. Unfortunately, this sounds like a big win for the bad actors, so one must take steps to protect yourself. It may seem that one has almost to remain paranoid to stay safe. It is not as bad as that, but one must remain vigilant and use the tools available for protection. Invest in security for all mobile devices. Consider the low cost of safety and the peace of mind it can give. It does not mean that one should start posting their social security number on social media, but it can help deter some threats. A home security system is designed to protect a home. However, all security companies know that a secured home must have a sign to alert people to the security system. The protection is for the house, but the sign makes criminals find an easier target. Phone security protection is no different. It does help protect the user, but it also offers bad actors a sign to go elsewhere.

References

1. S. L. Shah et al., TAMEC: trusted augmented mobile execution on cloud. Sci. Program. 1–8 (2021)
2. S. Rani et al., Threats and corrective measures for IoT security with observance of cybercrime: a survey. Wirel. Commun. Mob. Comput. 1–30 (2021)
3. D. Duarte, N. Snyder, *Mastering virtual teams* (Joseey-Bass, San Francisco, 2001)
4. D. P. Sakas, N. T. Giannakopoulos, Big data contribution in desktop and mobile devices comparison, regarding airlines' digital brand name effect. Big Data Cogn. Comput. 1–22 (2021)
5. R. Ruiz, Protecting mobile devices from cyber threats (ProQuest Dissertations Publishing, Utica, NY, 2021)
6. D. K. McGill, Expanding trust in mobile devices (2020)
7. Deloitte, Cyber crisis management: Readiness, response and recovery (2016), https://www2.deloitte.com/content/dam/Deloitte/global/Documents/Risk/gx-cm-cyber-pov.pdf
8. C. Crane, 80 Eye-Opening Cyber Security Statistics for 2019 (2019), https://www.thesslstore.com/blog/80-eye-opening-cyber-security-statistics-for-2019/#cyber-security-statistics-the-costs-of-cyber-security-attacks
9. Axim Global, If so much cybercrime is undetected and unreported, what's the answer? (2018), https://www.aximglobal.com/blog/if-so-much-cybercrime-is-undetected-and-unreported-whats-the-answer/
10. L. Rajaobelina, I. Brun, R. Line, C. Cloutier-Bilodeau, Not all elderly are the same: fostering trust through mobile banking service experience. Int. J. Bank Mark. 85–106 (2021)
11. C. Dorris, K. Winter, L. O'Hare, E. Lwonga, PROTOCOL: a systematic review of mobile device use in the primary school classroom and its impact on pupil literacy and numeracy attainment. Campbell Syst. Rev. 1–30 (2021)
12. S. Bhattacharya, A. Bashar, A. Srivastava, A. Singh, Nomophobia: no mobile phone PhoBIA (2019), https://www.ncbi.nlm.nih.gov/pmc/articles/PMC6510111/
13. K. Carver, V. Sritapan, C. Corbett, Establishing and maintaining trust in a mobile device. IT Prof. Technol. Solut. Enterp. 66–68 (2015)
14. J. Kahn, H. Abbas, J. Al-Muhtad, International workshop on cyber security and digital investigation (CSDI 2015) survey on mobile user's data privacy threats and defense mechanisms (2015), https://reader.elsevier.com/reader/sd/pii/S1877050915017044?token=62A6F62B330639F09F39AD2FB852BE2B33B58D78240190983DF4952B4EB7CBD086F2BAF87BC8669B51ACCE8E65ED2C9B
15. S. Ikeda, Over 813,000 apps removed from apple app store and google play in H1 2021 (2021), https://www.cpomagazine.com/data-privacy/over-813000-apps-removed-from-apple-app-store-and-google-play-in-h1-2021/
16. A. Abrams, Your cell phone is 10 times dirtier than a toilet seat. Here's what to do about it (2017), https://time.com/4908654/cell-phone-bacteria/#:~:text=Research%20has%20varied%20on%20just,bacteria%20than%20most%20toilet%20seats
17. N.H. Tanner, *Cybersecurity blue team toolkit* (Wiley, Indianapolis, 2019)

Robert Lee Gordon's academic background includes a Bachelor of Arts degree in History from UCLA, a Master of Business Administration, and a Doctorate of Management and Organizational Leadership from the University of Phoenix. Dr. Gordon also holds graduate certificates in Logistics Management, Project Management, Information Systems Security and is a certified professional coach. He has hundreds of published articles pertaining to supply chain management, strategic value-added purchasing, information technology (IT) and vendor relations, conflict in the virtual organization, and complexity. He has four books and several chapters covering Complexity, Project Management, IT security, Program Management, Contracting, Logistics, and Reverse Logistics.

Privacy-Leaking and Steganographic Threats in Wireless Connected Environments

Luca Caviglione

Abstract Wireless technologies, softwarization of hardware and the increasing diffusion of IoT nodes allow to access and control industrial settings, smart environments and a variety of remote services. This leads to a wireless connected world characterized by a mixed set of technologies handling various personal and sensitive data. Therefore, security and privacy are critical requirements, which should be pursued starting from the early stages of design. Unfortunately, the complex and composite nature of modern wireless deployments enables the creation of emerging and effective threats. For instance, the traffic of IoT nodes can be inspected to infer habits or to conduct reconnaissance campaigns, whereas exchanged digital artifacts could be exploited to conceal secret information or perform exfiltration of data. In this perspective, this chapter presents and outlines new threats targeting the technological ecosystems at the basis of the so-called wireless Internet. It also proposes design choices that should be considered to engineer more secure and private wireless environments.

1 Introduction

The disruptive impact of the Internet has been known from decades, but the advent of the COVID-19 pandemic magnified the importance of being able to remotely access hardware, software, and data from *anywhere* and at *anytime*. For instance, the availability of SOHO networks and the coverage of 4G/5G mobile communication technologies allowed many professionals to work from their homes. Even if not related only to wireless connectivity, the "stay at home" paradigm has brought to light several fragilities. For instance, the vaccination registration portal of the Lazio Region in Italy has been targeted by the RansomEXX and LockBit 2.0 ransomware.

L. Caviglione (✉)
Institute of Applied Mathematics and Information Technologies, Via de Marini 6,
Genova 16149, Italy
e-mail: luca.caviglione@cnr.it

A. K. Pathan (ed.), *Towards a Wireless Connected World: Achievements and New Technologies*, https://doi.org/10.1007/978-3-031-04321-5_2

A detailed analysis[1] revealed that the attack exploited the administration credentials of an employee working from home and then the infection spread through the VPN to the regional datacenter. Another example concerns novel malicious activities taking advantage of the shift from classrooms to online meetings. Specifically, threats like Zoombombing, spanning from innocuous pranks to organized disruption raids, revealed the limits of online conferencing when massively deployed or used for cases not intended in the original design. Nevertheless, the intensive use of the Internet during lockdown periods enlightened how many services are engineered around features rather than adequate security levels [1].

Summing up, the COVID-19 pandemic has confirmed the role of wireless technologies as the key driver to pursue the vision of a connected world. Furthermore, low-power and short-range communications enabled mobile devices to implement new applications and services for facing challenges difficult to forecast. As a paradigmatic example, the successful deployment of contact tracing applications was possible owing to the diffusion of the Bring Your Own Device (BYOD) model enabling a capillary adoption of personal smartphones for tasks ranging from work to education [2]. Moreover, wireless communications (i.e., cellular/IEEE 802.11 connectivity and Bluetooth) allowed both to sense and remotely transmit information for creating accurate snapshots of the diffusion of the coronavirus [3].

Unfortunately, this scenario can be a fertile ground for "classical" spear phishing and smishing attempts mimicking vaccination campaigns and exploiting human fears as well as for designing novel privacy-leaking attacks [4]. In this case, being able to engineer protocols with suitable built-in padding strategies, to not leak details on the number of sensed identifiers when syncing with the remote datacenter, is of prime importance, especially due to the nature of wireless connectivity. Similar issues are caused by the increasing "smartification" of homes and working environments. As possible examples, many houses, industrial or commercial settings regularly take advantage of wireless IoT nodes (e.g., smart lighting or sensors for precision agriculture), which can be enumerated and identified via scanning campaigns [5]. Furthermore, the traffic produced by smart speakers or voice assistants when contacting remote cloud facilities or orchestrating the various home appliances can be inspected to infer information about the habits of users [6].

As a consequence, the huge adoption of WLANs to guarantee connectivity, IoT and low-power wireless links to retrofit buildings or create innovative services (e.g., the deployment of IEEE 802.15.4 for Industry 4.0 purposes), and the increasing demand for mobility, lead to a complex attack surface and the birth of new threats, which should be investigated to not void the vision of a wireless connected world. In more detail, the wireless ecosystem could become the preferred aim for reconnaissance campaigns, especially due to the easy *side-channelization* of the various devices and links. At the same time, wireless connectivity and its adoption with edge/cloud computing result in a distributed and multi-domain computing continuum offering almost unbounded possibilities to hide information or to covertly exfiltrate data [7].

[1] U.S. Department of Health and Human Services, Analyst Note: https://www.hhs.gov/sites/default/files/lazio-ransomware-attack.pdf [Last Accessed: January 2022].

In this perspective, the goal of this chapter is to identify and outline threats and malicious schemes exploiting the technological ecosystem at the basis of the wireless Internet, with emphasis on emerging and unconventional attack techniques. Specifically, the contributions of this work are: (*i*) an assessment of the privacy and data leaking nature of modern wireless connected environments; (*ii*) the identification of complex, advanced attacks surfaces that can be implemented via information hiding and steganographic attacks; (*iii*) the definition of some precise development and engineering challenges that should be considered to not void the societal and economical opportunities of the wireless connected world.

The remainder of the chapter is structured as follows. Section 2 outlines emerging hazards that can endanger future wireless deployments, while Sect. 3 discusses their intrinsic leaking behaviors as well as ad-hoc steganographic threats. Section 4 introduces some design challenges that should be considered to make the wireless Internet more secure, while Sect. 5 deals with core guidelines. Lastly, Sect. 6 concludes the chapter and hints at future developments.

2 Wireless Connected World: Emerging Hazards

Among the several evolutions of modern malware, a major tendency deals with the increasing specialization of attacks, mainly to react against advancements in countermeasures and detection techniques. As a result of this "arms race", threats able to take advantage of specific features and usage patterns of wireless technologies are expected to emerge in the near future [8–10]. Unfortunately, precisely envisaging how attacks will evolve is a very hard exercise. At the same time, to anticipate such a rise, a mandatory task requires the accurate identification of the attack surface and technological/functional clusterization at the basis of the wireless Internet. To this aim, Fig. 1 depicts the three main domains that concur to the definition of the overall attack surface. Specifically, the first domain is denoted as *Mobile Domain* and considers nodes, hosts and appliances placed at the border of the network. The second domain is denoted as *IoT Domain* and delimits protocols, technologies and devices belonging to the IoT world or relevant to the smart-* paradigm. The third domain is denoted as *Network Domain* and focuses on the network infrastructure, including equipment responsible for connectivity.

In general, such domains are not disjoint and highly overlap. For instance, the advent of 5G connectivity makes the distinction between the Mobile Domain and the Network Domain less pronounced [11]. A similar consideration can be done for the IoT Domain. In fact, compared to the Mobile Domain, the separation is more in terms of features (e.g., low-energy connectivity, massive utilization of the cloud paradigm, and use of resource-constrained hardware) rather than protocols or functionalities. Besides, the wireless Internet is mostly build upon the tight interaction of nodes and individuals. Thus, both the BYOD and the Bring Your Own Technology paradigms cause the migration of data and attacks from an entity to another. Furthermore, barriers among personal, public and critical information are not clear,

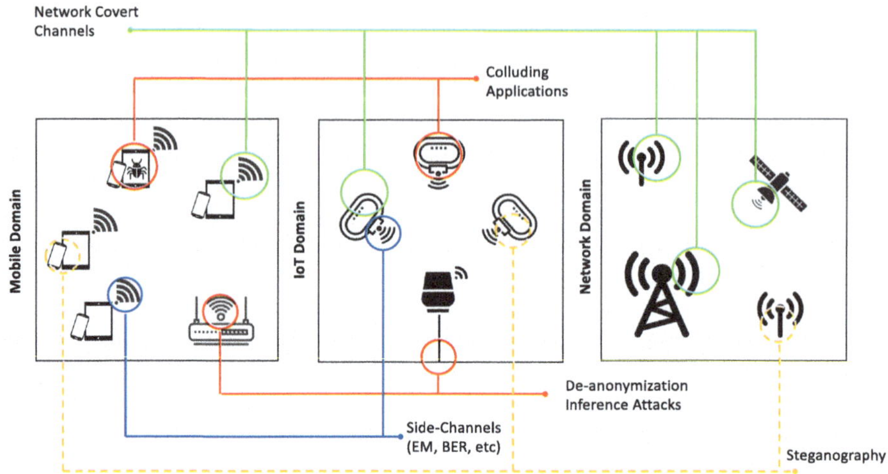

Fig. 1 Reference scenario for emerging privacy-leaking and steganographic threats

thus requiring to investigate partitioning schemes or countermeasures to prevent exfiltration [12]. Therefore, the wireless connected world appears as technologically balkanized, with an attack surface difficult to contour in a precise manner. Additionally, wireless devices concentrate multiple functionalities, e.g., electronic payments, smart transportation, and multimedia capabilities, making them a critical asset for the weaponization of attacks to break into corporate networks or highly-secure perimeters. Among the others, the two most prominently family of threats that should be considered dangerous and effective are:

- **Privacy and Data Leaking Attacks**: they represent malicious activities to implement various cyber reconnaissance techniques or to infer private information [13]. The ultimate goals of such attacks are the creation of complex campaigns for exploiting weaknesses of the victim, and the gathering of information to endanger the physical space of users.
- **Information Hiding and Steganographic Attacks**: threats exploiting information hiding and malware endowed with steganographic capabilities have been increasingly observed in the Internet such that the term *stegomalware* (i.e., steganography + malware) has become popular [8]. One of the main use of such techniques concerns the creation of network covert channels, especially to bypass security mechanisms or exfiltrate information [7]. In this vein, wireless communications can offer novel opportunities and mobile devices could lead to an almost unbounded variety of artifacts that can be used to conceal data.

We point out that, the technological maturity of wireless settings, including the bandwidth availability and the computing/storage resources of mobile nodes, make them also susceptible to the various threats already observed for the wired Internet. As a consequence, engineering mixed wired/wireless deployments requires an additional effort, especially to consider potential hazards due to the presence of mobile and wireless entities [14]. The next section will discus the most relevant threats and some synecdochical templates.

3 Privacy Leaking and Steganographic Threats

This section introduces emerging threats expected to focus on the wireless Internet. Specifically, Sect. 3.1 reviews attacks aimed at leaking information or endanger privacy of users, while Sect. 3.2 discusses steganographic hazards.

3.1 Privacy and Data Leaking of Wireless Ecosystems

As hinted, mobile and wireless devices are characterized by two clashing features. On one hand, they centralize personal data and applications for work, entertainment and education duties. On the other hand, they can be the objective of several information-leaking campaigns. As an example of the vulnerabilities caused by side-channels or the leaking nature of devices at the basis of the wireless revolution, [15] discusses smudge attacks arising by the widespread use of touch screens. Specifically, authors showcase how smudges can lead to information for guessing passwords, thus making the device a possible entry point for more complex raids.

Another interesting viewpoint to be considered is the massive deployment of smart speakers or voice-based assistant acting as digital hubs of the wireless world. A vast corpus of works concerning the tight relation between physical security and de-anonymization attacks launched against traffic of smart speakers has emerged (see, e.g., [6] and the references therein). In essence, even if encrypted, traffic produced by various appliances, as well as flows exchanged between the speaker and the remote backend, can be processed via AI-capable frameworks to guess the vocal commands (e.g., turn on/off lights), infer habits (e.g., sleep patterns), or when the house is empty. Risks can be magnified by the poor degree of knowledge of mechanisms and practices ruling security, including defective configurations. Moreover, the "work-out-of-the-box" flavor, jointly with the availability of many cost-effective IoT nodes (temperature sensors, smart lights and switches, just to mention some), lead to an hardware chain difficult to control and assess. Thus, future wireless ecosystems can be plagued with backdoors in equipments, incompatible privacy requirements, a technological supply chain hard to control in complex industrial settings and critical infrastructures, as well as a non-negligible loss of digital sovereignty.[2]

[2] EPRS Ideas Paper, Towards a more resilient EU: "Digital Sovereignty for Europe". Available online: https://www.europarl.europa.eu/RegData/etudes/BRIE/2020/651992/EPRS_BRI(2020)651992_EN.pdf [Last Accessed: January 2022].

Concerning specific threats targeting the wireless segment, apart classical techniques for spoofing identities, endangering lower layers of the stack, and de-authenticating nodes [14], side-channels-based attacks should be taken into a high account. For instance, choices in the design of the hardware can make feasible advanced threat models. This is the case of mixed-signal circuits where digital and analog circuitry reside on the same die, which is a common approach in many IoT and automotive products [16]. As a consequence of this engineering, data processing can leak information via EM emissions, thus enabling the remote collection of sensitive bits, such as cryptographic keys. Usually, the prime countermeasure resides in the hardware itself (e.g., shielding and optimized placement of chips). Yet, also the software could play a role: sensitive computations should be moved to the CPU or handled via suitable APIs for injecting patterns of instructions to add noise to EM emissions.

Another attack scheme concerns the use of channel state information that can be inferred from IEEE 802.11 hotspots. As an example, keystrokes of users can be inferred by correlating ICMP traffic and state information [17]. Such a scheme could be difficult to detect, especially due to the presence of a multitude of hotspots, and the habit of letting devices to join wireless networks in an automatic manner. Also in this case, ad-hoc security layers should be envisaged, especially to: randomize inputs of users, enforce policies for configuring the devices, and add artificial traffic for preventing exploitable correlations.

Lastly, "legacy" approaches considering encryption as the sufficient requirement to provide security should be regarded with a critical eye when designing the future wireless connected world. In fact, simple statistics, such as the frequency and the size of packets exchanged via IEEE 802.11 links, can be used to predict queries issued to services like Google, YouTube and Wikipedia [18]. Apart the aforementioned countermeasures at the network level, future applications should be wireless-aware or wireless-secure-by-design. Thus, proper padding, randomization, data interleaving, and buffering disciplines, should be considered from the very early engineering phases of various protocols and APIs.

3.2 Information Hiding and Steganography "Over the Air"

A trend characterizing the recent evolution of malicious software concerns the adoption of information hiding and steganography to conceal offensive routines, implement hidden and parasitic communication paths to exfiltrate data or orchestrate nodes of a botnet, retrieve additional payloads in a stealthy manner, and bypass security mechanisms [7, 19]. The wireless ecosystem is definitely a prime target for the increasing wave of information-hiding-capable threats since it enables to target at least two different core entites:

1. **devices**: modern mobile devices are endowed with a variety of sensors, as well as multimedia and computing capabilities. Their utilization via a BYOD paradigm

makes them a prime objective for exfiltration attempts or steganographic attacks [9]. Similar considerations can be done for IoT nodes or building automation and control networking, which often rely upon wireless connectivity for retrofitting purposes [20].

2. **wireless technologies**: the ubiquitous availability of wireless communications and the increasing softwarization of radio access networks open up to a wide-range of steganographic manipulations [21]. The progressive diffusions of infrastructures based on a "cable-free" paradigm, e.g., in developing countries via satellite communications, lead to a variety of protocols and complex cross-layers interactions that can be exploited for creating a plethora of effective covert channels [22]. Nevertheless, the wireless Internet is often at the basis of disaster relief operations or adopted to provide connectivity to industrial buildings, thus evaluating vulnerabilities exploitable for advanced persistent threats is now an essential step.

As regards devices, even if not strictly targeting the wireless link, a major offensive paradigm exploits the growing diffusion of sandbox-like security mechanisms of mobile devices, including IoT nodes. In essence, processes are confined inside secure execution enclaves, which prevent the unauthorized exchange of data or operations on shared resources (e.g., memory). To bypass such a scheme, applications can collude via local covert channels built through the manipulation of specific hardware or software resources [23]. For instance, information can be encoded with ad-hoc patterns of operations on the CPU: the overall load can be modulated to contain arbitrary information, e.g., a load exceeding an agreed threshold indicates a `1`. Indeed, such channels are fragile and characterized by a low-bandwidth, but proven to be effective to endanger privacy and to exfiltrate sensitive information, such as login credentials [9]. Moreover, wireless services increasingly rely on containerized applications or are based upon a technological posture implemented through virtual machines. Therefore, information hiding schemes can be used to let a container inspect the guest node device/host. Containers can then collude to understand whether they are running on the same hardware entity, e.g., to map the physical infrastructure [24].

Concerning techniques to stealthily exfiltrate data through a wireless link, they strictly depend on the exploited carrier (i.e., the place where the information is hidden) and can be applied to almost every layer of the ISO/OSI stack, as it happens for the wired case [22]. However, wireless communications offer additional features that can be abused to conceal data. A typical idea is to inject arbitrary content in the padding frames used in the IEEE 802.11 standard [25]. This approach could have some limitations when applied to practical settings. Yet, the exploitation of physical layer features, like camouflage subcarriers of the IEEE 802.11n spectrum mask, can lead to effective hidden communication paths nested within the wireless signal [26]. Indeed, complex access control schemes, signaling protocols, algorithms for managing the shared media, and handover mechanisms offer a variety of carriers that can be manipulated by an attacker to leak data. To this extent, the development of a wireless infrastructure or the creation of standards should be checked against the risk of covert channels.

Lastly, the wireless Internet will be undoubtedly the place where the majority of machine-to-machine communication happens. In this vein, protocols like the Message Queuing Telemetry Transport offer a variety of opportunities to implement parasitic communication attempts [27]. Messages can be arranged in patterns or interleaved with timing intervals to encode information (e.g., a delayed communication denotes a `1`) as well as optional fields can be overwritten to store secrets. Furthermore, the internals of the IEEE 802.15.4 protocol can be exploited by an attacker to hide, in an energy-efficient manner, additional information [28]. This could lead to exfiltration attempts difficult to detect, especially by using method-agnostic indicators, like signatures in power drains [29].

4 Development and Engineering Challenges

As discussed, the complexity of modern scenarios is exacerbated by the used variety of hardware and software components (e.g., resource-constrained nodes and containerized applications) and the presence of a wide-range of services with different performance and functional requirements. Hence, monitoring and inspecting the traffic produced by a device, assessing the overall network conditions, evaluating resource utilization or hardware-specific metrics, such as the quality of the link or the transmission power, usually require platform/vendor-specific approaches. Moreover, the multifaceted nature of threats often demands for specialized layers able to gather information from the entire protocol stack or from different portions of the guest operating system, for instance to spot privacy leakages or resource-draining attacks [30].

Indeed, the wireless connected world is a technological split space. Such an aspect should be evaluated jointly with the presence of sensitive data and the possibility of targeting almost every hardware/software entity with information-hiding-capable attacks [31]. Even if security of wireless applications and ecosystems has been largely investigated in the literature (see [14] and [32] for recent reviews) two increasing trends have to be considered. The first concerns the delivery of software via immutable containers or the implementation of services through virtualization. The second deals with the high diversification of vectors and techniques capable of endangering almost all the entities of a given deployment. Therefore, the majority of the approaches needs a rethinking, especially to face the challenges discussed in this chapter [33]. Among the various techniques, a promising idea leverages *code layering* and *code augmentations*. Such features permit to instrument at run-time containers used to build wireless applications via a Platform-as-a-Service paradigm, as well as to extend functionalities of the Linux kernel to monitor the underlying hardware and software [34]. Nevertheless, both approaches may prevent interrupting the service or changing vast portions of the software stack, thus leading to a decoupled design, i.e., a general inspection part and a threat-/service-specific logic. Section 4.1 will present design and architectural choices to be considered.

4.1 Code Augmentation for Wireless Ecosystems

The ability of designing and engineering wireless security in a stratified manner, with strata that can be independently modified, is surely a major challenge to be addressed for pursuing the vision of a mobile and always-connected world. To face this task, we propose as a reference example, to take advantage of the extended Berkeley Packet Filter (eBPF), which enables to inject at run-time code in the Linux kernel by offering a virtual-machine-like environment (see, e.g., [35] for further details and its usage for privacy-oriented applications). To prevent additional hazards, code is verified and subject to constraints in terms of loops, size of the stack and complexity, just to mention the most important. Concerning the suitability of eBPF to face the aforementioned challenges, we point out that it was initially designed for monitoring system performances, but nowadays has been increasingly adopted to inspect various behaviors of hardware and software components [36, 37].

Figure 2 depicts the reference layered architecture that we propose to address the various challenges characterizing the wireless Internet. Specifically, it is composed of two main parts:

- **Kernel Space**: it contains the kernel and the various hardware-/platform-specific drivers. Suitable abstraction is granted by the Linux kernel. Relying on Linux is not a limitation, since it is at the basis of many OSes deployed in mobile devices, IoT nodes, servers and datacenters, gaming consoles and appliances.[3] The kernel is augmented with the eBPF programs needed to inspect the various behaviors of the hosting node/network.
- **User Space**: it contains the software needed to implement the wireless service, e.g., containers, virtualized network functions, as well as security and task-specific services. Userland processes can retrieve the information collected by the eBPF via a mechanism based on *maps*, i.e., ad-hoc shared memory areas.

As a consequence, challenges arising from the fast-evolving nature of applications and services relying on wireless connectivity can be be easily faced via ad-hoc *eBPF programs*. Such layers can be created to inspect a precise behavior of the host. For instance, if a service is suspected to be the target of filesystem-based threats, the kernel could be instrumented to trace and report via eBPF the use of `__x64_sys_chmod` syscalls (see [38] for the detection of local covert channels based on the alteration of files permissions). Another example concerns cryptolockers: in this case, the volume of operations performed on block devices (collectable via `kprobe` events) can help to early detect massive encryption of files and partially neutralize an attack.

Concerning possible entities running in user-space, main reference blocks are:

- **Containerized Software**: as hinted, many modern services, especially those at the basis of 5G, are delivered through containerized software and virtualized frameworks, which cannot be inspected or altered. In this vein, adding a suitable eBPF

[3] Owing to its flexibility, the port of the eBPF technology on Windows has started with the official support from Microsoft, see: https://github.com/microsoft/ebpf-for-windows. [Last Accessed: January 2022].

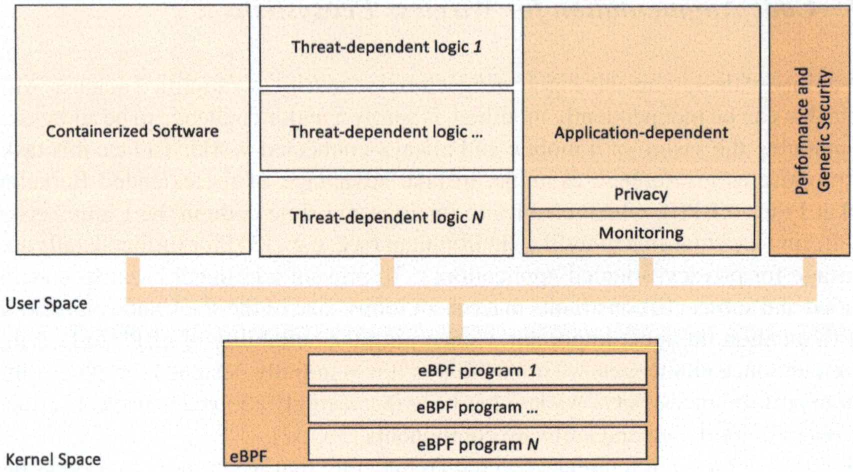

Fig. 2 Reference security architecture for future wireless ecosystems

program or an inspection layer in the OS allows to collect information to understand the correct evolution of a service, e.g., in terms of anomalous usage of resources.

- **Threat-dependent Logic**: by using the information gathered via in-kernel inspection, vendors can easily implement code to handle specific threats. For instance, eBPF proven to be effective to instrument a node for deploying IDS-like capabilities and track traffic with a high granularity [39]. At the same time, code augmentation has demonstrated its effectiveness in the preparation of datasets for machine learning frameworks [40], which are typically used to counteract the lack of general solutions against network covert channels [8].
- **Application-dependent**: applications are often provided in a monolithic manner or implemented via SDKs distributed by hardware and software vendors. Therefore, code layering can be used during the debug phase as well as to implement generic privacy and monitoring services. For instance, eBPF has shown to be a valuable tool to identify processes exchanging data with malicious servers [41].
- **Performance and Generic Security**: the ability of tracing each packet/syscall can be exploited to reveal processes or threads leaking data or implementing a colluding applications scheme [23, 24]. The user space can also team with specific programs/layers for assessing the performance of the hardware, e.g., to monitor the energy consumption in IoT nodes.

However, engineering a full-featured solution to simultaneously mitigate different threats and enforce security and privacy requirements is not a simple task. To this extent, the suggested code layering design can be viewed as a convenient way to create an array of *microservices* that can be orchestrated or chained to accomplish specific privacy/security tasks [42]. In Sect. 5, we provide some directions to help in the engineering phase of code-augmentation-based solutions for wireless ecosystems.

5 On the Engineering of a (more) Secure Wireless Internet

Wireless-capable devices are often endowed with limited computing resources or prone to battery depletion hazards, thus requiring to precisely evaluate the impact of countermeasures to not reduce the performance of the host/device [29]. Portability also plays a major role, since well-tested security mechanisms should be reused to flat costs and the risk of adding new vulnerabilities. In this perspective, Table 1 provides a "checklist" to guide the engineering towards a more secure wireless Internet, while considering the emerging privacy leaks and steganographic threats.

As shown, the table contains some indications on the various threats targeting wireless nodes and appliances, the required instrumentation (i.e., *what* has to be inspected to detect or mitigate a well-defined class of threats), the portability of the approach (i.e., if the eBPF filter or the code layering strategy *can be reused* for similar hardware or software settings), and the placement (i.e., *where* the architecture discussed in Fig. 2 has to be located within the overall ecosystem).

Possible design guidelines to be considered are:

Table 1 Main design and engineering considerations for developing countermeasures against emerging privacy and security issues of wireless frameworks

Threat	Instrumentation	Portability	Placement
Traffic de-anonymization	Protocol stack	Service dependent	Node/network
Device enumeration	Air interface	Device dependent	Node
Fingerprinting	Per-service	Service dependent	Node
Ransomware	Filesystems	Through block abstraction	Node
Screaming channels	Device driver	Implementation dependent	Node
EM/WiFi channels	Network	Attack dependent	Node/Network
Cryptominers	Threads/System Load	OS dependent	Node
Air-gapped channels	Multiple HW features	Limited to a specific technology	Node
Timing channels	Protocol stack	Protocol independent	Node/Network
Storage channels	Protocol stack	Protocol dependent	Node/Network
Web-based channels	Application	Threat dependent	Network
Steganography	Multiple SW features	Device independent	Network
Colluding Applications	All shared resources	Device dependent	Node
Cache-based channels	CPU	Implementation dependent	Node

- **Evaluate the effort for the instrumentation**: as shown, each threat may require to instrument different portions of the device, including various layers of the protocol stack or specific sub-systems. In general, the tighter the interaction with the hardware, the higher the benefits of exploiting some layering/augmentation techniques to make the design of security mechanisms more abstract. For instance, the detection of covert channels may happen in userland owing to data gathered by eBPF programs tweaked for a specific protocol or platform [34].
- **Pursue portability**: with portability we intend the ability of using the same layering strategy for similar security and privacy hazards. In this case, the user+kernel space implementation can have strong dependencies both with the threat (i.e., a specific signature to exploit) and the targeted technology (i.e., a specific hardware component). For instance, cache-based channels often take advantage of the multi-core nature of CPUs deployed in smart devices or wireless access points.[4] Indeed, the literature abounds of techniques to tame side- and covert-channels exploiting caches, CPUs and virtual machines (see, [43] and [44], respectively). A possible approach can leverage in-kernel probes tailored for a specific platform and general, reusable security layers that can be distributed to users. We point out that, this can also allow to deliver containerized security frameworks or generic over-the-air updates to mobile devices or network nodes.
- **Carefully evaluate the placement**: understanding where to deploy the proposed framework is subject to various tradeoffs. For instance, if a tight coupling is needed, the instrumentation should be done within the node to inspect. However, this may cause excessive overheads or impair the levels of Quality of Experience. Thus, a possible approach is to shift the userland part at the border of the network for the most resource-intensive tasks. Similarly, when the hazard is expected to aim a group of nodes (see, e.g., the case of device enumeration attacks of web-based channels) a good design choice considers deploying the framework via a *Something-as-a-Service* paradigm. Despite scalability properties, a carefully-evaluated placement strategy may also lead to additional benefits. In particular, threat-dependent or general security services can run in the core network or "outside" the devices, whereas instrumentation can be done close to the hardware or the software artifact to protect. In this case, solutions like eBPF can take advantage of comprehensive toolchains to ease the deployment and development phases [45].
- **Remedy to "immutable" choices**: as discussed, poor hardware design or APIs without optimal isolation features may lead to exploitable behaviors. In general, such two aspects cannot be fixed *a posteriori*, e.g., due to the immutable nature of the hardware or the unavailability of source code. Thus, the possibility of stacking layers can provide effective fixes, for instance to inject additional noisy patterns, provide padding or align data structures, as well as to sanitize traffic by overwriting protocol fields or deploying buffering disciplines.

[4] For an example of a covert channel exploiting a shared register in a commercial multi-core CPU system, see the M1RACLES - CVE-2021-30747 targeting Apple Silicon. Even if the vulnerability should not be considered as dangerous, it can open up to various privacy-leaking attacks. Moreover, this highlights the importance of knowing and considering this class of threats, which are slowly emerging. Available online: https://m1racles.com [Last Accessed: January 2022].

5.1 Overall Overview

To the aim of providing security and privacy requirements in the wireless connected world, Fig. 3 showcases the overall overview of the proposed framework template. As depicted, code augmentation for data gathering and instrumentation can be placed in three different layers of the network, namely within end nodes, access points or in the radio access network, and in the core network. In other words, eBPF (or similar techniques) can be used to augment the capabilities of the lower layers of different entities composing the scenario under consideration.

In some cases, the threat to be addressed or the hosting device could account for specific constraints. For instance, revealing anomalous activities on a register shared among different execution cores requires to instrument the host and cannot be done remotely. Instead, when threats are not physically bounded or confined to a single node, information can be gathered in other parts of the network. As an example, network covert channels built via the manipulation of traffic behaviors can be detected and neutralized at the border of the network: this can offload hosts and protect a vast user population with a unique appliance. Yet, data can be (partially) gathered in end nodes without causing overheads to endow the decision maker with suitable measurements. For the case of side and air-gapped channels, the prime line of defense is typically implemented at the hardware level. However, gathering analytics about the experienced bit error rate, frame error rate, S/N ratio, and low-level EM emissions directly from the access point or the wireless router may lead to spot the privacy-leaking attempt [17].

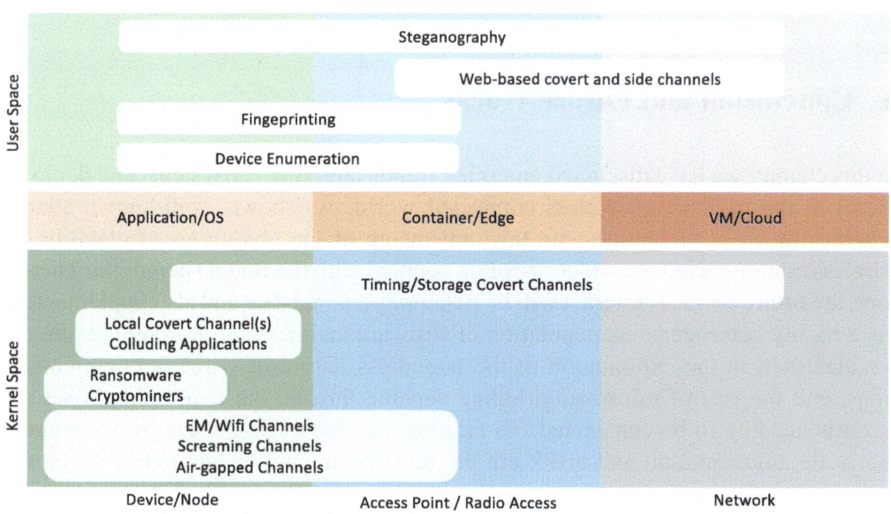

Fig. 3 Overall overview of the proposed security paradigm when deployed in a wireless ecosystem

As regards the user space, it groups the various services/applications in charge of enforcing privacy and security requirements or mitigate specific threats. Such high-level counterparts can operate in a standalone manner (e.g., by implementing application layer proxies, access control policies or personal security tools) or exploit low-level measurements gathered by specific eBPF programs (or other layering approaches). Applications can run locally or in a distributed manner. As an example, techniques for mitigating steganographic and information-hiding-capable threats often require to inspect different carriers, including digital media, thus they can both run locally or in suitable proxies [46]. Similarly, protections against network-oriented attacks like fingerprinting and enumeration can require data generated in the user space, thus preventing the need of in-kernel measurements.

As shown in Fig. 3, the proposed paradigm can be implemented within different architectural layers. Specifically, security services can be deployed in the host or scaled up according to the growing availability of resources. Despite this, the approach requires to develop some "glue" middleware to allow the collection from the user space of information gathered by in-kernel probes. Similarly, additional functionalities could be needed to move data towards a decision maker. In this perspective, toolkits[5] able to collect information coming from the various entities composing the network, provide some filtering and processing capabilities, and then implement some form of advanced detection, should be considered critical for the successful implementation of security mechanisms in future wireless networks.

Lastly, we point out that the proposed paradigm is not in contrast with the layering of Fig. 1, which is more focused on elaborating where attacks happen. Instead, both abstract templates should be considered when designing future frameworks to endow the wireless Internet with proper security and privacy requirements.

6 Conclusion and Future Works

In this chapter we have discussed emerging trends targeting ecosystems and deployments at the basis of a wireless connected world. As shown, well-known attack schemes or novel techniques can take advantage of the ubiquitous availability of wireless communications, which are often coupled with the BYOD paradigm. Therefore, the future wireless Internet will be richer of personal data and also implemented via a highly heterogeneous population of software and devices. Among the others, we identified in the exploitation of the boundless nature of wireless communications, and the rise of information-hiding-capable threats, the two main classes of hazards needing to be considered. To face such a challenging scenario, we argued that code augmentation and eBPF are the most promising technological enablers.

[5] See, as a paradigmatic example, the toolkit envisaged in SIMARGL—Secure Intelligent Methods for Advanced Recognition of Malware and Stegomalware: https://simargl.eu [Last Accessed: January 2022].

Accordingly, we introduced some design and engineering guidelines to develop both threat-specific and portable detection and mitigation techniques.

Future works aim at refining prototypal implementations, which demonstrated to be effective in spotting threats leveraging steganographic techniques and covert channels (see [34, 38, 40, 47] for details). Since overheads introduced by the layering approach are very limited,[6] we plan its adoption in a variety of nodes with mixed computing and storage requirements. In parallel, the development of some toolkits or services to orchestrate the various layers is a relevant part of our ongoing research. Lastly, the development of formal frameworks to understand the limitation of the proposed idea, as well as its functional requirements, is another key research action.

Acknowledgements This work was partially supported by the European Commission and the Horizon 2020 Program within the framework of Project SIMARGL (Secure Intelligent Methods for Advanced RecoGnition of Malware and Stegomalware), Grant Agreement No. 833042.

References

1. I.A. Secara, Zoombombing-the end-to-end fallacy. Netw. Secur. **2020**(8), 13–17 (2020)
2. N. Ahmed, R.A. Michelin, W. Xue, S. Ruj, R. Malaney, S.S. Kanhere, A. Seneviratne, W. Hu, H. Janicke, S.K. Jha, A survey of COVID-19 contact tracing apps. IEEE access **8**, 134577–134601 (2020)
3. M.J.M. Chowdhury, M.S. Ferdous, K. Biswas, N. Chowdhury, V. Muthukkumarasamy, Covid-19 contact tracing: challenges and future directions. IEEE Access **8**, 225703–225729 (2020)
4. B. Pranggono, A. Arabo, Covid-19 pandemic cybersecurity issues. Internet Technol. Lett. **4**(2), e247 (2021)
5. M. Miettinen, S. Marchal, I. Hafeez, N. Asokan, A.R. Sadeghi, S. Tarkoma, IoT sentinel: automated device-type identification for security enforcement in IoT, in *2017 IEEE 37th International Conference on Distributed Computing Systems (ICDCS)* (IEEE, 2017), pp. 2177–2184
6. D. Caputo, L. Verderame, A. Ranieri, A. Merlo, L. Caviglione, Fine-hearing Google Home: why silence will not protect your privacy. J. Wirel. Mob. Netw. Ubiquit. Comput. Dependable Appl. **11**(1), 35–53, e247 (2020)
7. W. Mazurczyk, L. Caviglione, Information hiding as a challenge for malware detection. IEEE Secur. Priv. **13**(2), 89–93, e247 (2015)
8. L. Caviglione, M. Choraś, I. Corona, A. Janicki, W. Mazurczyk, M. Pawlicki, K. Wasielewska, Tight arms race: overview of current malware threats and trends in their detection. IEEE Access (2020)
9. W. Mazurczyk, L. Caviglione, Steganography in modern smartphones and mitigation techniques. IEEE Commun. Surv. & Tutor. **17**(1), 334–357, e247 (2014)
10. S. Wendzel, S. Zander, B. Fechner, C. Herdin, Pattern-based survey and categorization of network covert channel techniques. ACM Comput. Surv. (CSUR) **47**(3), 1–26, e247 (2015)
11. Y. Huo, X. Dong, W. Xu, M. Yuen, Enabling multi-functional 5g and beyond user equipment: a survey and tutorial. IEEE Access **7**, 116975–117008, e247 (2019)
12. K.W. Miller, J. Voas, G.F. Hurlburt, BYOD: security and privacy considerations. It Prof. **14**(5), 53–55, e247 (2012)

---

[6] In our preliminary trials, for the case of spotting the presence of network covert channels targeting IPv6 traffic, packets processed with a code layering scheme based on eBPF are delayed only of few nanoseconds.

13. W. Mazurczyk, L. Caviglione, Cyber reconnaissance techniques. Commun. ACM **64**(3), 86–95, e247 (2021)
14. Y. Zou, J. Zhu, X. Wang, L. Hanzo, A survey on wireless security: technical challenges, recent advances, and future trends. Proc. IEEE **104**(9), 1727–1765, e247 (2016)
15. A.J. Aviv, K. Gibson, E. Mossop, M. Blaze, J.M. Smith, Smudge attacks on smartphone touch screens, in *Proceedings of the 4th USENIX conference on Offensive technologies* (2010), pp. 1–7
16. G. Camurati, S. Poeplau, M. Muench, T. Hayes, A. Francillon, Screaming channels: when electromagnetic side channels meet radio transceivers, in *Proceedings of the 2018 ACM SIGSAC Conference on Computer and Communications Security* (2018), pp. 163–177
17. M. Li, Y. Meng, J. Liu, H. Zhu, X. Liang, Y. Liu, N. Ruan, When CSI meets public WiFi: inferring your mobile phone password via WiFi signals, in *Proceedings of the 2016 ACM SIGSAC Conference on Computer and Communications Security* (2016), pp. 1068–1079
18. S.A. Sharma, B.L. Menezes, Implementing side-channel attacks on suggest boxes in web applications, in *Proceedings of the First International Conference on Security of Internet of Things* (2012), pp. 57–62
19. K. Cabaj, L. Caviglione, W. Mazurczyk, S. Wendzel, A. Woodward, S. Zander, The new threats of information hiding: the road ahead. IT Prof. **20**(3), 31–39, e247 (2018)
20. J. Kaur, J. Tonejc, S. Wendzel, M. Meier, Securing bacneâĂŽs pitfalls, in *IFIP International Information Security and Privacy Conference* (Springer, 2015), pp. 616–629
21. L. Bonati, S. D'Oro, F. Restuccia, S. Basagni, T. Melodia, SteaLTE: private 5g cellular connectivity as a service with full-stack wireless steganography. arXiv preprint arXiv:2102.05606 (2021)
22. B. Carrara, C. Adams, Out-of-band covert channels-a survey. ACM Comput. Surv. (CSUR) **49**(2), 1–36, e247 (2016)
23. C. Marforio, H. Ritzdorf, A. Francillon, S. Capkun, Analysis of the communication between colluding applications on modern smartphones, in *Proceedings of the 28th Annual Computer Security Applications Conference* (2012), pp. 51–60
24. X. Gao, Z. Gu, M. Kayaalp, D. Pendarakis, H. Wang, Containerleaks: emerging security threats of information leakages in container clouds, in *2017 47th Annual IEEE/IFIP International Conference on Dependable Systems and Networks (DSN)* (IEEE, 2017), pp. 237–248
25. K. Szczypiorski, W. Mazurczyk, Steganography in IEEE 802.11 OFDM symbols. Secur. Commun. Netw. **9**(2), 118–129 (2016)
26. J. Classen, M. Schulz, M. Hollick, Practical covert channels for WiFi systems, in *2015 IEEE Conference on Communications and Network Security (CNS)* (IEEE, 2015), pp. 209–217
27. A. Mileva, A. Velinov, L. Hartmann, S. Wendzel, W. Mazurczyk, Comprehensive analysis of MQTT 5.0 susceptibility to network covert channels. Comput. Secur. **104**, 102207 (2021)
28. D. Martins, H. Guyennet, Steganography in MAC layers of 802.15.4 protocol for securing wireless sensor networks, in *2010 International Conference on Multimedia Information Networking and Security* (IEEE, 2010), pp. 824–828
29. L. Caviglione, M. Gaggero, E. Cambiaso, M. Aiello, Measuring the energy consumption of cyber security. IEEE Commun. Mag. **55**(7), 58–63, e247 (2017)
30. L. Catuogno, C. Galdi, N. Pasquino, Measuring the effectiveness of containerization to prevent power draining attacks, in *2017 IEEE International Workshop on Measurement and Networking (M&N)* (IEEE, 2017), pp. 1–6
31. S. Wendzel, L. Caviglione, W. Mazurczyk, A. Mileva, J. Dittmann, C. Krätzer, K. Lamshöft, C. Vielhauer, L. Hartmann, J. Keller, T. Neubert, A revised taxonomy of steganography embedding patterns, in *The 16th International Conference on Availability, Reliability and Security, ARES 2021. Association for Computing Machinery, New York, NY, USA* (2021)
32. L. Chettri, R. Bera, A comprehensive survey on internet of things (iot) toward 5g wireless systems. IEEE Internet Things J. **7**(1), 16–32, e247 (2019)
33. M. Repetto, A. Carrega, R. Rapuzzi, An architecture to manage security operations for digital service chains. Future Gener. Comput. Syst. **115**, 251–266, e247 (2021)

34. L. Caviglione, W. Mazurczyk, M. Repetto, A. Schaffhauser, M. Zuppelli, Kernel-level tracing for detecting stegomalware and covert channels in Linux environments. Comput. Netw. 108010 (2021)
35. S. Rivera, V.K. Gurbani, S. Lagraa, A.K. Iannillo, R. State, Leveraging eBPF to preserve user privacy for DNS, DoT, and DoH queries, in *Proceedings of the 15th International Conference on Availability, Reliability and Security* (2020), pp. 1–10
36. S. Miano, M. Bertrone, F. Risso, M. Tumolo, Creating complex network service with eBPF: experience and lessons learned, in *Proceedings of IEEE High Perform. Switching Routing (HPSR)* (Bucharest, Romania, 2018), pp. 1–8
37. S. Miano, M. Bertrone, F. Risso, M., Vásquez Bernal, Y. Lu, J. Pi, Securing linux with a faster and scalable iptables. ACM SIGCOMM Comput. Commun. Rev. **49**(3) (2019)
38. A. Carrega, L. Caviglione, M. Repetto, M. Zuppelli, Programmable data gathering for detecting stegomalware, in *Proceedings of the 2nd International Workshop on Cyber-Security Threats, Trust and Privacy Management in Software-defined and Virtualized Infrastructures (SecSoft)* (IEEE, 2020)
39. M. Bachl, J. Fabini, T. Zseby, A flow-based IDS using machine learning in eBPF. arXiv preprint arXiv:2102.09980 (2021)
40. M. Repetto, L. Caviglione, M. Zuppelli, bccstego: a framework for investigating network covert channels, in *The 16th International Conference on Availability, Reliability and Security* (2021), pp. 1–7
41. L. Deri, S. Sabella, S. Mainardi, P. Degano, R. Zunino, Combining system visibility and security using eBPF, in *ITASEC* (2019)
42. M. Alam, J. Rufino, J. Ferreira, S.H. Ahmed, N. Shah, Y. Chen, Orchestration of microservices for IoT using Docker and edge computing. IEEE Commun. Mag. **56**(9), 118–123, e3134 (2018). https://doi.org/10.1109/MCOM.2018.1701233
43. Y. Lyu, P. Mishra, A survey of side-channel attacks on caches and countermeasures. J. Hardware Syst. Secur. **2**(1), 33–50, e3134 (2018)
44. J. Betz, D. Westhoff, G. Müller, Survey on covert channels in virtual machines and cloud computing. Trans. Emerg. Telecommun. Technol. **28**(6), e3134 (2017)
45. T. Nam, J. Kim, Open-source IO visor eBPF-based packet tracing on multiple network interfaces of Linux boxes, in *2017 International Conference on Information and Communication Technology Convergence (ICTC)* (2017), pp. 324–326
46. J. Blasco, J.C. Hernandez-Castro, J.M. de Fuentes, B. Ramos, A framework for avoiding steganography usage over HTTP. J. Netw. Comput. Appl. **35**(1), 491–501, e247 (2012)
47. L. Caviglione, M. Zuppelli, W. Mazurczyk, A. Schaffhauser, M. Repetto, Code augmentation for detecting covert channels targeting the IPv6 Flow Label, in *2021 IEEE 7th International Conference on Network Softwarization (NetSoft)* (2021), pp. 450–456

Luca Caviglione is a Senior Research Scientist at the Institute for Applied Mathematics and Information Technologies of the National Research Council of Italy. He holds a Ph.D. in Electronic and Computer Engineering from the University of Genoa, Italy. His research interests include optimization of large-scale computing frameworks, traffic analysis, and network security. He is an author or co-author of more than 150 academic publications, and several patents in the field of p2p and energy-aware computing. He has been involved in Research Projects and Network of Excellences funded by the ESA, the EU and the MIUR. He is a Work Group Leader of the Italian IPv6 Task Force, a contract Professor in the field of networking/security and a Professional Engineer. He is the head of the IMATI Research Unit of the National Inter-University Consortium for Telecommunications and part of the Steering Committee of the Criminal Use of Information Hiding initiative.

Anti-Phishing Approaches in the Era of the Internet of Things

Mallika Boyapati, Bala Chaitanya Gutta, Md. Zakirul Alam Bhuiyan, and Junggab Son

Abstract In today's Internet era, Internet of Things (IoT) based products and applications are adopted by users for many different purposes like shopping, managing finances, smart home security hubs, etc. Some of them are implemented as web page applications hosted over the Internet, which essentially inherits existing threats and attacks on them. One of the most common security attacks on web page applications is phishing. Phishing is a social engineering attack in which an adversary tries to steal users' sensitive information including credentials by tricking them into believing the user is on a legitimate web page. Adversaries tend to adopt new and sophisticated ways to forge the web page designs in a crafty way and trick users to visit the malicious links. The crafty phishing web pages are used as a medium to carry out the art of phishing even for IoT-based applications. This chapter focuses on the state-of-the-art technologies that can be utilized to defend against phishing attacks in the era of IoT. Specifically, technologies to detect web page, zero-day, and adversarial phishing attacks, including their features, are introduced and discussed.

1 Introduction

With the advancement of devices, the security issue became one of the biggest concerns in the Internet of Things (IoT) environment. IoT devices are getting more powerful with additional abilities to control systems, and thus they can be exploited to collect a lot of data correlated to users' activities, such as locking and unlocking

M. Boyapati · B. C. Gutta · J. Son (✉)
Information and Intelligent Security Lab., Department of Computer Science,
Kennesaw State University, Marietta, GA 30060, USA
e-mail: json@kennesaw.edu

M. Boyapati
e-mail: mboyapat@students.kennesaw.edu

Md. Z. A. Bhuiyan
Department of Computer and Information Sciences, Fordham University,
Bronx, NY 10023, USA
e-mail: mbhuiyan3@fordham.edu

A. K. Pathan (ed.), *Towards a Wireless Connected World: Achievements and New Technologies*, https://doi.org/10.1007/978-3-031-04321-5_3

the doors, controlling temperatures, starting up cars, etc., only with touching a button on the smart devices. It is worth noting that users can perform these activities by accessing a web page from a browser on smart devices (web-based IoT).

A practical example of the web-based IoT is the design and development of security surveillance systems at home [1]. These surveillance systems are built using Single Board Computers (SBC) that connect to WiFi. WiFi-enabled surveillance systems process the sensor data and send it to the controller sections. The controllers enable the devices that capture the live feed and notify the users on the email. The user can click the link provided in the email to check the live video. The user will be able to respond using the web page link based on a situation. Another example is Bank of America,[1] a leading financial organization. They implemented a new, voice-based command feature to their web page applications. Customers can control their financial transactions with a voice command to Google Home or Alexa. Also, they implemented a bank locker security system connected to IoT applications. The locker can be operated with their web application launched on a web browser on a user's wireless devices [2]. Not only the financial and retail sectors, but most sectors, in general, are also taking a step further and introducing IoT applications while connecting via web pages. Although the examples seem great, the security of these devices, applications, and user data is a real concern and an unavoidable consideration [3]. Of these security attacks, one of the major concerns is web page phishing attacks.

In recent years, the advent of phishing attacks that use web pages as a medium has increased. In most cases, an adversary uses a phishing page directly or uses a phishing application launched on web pages. These phishing pages can be run on any of the devices, which means any IoT devices connected to the Internet through a web browser that can host or open up web pages/applications are prone to web page phishing attacks. Moreover, a phishing kit that makes the attack easy is available from the dark web, and the number of purchases is increasing day by day.[2] Benign users who do not know about the attack are always susceptible to being affected which results in billions of dollars are being transferred to fraudulent accounts where the hacker can lure the user for the credentials. Hackers have always been busy, and the threats are growing every day. Therefore, IoT applications should be well-prepared against phishing attacks in order to protect users as well as systems from being damaged.

The damage caused by phishing is worth noticing these days. The phisher might transfer funds, make purchases, exploit the actual legitimate user accounts from the harvested credentials. Also, there is a chance that the phisher sells the information to criminals. It is common that the phisher shuts down the web page once a victim is caught in order to make tracking it difficult. Even though the site is not shut down, if the web page is hosted on a server outside the country, it will not be an easy task to catch and prosecute the phishers. This served as a motivation to many research

[1] https://www.bankofamerica.com/ (Last Accessed: January 25, 2022).

[2] https://www.globalsign.com/en/blog/warning-advanced-phishing-kits-now-available-on-the-dark-web (Last Accessed: January 25, 2020).

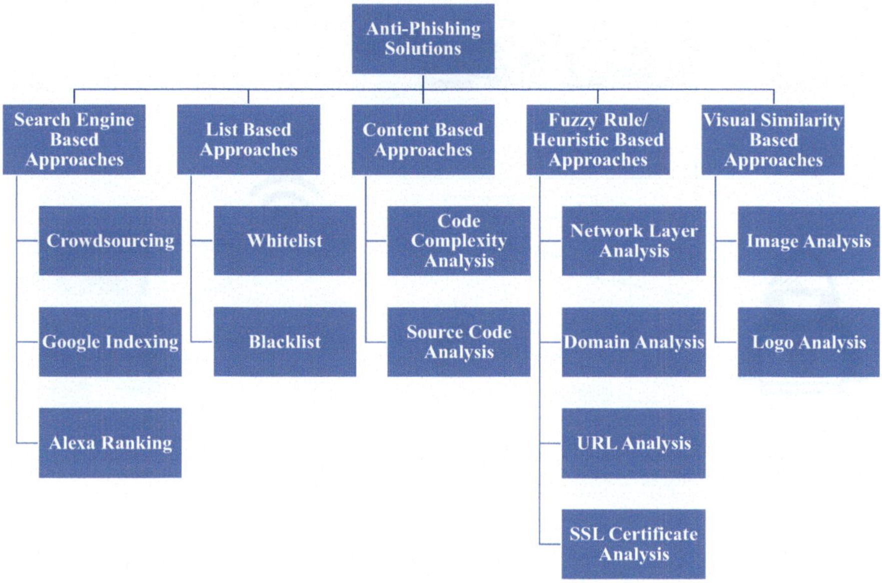

Fig. 1 Categorization of anti-phishing solutions

schemes that can effectively detect phishing pages and warn the users before any phishing attacks occur.

This chapter focuses on the cutting-edge approaches that can be employed for web phishing detection in IoT environments. Although adversaries try to replicate the web page applications as legitimate as possible, there are a lot of differences that are being observed between legitimate and phishing web pages. After carefully analyzing extant anti-phishing solutions, we categorized them into five classes: Search engine-based, List-based, Content-based, Fuzzy role/heuristic-based, and Visual similarity-based approaches, as shown in Fig. 1, each of which is introduced and discussed in the following sections.

2 Phishing Attacks in IoT Environments

As shown in the Fig. 2, a phishing attack generally has three major life cycles:

1. Phish: An adversary who creates a phishing page, sends it to users via email, or embeds a bad URL in some malicious pages. This is the "Phish" phase.
2. Bite: When a user unknowingly clicks the URL, it is called the "Bite" phase.

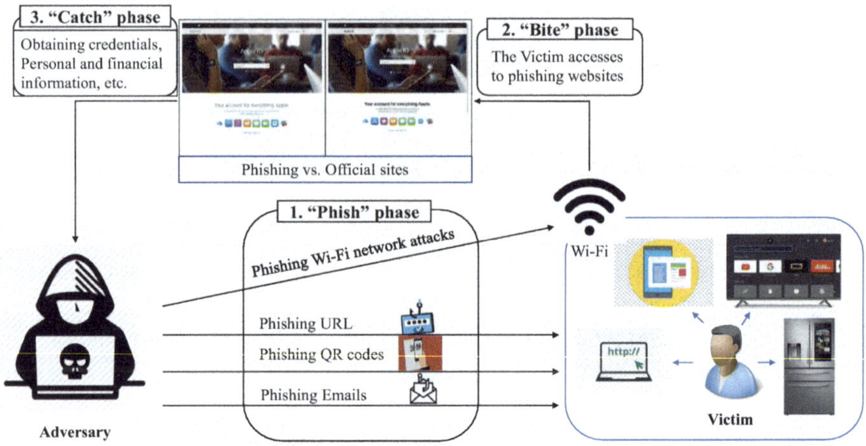

Fig. 2 Feasible phishing scenarios in IoT environments

3. Catch: There can be many reasons why the user might be convinced that the web page is legitimate, such as visual similarity, content similarity, etc. After getting convinced, if the user provides his credentials of the legitimate site, the phisher will be able to capture them. This is the "Catch" phase.

Adversaries create clone web pages to spoof users. They try to come up with new ways in sending the created phishing webpages to users and this is called the "Phish" phase. Due to the recent advances in web-access methods, they can adopt diverse ways to distribute phishing websites. To name a few, Hypertexts, QR codes, open unprotected Wi-Fi networks, etc., can be an option for a medium that can be opened from users' smartphones, IoT devices, and any other types of devices that are equipped with a web browser. As per the Phishing statistics survey, about 6.95 million phishing pages were created in 2020.[3]

These days, free Wi-Fi is something that people can leverage at public places without considering the security consequences. However, the adversaries look for targets who access the open unprotected Wi-Fi networks. Password-protected Wi-Fi networks are usually secured by Wi-Fi Protected Access II (WPA2) protocol which leverages a very strong cryptographic hash function. However, if the Wi-Fi is tampered with or provided by an adversary, those security features can be easily disarmed. The adversary can add a clickbait that has a link to a phishing website or just redirect users' connection to the website.

The "Bite" phase in which a user is connected to a phishing website is more serious and pervasive in IoT environments. Participating devices are generally resourced constraint and battery-powered, and thus, executing an anti-phishing solution in the background might not be feasible. These days, a service provider provides more

[3] https://www.csoonline.com/article/3634869/top-cybersecurity-statistics-trends-and-facts.html (Last Accessed: January 25, 2022).

easier and convenient ways to access the application, such as URL integrated photos and texts, QR codes, etc., which eventually makes it users more difficult to realize which website they are connecting to.

There are many reasons why the user might be convinced that the phishing web page is legitimate, such as visual and content similarities, etc. Once a user is convinced, he/she will doubtlessly reveal a login credential, credit information, and/or personal information. This is called the "Catch" phase which the adversary collects valuable data for monetary profits and where the user becomes a victim. According to the article published in one of the cyber Magazines, about 5,200 phishing attacks occurred on IoT devices every month in 2021.[4] These attacks have the damaging effect severe on financial loss and leakage of personal information to users and productivity, reputation, and the loss of data to the business.[5] Hence, phishing webpage detection is still of utmost importance in the era of the wireless connected world to prevent users using wireless browsers from being a victim of phishing attacks.

3 Search Engine Based Approaches

Analysis of results retrieved by querying the URL of the web page on popular search engines is the key idea of Search Engine based approaches. The classification of web pages is made based on the results given by the search engine. Search engines usually return results based on web page rankings. Legitimate web pages in general will have the highest rankings. Also, genuine websites have a longer life span. Unlike phishing pages, the index values grow with time along with the public reputation of the web page. The search engines, when queried with a URL, return the most indexed and reputed values as results. This makes it easier for the developer to develop an application that queries the URL entered on any browser and compares the results. This approach can take a longer time for the algorithm to run as they are dependent on results generated by third-party services (Fig. 3).

Effectiveness of the detection of zero-day phishing web pages is one of the major advantages of search engine based approaches. This approach is easy to implement and deploy the application with feasibility and is a great algorithmic approach for detecting phishing on highly-targeted web pages. It works very well when legitimate web page owners get indexing and ranking by search engines like Google, Alexa, etc. By using a crawler in the algorithm, attacks in the wild can also be identified by this approach [4].

Search-based algorithms can be implemented using certain packages and methods with Python. The algorithm takes URL as input and gives web page classification as output. The overall procedure is as follows:

---

[4] https://securityboulevard.com/2021/09/cyber-threats-haunting-iot-devices-in-2021/ (Last Accessed: January 25, 2022).

[5] https://www.stage2data.com/what-damage-can-phishing-cause-to-your-business/ (Last Accessed: January 25, 2022).

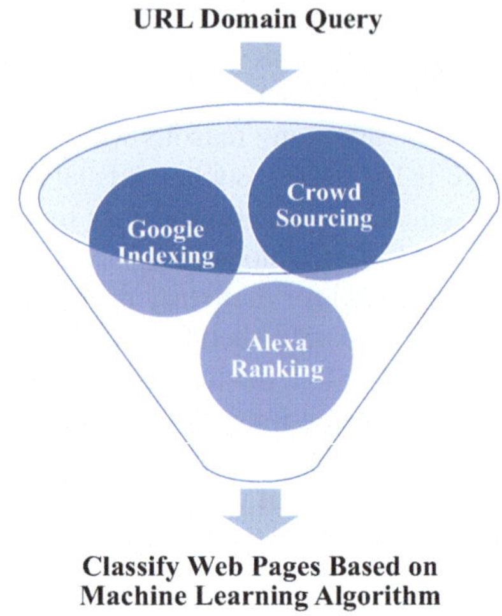

Fig. 3 Search engine based approach

1. URL is given as input to the algorithm
2. Domain name of the web page can be extracted by parsing the URL. Title of the web page can be extracted using BeautifulSoup package and OPenURL.title method.
3. A query can be generated with the domain and title of the web page.
4. Query crawling can be done by using any of the search engines like Google or Alexa. Crowd sourcing can also be applied in this step.
5. Store the first "K" results to an array. The value of K can be considered as a threshold. This can be decided based on the training cases and examples used. Usually, assigning it to 3 will be the ideal scenario.
6. Compare the Input URL with the stored URL array. If the Input URL matches to one of the elements of the URL array, then the web page is considered as legitimate. On the contrary, it is considered as phishing based on the ranking feature.
7. The class label for the web page is returned as output.

There are a few outliers for this approach. The legitimate web pages that are launched recently and do not have good domain age or the rankings will be marked as phishing. This can be overlooked as search engines usually update their indexes and ranking every 5 d. The ranking results of legitimate web pages usually maintain a trend from their launch until they reach their optimal ranking. The phishing pages that are fooling the users for a month might have ranking and index values. There is a chance that the system will not be able to discover them as the hacker might try a domain name very similar to legitimate web pages.

3.1 Crowd Sourcing

Crowdsourcing is a technique that uses public-reported phishing datasets. A few crowdsourcing phishing web pages are OpenPhish, PhishTank, etc. that take input from other peers to update the public datasets. The reports from the users are validated before entering them as records of the dataset. web page URL's present in the datasets are retrieved when a URL is queried in the search engines. Querying the search engine with the input URL might generate the top results as one of these crowdsourcing pages, which means that the web page is already marked as phishing.

Although crowdsourcing is a very effective method to detect the phishing web pages detected previously, some pages might not be marked as phishing on those pages. New pages keep emerging frequently making the users prone to zero-day phishing attacks. These attacks can be prevented using Google indexing or Alexa Ranking.

3.2 Google Indexing

The search engine-based approach that uses Google as the search engine is the Google indexing approach. The Google database has many legitimate sites verified. The web pages verified are given an index value. Querying this database with a URL or any keywords in the domain name returns the legitimate pages and their index values. If the web page is not retrieved by Google index, it means the web page is quite new, and it is not yet verified by Google. There is a high probability that the unverified web pages are zero-day phishing web pages.

Crawling is an approach used by the Google search engine to verify the website's trustworthiness and then index them. Search engines usually crawl and cache many new websites per second in the wild. When a URL is crawled, Google will be searching for the most relevant content by trying to parse the URL. The search results are ordered by a factor of relevancy. Usually, the web pages displayed at the top of the search query results are given the highest indexing. The index value plays a major role in classifying the web page as legitimate or phishing. Google indexing runs based on the "Web of trust." It relies on the ratings given by the users. Each of the URLs can be searched on Google for its index value to determine its trustworthiness.

3.3 Alexa Ranking

The search engine-based approach that uses Alexa as a search engine is the Alexa ranking approach. Alexa usually prefers to rank web pages only when a few metrics are being satisfied. For example, Alexa takes user-friendliness and relevance as some metrics to rank the web pages. Web site traffic is one of the important metrics Alexa uses to rank a web page. The web pages with high traffic will have high rankings. The interesting feature about Alexa ranking is that they only rank web page domains but not sub domains. The domain name can be parsed from the URL and each domain can be searched for Alexa ranking. When using only Alexa ranking, the web pages having high traffic will show up as top-ranked web pages. When a URL is queried using Alexa, the top-ranked results returned can be compared with the URL and a decision on the classification label can be made by the algorithm.

4 List Based Approaches

The key idea of List-based approaches is searching for the URL presence in Blacklists and Whitelists to classify web pages as legitimate or phishing. Input URL is compared with both Blacklist and Whitelist datasets for verifying its presence. Blacklists contain the phishing URL datasets, and Whitelists contain the legitimate URL datasets.

List-based approaches can be very advantageous in the scenarios where the adversaries maintain the same URL but change the IP address and service location. The phishing web pages that have a good lifespan and are blacklisted can be easily detected by this approach (Fig. 4).

A few outliers of List-based approaches are the web pages not listed on the Whitelist and Blacklist databases. Most of the zero-day phishing web pages have a very short lifespan indicating a high probability of blacklist missing a lot of phishing pages. The problem arises when a phishing page not stored on the blacklist becomes active after being idle for a certain period. One solution is to update the blacklist frequently even though the web page is inactive.

Few other outliers of this approach are web pages that are not on the whitelist but are legitimate. Newly emerging companies hosting their applications on recently purchased domains are one of the examples. As these domains have a less life span, they might not be on legitimate dataset lists. The problem arises when the algorithm compares the web pages with a legitimate dataset and classifies them as phishing web pages. Updating the whitelists frequently can be a solution to this problem.

List-based approach algorithms can be implemented by the following steps and using a few methods in python libraries. Along with the URL, the algorithm has a

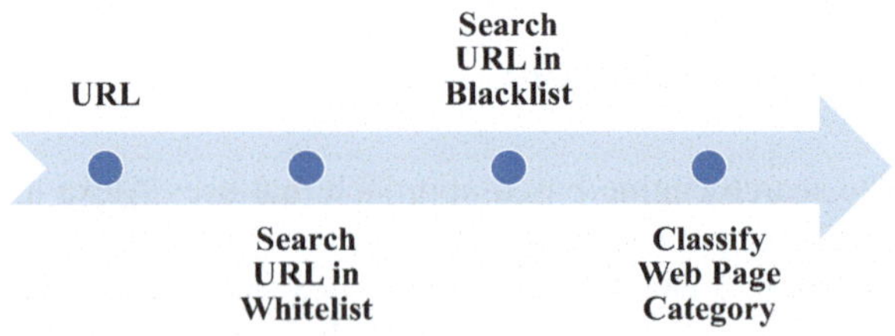

Fig. 4 List based approach

whitelist and blacklist dataset as input. The output would be the classification label for the web pages.

1. URL is given as input to the algorithm.
2. Read the Whitelist Dataset and the Blacklist Dataset.
3. Retrieve IP address and domain name for the URL. These can be retrieved using dns.resolver.query method using python.
4. Initialize a query with both IP and Domain name.
5. Search Blacklist and Whitelist databases for either Domain name or IP.
6. If the query returns a positive existence in the whitelist, the web page can be considered as legitimate, and if the query returns positive existence in blacklist, the web page can be considered phishing.
7. The class label for the web page can be returned as output.

4.1 Whitelist

A whitelist is a database of all legitimate web page URLs, Internet Protocol (IP) addresses, Domain Name servers (DNS), Internet Service Provider locations, etc. The Whitelisting approach is an approach that seeks to detect the legitimate web pages. Whitelist datasets that perform a server-side validation rather than just checking for the SSL certificates for the benign nature of the web pages are reliable. Opting a reliable whitelist is a key to improving the web page classification accuracy of the algorithm. One issue that is unavoidable with this approach is the time frame taken for URL validation before it's added to the lists. If the timeframe is too long, there is a chance of misclassification of benign web pages as phishing. The web page hosts should validate the web pages to make sure they are not outliers in this approach.

4.2 Blacklist

A blacklist is a database on the web to store and update the detected phishing web pages active for a few days. The dataset is updated frequently in a particular period. It consists of domain names and IP addresses of all suspicious web pages. Usage of DNS-based blacklists is a promising way to get efficient results for the algorithm.

Blacklists usually hold the website URLs of phishing web pages, and parts of the URL names are stored in these lists. If the URL's key words are parsed and searched in the blacklist with a query, the phishing pages can be detected. Although the web page not found on the Blacklist is considered legitimate, the important factor that needs to be considered is effectiveness of the coverage of the list. The Blacklists can be downloaded and stored on the client's computer or can be searched on the web. Also, it is important to note to consider the quality of the blacklist. Usually the phishing web pages are short-lived, and most of the damage would be done before the phishing web page is flagged red. Therefore, another factor that needs to be considered is the timeframe before an actual URL is included in the blacklist.

5 Content Based Approaches

The key idea of content-based approaches is to consider the web page's content for classifying the web pages. The content of the web pages is scraped and downloaded from a URL. Web scraping is employed to ensure required information collection from the Internet [5]. The scraped information is analyzed to detect any malicious links included in the contents or scripts. With the analysis results, certain features of web page's content are collected and used for web page classification as legitimate or phishing. The features utilized rely on the content that is being hosted on the web page, which makes this approach a content-based approach. This approach is independent of using third-party services like search engines, lists, etc.

Content-based approaches are very useful in detecting phishing web pages that have basic functionalities and look like legitimate web pages. The functionalities include dynamic nature of legitimate web pages unlike phishing. Phishing web pages are created by scraping the HTML and XML codes of legitimate web applications where dumping the internal files is quite hard and can be done only by web developing professionals. The similarity between the contents of legitimate and phishing web pages are calculated and compared with a threshold value. This approach is very efficient in detecting the phishing web pages emerging on a particular target brand.

A few outliers of the content-based approaches are the phishing pages having similar content and functionalities as legitimate pages. There is a tendency that such web pages make use of spaces or special characters at the end of paragraphs. The usage of special characters at the end of sentences and paragraphs can be detected by similarity indexes. Also, the domain and SSL certificates might not match these phishing web pages. Other outliers would be highly dynamic web pages developed with good programming practices. These pages might be developed by a web-developing professional with complex coding techniques. These web pages are hard to be detected as phishing web pages (Fig. 5).

Content-based algorithms can be implemented by the following steps and employing certain methods. These algorithms take URL input and return web page classification as phishing or legitimate as their output.

Fig. 5 Content based approach

1. URL is given as input to algorithm.
2. Extract the source code of the web page by employing web-scraping methods. The urllib2.urlopen.read() is one web-scraping method using python.
3. Construct a segmentation set by parsing the contents scraped from the URL using source code parser. One of the examples of such methods is the HTML parser. The method that can be used to parse the source code will be requests.get(url, 'code.parser').
4. From each segment, count all the required code blocks to determine each feature used. List of features that can be used in the algorithm are stated in Tables 1 and 2. Set a standard rule matrix for each feature. For example, separating the scripting libraries for easy maintainability indicates 1, or it is a 0 value. The usage of dynamic contents on a web page is considered 1, and not using them is considered 0. The 1 and 0 in the tables can be used to set rule matrix.
5. Set a general threshold value each page has to meet to be legitimate based on the number of features.
6. Define a page threshold variable and initialize its value to zero.
7. For each segment, determine if each web page follows good programming practices by evaluating each feature.
8. If the web page follows a rule from the rule matrix, increment the page threshold by 1.
9. Finally, if the page threshold value exceeds the general threshold value set then the web page is considered as legitimate. If not, the web page can be considered phishing.
10. The classification results are returned as output of the algorithm.

5.1 Source Code Analysis

In source code analysis, the features that relate to the source code to classify web pages are gathered for analysis and fed into the machine learning algorithms to build a classification model. This technique depends on the basic fact that the contents of the source code are different from the phishing and the legitimate web pages. The features can be extracted from the scripting library tags used in the source code. The legitimate web pages usually do not follow extreme and good programming practices as their main goal is to catch user information and disappear. Developing a full-fledged web page takes a lot of time and coding skills. The legitimate web pages usually have a lot of code libraries used to implement some functionalities like dynamic contents, cookies, cache controls, etc.

The front-end source code utilized to host the web page can be extracted and analyzed by parsing it using machine learning techniques [6]. The keywords, misspellings, punctuation marks, reference links, etc. are separated from the parsed information. Features extracted from the source code used in web page classification

Table 1 Features extracted from source code for phishing detection

Feature	Presence of feature	Absence of feature
Server and cache controls [7]	Legitimate	Phishing
Lot of page events [8]	Phishing	Legitimate
Abnormalities found in script content [9]	Phishing	Legitimate
Lot of I-frames [10]	Phishing	Legitimate
Set replacement text of I-frames to None [10]	Phishing	Legitimate
Lot of redirect links [8]	Phishing	Legitimate
Number of scripting libraries used [11]	Legitimate	Phishing
Less functionality for get and post requests [12]	Phishing	Legitimate

are listed in Table 1. The table explains the impact of presence and absence of the certain feature on the web page. This information can be used to set rules in a rule matrix.

Each of the features that can be considered from the analysis of the source code can be set with certain rules with a logic behind each rule that can be employed in web page classification.

Usual source code characteristics observed on legitimate web pages are:

- Providing cache controls and cookies to the browser to prevent the browser from sending multiple requests within a time frame. These features can be extracted from the HTTP headers.
- User actions or page events are specified only when needed.
- Observation of good coding practices.
- The pages do not redirect to some phishing links when the mouse hovers over blank spaces.

Unlike legitimate web pages, phishing web page source code characteristics are as follows:

- Does not have cookie and cache controls.
- User actions are specified when they are not really required.
- Good coding practices will not be observed as adversaries tend to dump code from the legitimate target web page applications.
- The pages allow many redirection links when the mouse hovers over blank spaces.
- Display content in the Iframes are specified as None to hide the abnormal texts and links embedded. Any visibility of Java script codes within an Iframe is also normal. Presence of a lot of Iframes on the web page is normal in phishing web pages.
- Misspellings in web page texts. These can be detected by employing text-mining techniques.
- Presence of redirection link that does not match domain or subdomain names.

- Libraries that generate a lot of JavaScript popups.
- Post requests sending the data to server and fetching no other major functionalities. For e.g.: there can be a login form or a payment form that posts the requests, but users end up landing on the same page. The adversary will be able to get the credentials though.

5.2 Code Complexity Analysis

Most of the legitimate websites follow good programming practices to secure their web pages from client-side attacks like CSRF, XSS, etc. They also use security techniques that prevent hackers from scraping the web page contents. Employing such coding practices ensures that the adversary cannot scrape the web page contents and dump them into the phishing web site. Also, security techniques that prevent web page sniffing can be observed on a legitimate web page, which prevents others from checking for web page user traffic. User activities are usually monitored. In addition to all these techniques, the programmers for legitimate web pages would be professional and will be incorporating good coding practices by separating HTML codes from Java scripts unlike phishing web pages. This improves maintainability of the web pages. Unlike legitimate web pages, phishing web pages' main intent would usually be considering the credential stealing from the users. They do not need to incorporate the user activity tracking codes for this. So, a HTML page with similar content can do their job instead of complex code structures.

The code complexity analysis approach mainly focuses on the fact that adversaries try to recreate the targeted pages with fewer functionalities as they might not be highly skilled or professional web developers. The machine learning algorithm can make use of code complexity features to determine if the web page is phishing. A few of those features are listed in Table 2.

Usually, phishing web pages tend to have very low inline block counts contrary to legitimate pages. Legitimate web pages usually tend to have more of the external blocks in the code. Most of the legitimate web applications try to implement dynamic changes without refreshing the page for every single user action, unlike phishing pages. The higher the count of total decisions that need to be taken dynamically,

Table 2 Features extracted from code complexity analysis

Feature	Presence of feature	Absence of feature
Less inline blocks of code [13]	Phishing	Legitimate
Less external blocks of code [13]	Phishing	Legitimate
Less landing page variants [13]	Phishing	Legitimate
Lot of cyclomatic code complexity [13]	Legitimate	Phishing
Lot of LOC in external blocks [13]	Legitimate	Phishing

the higher the cyclomatic complexity. Phishing pages tend to have less cyclomatic complexity. The number of lines of code other than the library in the external block speaks about the effort invested by the developer to build the web pages. Legitimate sites usually tend to have more lines and complexity in their hosting web pages.

6 Fuzzy Rule/Heuristics Based Approaches

Extraction of a set of features and imposing If-then rules to classify the web pages as phishing or legitimate is the key idea of fuzzy rules heuristic based approaches [14]. Although setting all the rules and gathering the member functions is a difficult task and takes longer time, this approach can be very helpful in processing the ambiguous variables. Fuzzy logic always has an interpretation that the possible logic is infinite. For applying the fuzzy logic systems, the input URL should be converted to fuzzy input and then specify a member function. This function will have the meanings of the terms applied in the rules defined.

A heuristics-based approach is very efficient in the detection of phishing web pages that do not follow all the heuristics or the defined set of rules. Also, the approach is very effective in the detection of phishing web pages that try to dump the source code and content from the legitimate pages of target brands.

A few outliers of the heuristics-based approach are the web pages that might have good traffic at some points when users are at "bite" phase. Also, the web pages that have changed their domain name recently have a high probability of getting classified as phishing (Fig. 6).

Fuzzy/Heuristics based algorithms can be created by the following steps and implementing a few if-then rules. The algorithm takes URL as its input and gives web page classification label as output.

1. Take Input URL.
2. Extract features from URL. This can be done by using URL parser. The features that can be extracted form URL are mentioned in Table 3.
3. Open SSL certificates and extract features from SSL certificates. The features from SSL certificates can be extracted using $OpenSSL.crypto$ library and $cert.getcomponents$ methods with python. The features that can be extracted for analysis purposes are mentioned in Table 5.
4. Extract network layer features. This can be done by using $driver.get_log$ method with python. The components that can be considered as features are being described in Table 4.
5. Extract domain features. This can be done by using $gettld$, $.fld$, $.subdomain$, $.parsedurl$ methods with python. Features that can be considered are given in Table 6.
6. Calculate the values of heuristics by constructing the if-then rules from all the features.
7. Calculate fuzzy values from the heuristics using membership functions.

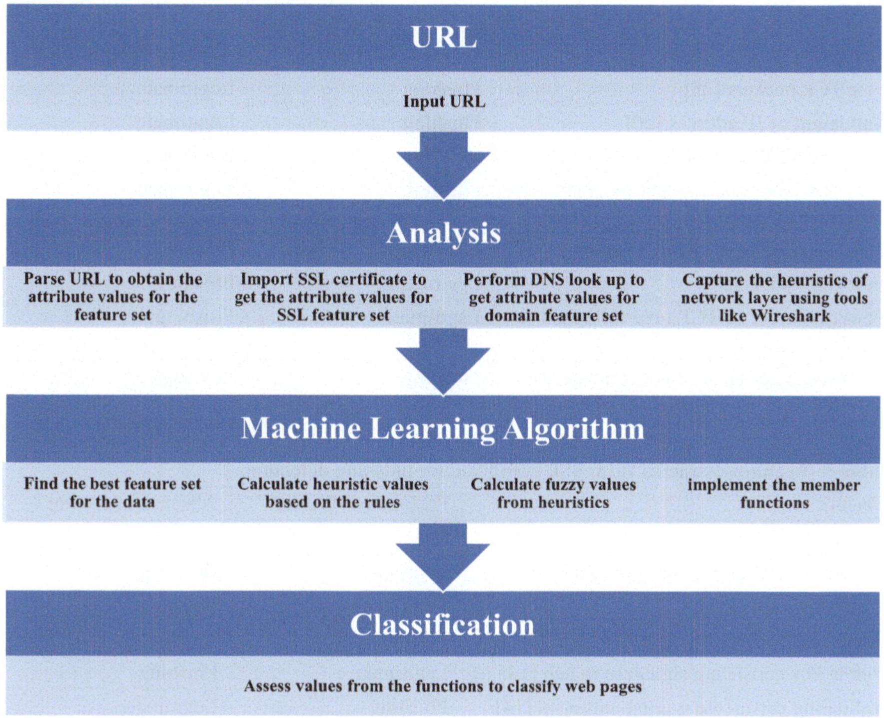

Fig. 6 Fuzzy Rule/ Heuristics Based Approaches

Table 3 Features extracted from URL for phishing detection

Feature	Presence of feature	Absence of feature
Long length of URL [15]	Phishing	Legitimate
Protocol using HTTPS [16]	Legitimate	Phishing
Lot of special characters [17]	Phishing	Legitimate
Lot of subdomains [18]	Phishing	Legitimate
Usage of URL tokens [19]	Legitimate	Phishing
URL redirection [20]	Phishing	Legitimate
High frequency of entropy values [21]	Phishing	Legitimate
Security sensitive keywords presence in URL [22]	Phishing	Legitimate
Domain name matches brand domain [23]	Legitimate	Phishing
Encoded host name [24]	Phishing	Legitimate

Table 4 Features extracted from network layer for phishing detection

Feature	Presence of feature	Absence of feature
Few TCP packets [25]	Phishing	Legitimate
Validation of IP address [26]	Phishing	Legitimate
Few remote IP addresses [27]	Phishing	Legitimate
Very few DNS queries sent by crawler for DNS record [28]	Phishing	Legitimate
Low response time of DNS server	Legitimate	Phishing
Few UDP packets	Legitimate	Phishing
Fewer Number of TCP urgent flagged packets	Legitimate	Phishing
Lower average local/remote packet rate	Phishing	Legitimate

Table 5 Features extracted from SSL certificate for phishing detection

Feature	Presence of feature	Absence of feature
SSL certificate [29]	Legitimate	Phishing
Certificate owner matches domain [30]	Legitimate	Phishing
Authority name [31]	Legitimate	Phishing
Abnormal cookie presence [16]	Phishing	Legitimate
Public key certificate signature match [14]	Legitimate	Phishing
Validating certificate issuing authority [14]	Phishing	Legitimate

Table 6 Features extracted from domain of web page for phishing detection

Feature	Presence of feature	Absence of feature
DNS server domain name match [32]	Legitimate	Phishing
Availability of WHOIS registration info [33]	Legitimate	Phishing
Frequent domain name update [34]	Phishing	Legitimate
IP prefix out of range [35]	Phishing	Legitimate
IP belongs to different country [36]	Phishing	Legitimate
Presence of PTR record	Legitimate	Phishing

8. Set a general heuristic threshold for classifying the web pages based on number of if-then rules.
9. define a page heuristics threshold variable and initiate it with a 0 value.
10. Assess the results corresponding to each of the rule-based degrees for the input URL.

11. Based on the if-then rules increment, the page heuristics variable when the web page satisfies a condition to be legitimate.
12. Implement a defuzzification process to classify the web pages.
13. If the page heuristic variable is less than the set general threshold, the web page is considered phishing. If not, the web page is considered legitimate.
14. Return the classification labels as the algorithm output.

6.1 URL Analysis

The goal of an adversary is to make sure the user is redirected to the web page without his knowledge. There could be many ways to trick users into visiting the malicious links. One among those techniques is to embed the URL in the email text or insert a valid URL link into one of the malicious links. Some phishing web pages have a replaced domain name by adding a special character to it, and this technique is called Homograph spoofing. Misspellings introduced in a domain name are called typo squatting. Some pages use domain names that sound like legitimate target web pages, and this technique is called sound squatting. One other important technique used by the adversaries is combo squatting. This implies they add meaningful structure in the URL where a user can easily be tricked (For e.g.: Facebook-support.com). This is very misleading as the user might think it is a real Facebook support page. Although, it is a phishing web page, by careful analysis of the URL, the malicious links can be detected. The users that do not check for domain names, spellings, meanings etc. can be easily tricked with such techniques.

URL is unique for a web page and multiple web pages cannot be hosted on one URL. So, we can take advantage of the structural properties of the URL to classify if the web page is legitimate or phishing. The structure of the URL would be protocol followed by subdomain, domain name, top level domain, path, and anchor. Each part of the URL has a meaningful significance, and properties that can be leveraged. The features extracted from URL are listed and described in Table 3.

Lengthy URLs generally tend to be phishing pages. The if-then rule can be formulated in such a way that if the length exceeds more than a certain threshold of characters, then doubt phishing. By analyzing the protocol of a URL, evidence of HTTPS usage can be collected. There is a tendency that if the web page is not hosted on the secure communication layer (HTTPS), the web page has a higher probability of being a phishing page. Generally, the URL with more than 2 or 3 reserved characters can be doubted as phishing. Some URLs might have a lot of special characters when they use foreign languages to use character encoding. Therefore, this feature can be considered but not taken as a base to classify the web pages. The structure of the URL can allocate a top-level domain, domain name, and subdomains. The subdomains within a domain name can be anything meaningful. It can be either specification of language or category of the domain. The presence of more than two subdomains for one page in the URL is a thinker. Legitimate sites tend to organize their URL structure meaningfully. Basic structure of URL has information regarding the subdomains.

URL token is a parameter used when a web server would like to identify the user and communicate some sensitive details securely. The bag of words approach can be used to identify tokens in URL. Adversaries tend to obtain details entered in the input fields with a high probability of having no prior knowledge of web development. Use of secure communication channels and tokens is not usually observed in phishing web pages. URL redirection is done by embedding redirection instructions in URL. The legitimate web pages do not use redirection unless there is a category the user specifically wants them to be redirected. So, presence of URL redirection can be considered phishing web pages.

For legitimate web pages, the entropy value frequency is high in a particular range. Beyond that range, if the web pages have entropy values, the pages can be considered phishing. In other words, legitimate web pages have lower scores for their domains. So, the frequency of the characters can be considered to take the overlap in the dataset. The sensitive information from the users should not be passed through URL using the parameters. Confidentiality of user data is ensured by legitimate web pages. So, the web pages with user information in URL can be considered phishing. Presence of a brand name does not indicate the page is legitimate. By making sure the brand name is the same as a domain name and exactly matches, the web page can be considered legitimate. If not, a web page can be considered phishing without any second thoughts. The legitimate pages generally tend to use a host name directly instead of encoding it in the URL. The pages having encoded host names can be considered phishing.

6.2 Network Layer Analysis

Any web page that is hosted on a server will have a network layer for establishing a communication between client and server ends. The network layer allows exchange of the data packets between the connected ends and decides the physical path used for data transmission. There are many details available on this layer about the source IP and the destination IP etc. The adversary can intrude into the network path to capture the packets and reroute them for sensitive information in the packets. To prevent an adversary from capturing the packets, the communication needs to be on a secure layer and encrypted.

There are many features in network layer that are considered crucial for phishing web page classification. For example, the number of packets transmitted from the IP can also be seen, which indicates the traffic to the web page. The more the users the web page has, the higher is the traffic. Different features from the network layer are analyzed that help to classify the web pages as legitimate or phishing are listed and described in Table 4.

Transmission control protocol (TCP) packets have every little detail about the sender IP address, user details, etc. It is a communication handshake that establishes connection before data transfer. The number of packets indicates the site traffic. If a browser sends a request to the server or a user performs any action on the web page, a TCP packet is generated. So, the higher the web page traffic, the fewer chances that the web page is phishing. If TCP packets are very low, then the web page might be considered as phishing. The IP address of the server from which web page is hosted is detected using NS lookup command. The IP address can be checked for the validity using Strtok() function and a few if then rules. The pages that do not get validated can be considered as phishing web pages. The number of remote IP addresses connected to the server indicates the traffic to the server. Legitimate web pages have lot of traffic. So, the connected remote IP addresses will be more for legitimate web pages. Phishing pages might not have remote IP addresses. The user with less knowledge on URLs will be searching them on Google. This step will make the crawler to search for the correct domain name. The legitimate web pages and target brands have a lot of DNS query requests from the crawler compared to phishing web pages.

DNS servers tend to send query results for legitimate web pages sooner compared to phishing web pages. Legitimate pages often searched can be usually retrieved sooner. Hence, response time for DNS server acts as an interesting indicator for doubting phishing. UDP packets consist of source and destination ports. These UDP packets do not need a connection: it broadcasts the data. The number of UDP packets should always be less. Generally, video streaming and gaming is transmitted using UDP since it is faster than TCP. On the contrast, many legitimate web pages tend to use TCP, hence why the number of UDP packets for a web page should always be less. The urgent flagged packets should be very less compared to other packets. If the ratio is greater than half, then the web page can be doubted as phishing. The urgency in the flagged packets indicate the sensitive information from or to a user. Usually, phishing pages tend to collect sensitive information from users. Legitimate pages tend to take the information and collect them. Local packet rate average should be high for the legitimate pages. Phishing pages do not tend to have high local packets or remote packets.

6.3 SSL Certificate Analysis

A SSL certificate is given for each web page to make the communications in a secured way, i.e. web pages on HTTPS ports have SSL certificates. The SSL certificate makes the TLS encryption possible, which will be very helpful for data transmission on a secured channel. Each of the web page should have this SSL certificate to ensure high security for the data transmission between the host and the user. If the data is transmitted without using a secured layer, there is a chance of stealing the information by the hackers. Most of the legitimate pages host their sites using SSL certificates. Also, the web page hosts should ensure that their web pages are hosted on a secure channel whenever the user is asked for user credentials or payment gateways or any other input fields. The data from each user is important to be secured. Companies hosting the legitimate web pages these days opt for SSL communication unless the web page has a pdf or policies or standard rules and is just published on the web without any backend functionalities.. Sometimes, the user should be very careful while downloading the files that are not hosted on the HTTP as there can be malware attached by adversary to it. When a user downloads such files, they might be the victims.

It is very easy for the adversary to obtain fake SSL certificates. So, SSL certificate presence alone is not sufficient to classify the web page as legitimate. The certificate has certain features that can be analyzed to verify its authenticity. Analyzed SSL certificate features are listed and described in Table 5.

Most of the web pages that do not have an SSL certificate are phishing. The Presence of an SSL certificate is key these days if users would like to transfer sensitive information to the web page host. SSL certificate makes sure the communication is done using secure layers. If the certificates "issued to" column, or the owner's name does not match with the domain name, then it is a phishing page. Usually, legitimate web pages make sure the organization's name matches the name of the certificate owner, organization name, and domain name. Phishing pages that target the brands can easily be detected by matching authority names and domain names.

Legitimate web pages never have abnormal cookies. Abnormal cookies refer to cookies that do not expire. Any cookie should act as per the standard guidelines determined by OWASP security organization.[6] If any cookie acts abnormal and stores sensitive data of customers when they are not on the web page, then it is a phishing web page. If the signature in a public key certificate matches the domain and senders name then the web page can be considered Legitimate. On the contrary, if there is a mismatch in validation, the web page can be considered as phishing web page. The authority that issued the certificate is also important. There are many fake certificate using companies. The trust certificate issuing authority can be verified. If the issuer is not in the top 'n' range, then the web page can be doubted as a phishing page.

6.4 Domain Analysis

Web page domain is an address the user can search for over the Internet. Domain names need to be selected in such a way that it is traceable by the users. Each time a user visits the web page, a Domain Name System(DNS) lookup is being performed. Also, each domain can have a lot of sub domains arranged in a hierarchical order based on the company structure and departments. Feature-oriented sub domains indicate the domain names that come under a domain and arranged in a systematic manner. This helps in identification of domains that are not phishing. These sub domains are usually customized by the company in a meaningful manner. When a host would like to change the domain, the organization should follow SEO practices to make

[6] https://owasp.org/ (Last Accessed: January 25, 2022).

sure there is no loss of traffic. Analyzed features associated with domain name for classifying web pages are listed and described in Table 6.

The DNS server name should be matched with the domain of the web page. Usually, if the DNS server names do not match the web page domain, the web page can be doubted as phishing. The brands that are most targeted will have the same domain as the organization name. The web pages whose information regarding the WHOIS registration is available can be considered as legitimate web pages. For phishing pages, this information will be missing. The definition of registration date should have the same age as the domain. The domain of the web page should always be almost as old as the organization. Domain updates are made when the settings of the domain need to be changed for a web page. Domain names change in rear cases, like when the business name changes. So, usually, change in domain name is rarely expected. So, frequent changes in domain names imply the web page is phishing. Legitimate pages usually do not change their domain names.

There are two IPv allocations available at this point, 4 and 6. This is also known as subnet mask. These values have ranges and formats. If the IP prefix is out of range, then the web page can be considered as phishing. Certain ranges or set of IP addresses belong to certain regions. If the IP address of the domain does not belong to the region in which the IP or the device is present, then doubt phishing for the web page. The algorithm warns users not to go ahead unless they know they are on the legitimate site that is hosted out of the region. For legitimate web pages, presence of PTR record is observed. Phishing pages usually do not maintain PTR records as they perform a reverse DNS lookup. This is one of the crucial features that can be checked for a web page to classify the web pages.

7 Visual Similarity Based Approaches

Adversaries try to recreate web pages similar to the target brands to scam the users and steal their confidential information. Small technical and minute details might not be considered by the adversaries, like logo similarity, when recreating the web pages. Algorithm can take advantage of these flaws to classify the web pages. A few examples of such flaws could be logos and image dimensions.

As soon as the URL is entered in a web browser, multiple screenshots of different regions of the web pages can be captured and compared with original target brand datasets to get the similarity indexes. The contours and the colors can be matched for similarity in the web pages. Hashes can be generated and matched from the screenshots.

A visual similarity-based approach uses screenshots of web pages to compare the similarity and classify the web pages. There are a lot of advantages to such approaches. Phishing web pages that have same structure and content but different URL or text snippets can be effectively identified. Phishing web pages that have the same content but different dimensions of logos or text than original web pages can be effectively captured. Phishing web pages with similar content but logos and images

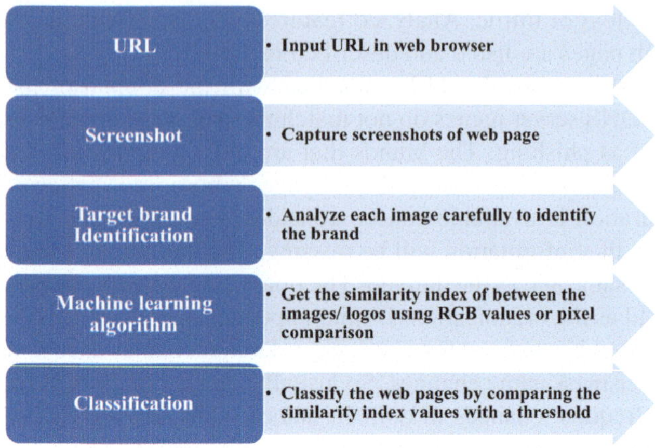

Fig. 7 Visual-Similarity based Approaches

downloaded from the Internet and have different color heuristics can be effectively detected.

A few outliers would be the web pages that dump the scraped source code from the original web page. These kind of phishing pages look like the original web pages. The web pages that use the favicons and logos from the original page without a lot of change in the dimensions. Detection of such phishing web pages, which have similar visual images, cannot be easily.

Visual similarity based approach algorithms can be implemented by the following steps. Each algorithm can take URL and a target brand data set and returns web page classification as output.

1. URL of the web page is given as input for algorithm.
2. Maintain a target brand web page screenshots dataset and read it.
3. Enter URL in the browser and capture the screenshot data for the web page.
4. Capture sub images of the web page. Cropped images for the logo, URL, titles, snippets and favicon can be captured.
5. Identify the target-brand by searching the target brand datasets for the similar logo.
6. The page similarity index between the identified target-brand and the captured images are calculated. The features can be extracted from Tables 7 and 8.
7. Set a general similarity threshold index for the web page to be classified as legitimate.
8. If the page similarity index is more than the set threshold value, then the web page is considered legitimate. If not, the page is considered phishing.
9. The web page classification label is returned as output for the algorithm.

Table 7 Features extracted from screenshot images of web page for phishing detection

Feature	Presence of feature	Absence of feature
The similarity of web page screenshot [37]	Legitimate	Phishing
The similarity in URL screenshot [38]	Legitimate	Phishing
Snapshot similarity of title and snippet image of text above URL [39]	Legitimate	Phishing

Table 8 Features extracted from logos for phishing detection

Feature	Presence of feature	Absence of feature
Embedded logo size variations [40]	Phishing	Legitimate
Favicon usage detection and similarity [41]	Legitimate	Phishing

7.1 Image Analysis

The screenshot of the complete website has a lot of information. Each of the parts of the web page can be captured as sub images. Each of the sub images has its own importance. The URL sub image of the web page can be taken as a screenshot and compared for accuracy in similarity with the target brand. The title and snippet are other features or sub images that can be considered. A profile can be created for each of the web pages that can store all the features concerned to the targeted brands. Each of the features can be compared to the identified target for the similarity index.

This approach deals with images and similarity and distance between the images, so the optimization techniques for the algorithm comes into picture. If proper optimization techniques are not used, each URL can take up a very long time to classify the web pages.

This approach takes snapshots of images and compares them with original web page screenshots that are already stored for target brands. A few of these features are listed and described in Table 7.

The URL is entered in a web browser and screenshots are captured. The screenshots of the web pages are compared to the screenshots of the benign web pages of the top brands. If the screenshots match, the web page is considered benign. If the similarity index is not greater than the similarity-threshold considered, then it is considered a phishing page. The similarity index of the web page is calculated using methods like gray scale pixel comparison, RGB values comparison etc.

The similarity in the URL can be detected by comparing the URL sub image with the legitimate web pages. Each URL that can be identified as similar with one of the pages in the benign dataset is considered as legitimate. On the other hand, if the similarity index of the web page is less than considered similarity threshold, it is considered a phishing web page. The titles and snippet screenshot of the web page is another sub image or feature and is compared to original benign web pages, often for phishing web pages. The similarity index is considered, and if the threshold exceeds the value, the web page is considered as legitimate. Often, the snippet dimensions of phishing web pages will not be the same.

7.2 Logo Analysis

Logo recognition is one of the object recognition tasks. Invariant features can scale for efficient phishing web page detection. Many adversaries might try to utilize logos downloaded from the Internet with minute differences. These differences can be identified when scaling techniques are used for logo analysis. The logos that do have and do not have the exact similarity can be detected with logo analysis techniques.

The logo of the page is usually seen by every user. Although the adversaries try to use the logo with similar dimensions as legitimate web pages, exact match as the legitimate pages is not possible. The original logos usually cannot be downloaded from the legitimate web pages and hence the logos used on the phishing pages might be downloaded from the Internet and embedded into the site. The logos can be compared for the dimensions, contour, and color match with the identified target brand to detect the phishing web pages. Logo analysis approach takes the screenshot of web pages and creates sub images of features from the screenshots to compare with legitimate web pages screenshot datasets. Few of these features are being listed and described below in Table 8.

The logo on the web page can be considered as a feature to classify the web pages. The logo is taken as a sub image screenshot and compared with the original logo. If the similarity index match, then the web page is considered benign. Legitimate web pages generally tend to use a favicon. Favicon of a web page can be captured as a sub image screenshot and compared to the complete benign dataset. If the dimensions and the RGB values completely match with no filters then the web page can be considered benign. If the generated favicon is not consistent with the domain, then the web page can be considered phishing.

8 Conclusion

Phishing web pages, fake but look like legitimate pages, can be critical to web-based IoT applications as well. Due to its diversity and variety, there is no perfect approach to detect all kinds of phishing web pages. Each approach has its own advantages and limitations, and thus, an adequate choice of phishing page features and technologies plays a vital role in order to develop an effective detection system. This chapter introduced and discussed state-of-the-art anti-phishing technologies that can be utilized in the era of the IoT, which of them are categorized into five classes in terms of similarity, such as (a) search engine-based, (b) list-based, (c) content-based, (d) fuzzy rule/heuristic-based, and (e) visual similarity-based approaches.

The results will give inspiration to researchers in this field so that they can think of a direction about how the features and technologies can be combined to obtain a better performance.

Acknowledgements This research was supported by the MSIT (Ministry of Science, ICT), Korea, under the High-Potential Individuals Global Training Program (2021-0-01547-001) supervised by the IITP (Institute for Information and Communications Technology Planning and Evaluation).

References

1. S. Sruthy, S.N. George, WiFi enabled home security surveillance system using raspberry pi and iot module, in *Proceedings of the 2017 IEEE International Conference on Signal Processing, Informatics, Communication and Energy Systems (SPICES)* (IEEE, 2017), pp. 1–6
2. A. Kumar, P. Sood, U. Gupta, Internet of things (IoT) for bank locker security system, in *Proceedings of the 6th International Conference on Signal Processing and Communication (ICSC)* (IEEE, 2020), pp. 315–318
3. J. Lau, B. Zimmerman, F. Schaub, Alexa, are you listening? privacy perceptions, concerns and privacy-seeking behaviors with smart speakers. Proc. ACM Hum.-Comput. Interact. **2**(CSCW) (2018)
4. C.-Y. Wu, C.-C. Kuo, C.-S. Yang, A phishing detection system based on machine learning, in *Proceedings of the International Conference on Intelligent Computing and its Emerging Applications (ICEA)* (IEEE, 2019), pp. 28–32
5. L. Barlow, G. Bendiab, S. Shiaeles, N. Savage, A novel approach to detect phishing attacks using binary visualisation and machine learning, in *Proceedings of the IEEE World Congress on Services (SERVICES)* (IEEE, 2020), pp. 177–182
6. C. Opara, B. Wei, Y. Chen, Htmlphish: enabling phishing web page detection by applying deep learning techniques on html analysis, in *Proceedings of the International Joint Conference on Neural Networks (IJCNN)* (IEEE, 2020), pp. 1–8
7. M. Chatterjee, A.-S. Namin, Detecting phishing websites through deep reinforcement learning, in *Proceedings of the 43rd IEEE Annual Computer Software and Applications Conference (COMPSAC)*, vol. 2 (IEEE, 2019), pp. 227–232
8. M.M. Vilas, K.P. Ghansham, S.P. Jaypralash, P. Shila, Detection of phishing website using machine learning approach, in *Proceedings of the 4th International Conference on Electrical, Electronics, Communication, Computer Technologies and Optimization Techniques (ICEECCOT)* (IEEE, 2019), pp. 384–389
9. F. Tajaddodianfar, J.W. Stokes, A. Gururajan, Texception: a character/word-level deep learning model for phishing url detection, in *Proceedings of the IEEE International Conference on Acoustics, Speech and Signal Processing (ICASSP)* (IEEE, 2020), pp. 2857–2861
10. S. Zaman, S.M.U. Deep, Z. Kawsar, M. Ashaduzzaman, A.I. Pritom, Phishing website detection using effective classifiers and feature selection techniques, in *Proceedings of the 2nd International Conference on Innovation in Engineering and Technology (ICIET)* (IEEE, 2019), pp. 1–6
11. A. Abuzuraiq, M. Alkasassbeh, M. Almseidin, Intelligent methods for accurately detecting phishing websites, in *Proceedings of the 11th International Conference on Information and Communication Systems (ICICS)* (IEEE, 2020), pp. 085–090
12. A. Alswailem, B. Alabdullah, N. Alrumayh, A. Alsedrani, Detecting phishing websites using machine learning, in *Proceedings of the 2nd International Conference on Computer Applications Information Security (ICCAIS)* (IEEE, 2019), pp. 1–6
13. A. Niakanlahiji, B.-T. Chu, E. Al-Shaer, PhishMon: a machine learning framework for detecting phishing webpages, in *Proceedings of the IEEE International Conference on Intelligence and Security Informatics (ISI)* (IEEE, 2018), pp. 220–225

14. S. Roopak, A.P. Vijayaraghavan, T. Thomas, On effectiveness of source code and SSL based features for phishing website detection, in *Proceedings of the 1st International Conference on Advanced Technologies in Intelligent Control, Environment, Computing Communication Engineering (ICATIECE)* (IEEE, 2019), pp. 172–175
15. J. Stobbs, B. Issac, S.M. Jacob, Phishing web page detection using optimised machine learning, in *Proceedings of the 19th IEEE International Conference on Trust, Security and Privacy in Computing and Communications (TrustCom)* (IEEE, 2020), pp. 483–490
16. Y. Sonmez, T. Tuncer, H. Gökal, E. Avcı, Phishing web sites features classification based on extreme learning machine, in *Proceedings of the 6th International Symposium on Digital Forensic and Security (ISDFS)* (IEEE, 2018), pp. 1–5
17. M. Korkmaz, E. Kocyigit, O.K. Sahingoz, B. Diri, Phishing web page detection using N-gram features extracted from URLs, in *Proceedings of the 3rd International Congress on Human-Computer Interaction, Optimization and Robotic Applications (HORA)* (IEEE, 2021), pp. 1–6
18. M.M. Yadollahi, F. Shoeleh, E. Serkani, A. Madani, H. Gharaee, An adaptive machine learning based approach for phishing detection using hybrid features, in *Proceedings of the 5th International Conference on Web Research (ICWR)* (IEEE, 2019), pp. 281–286
19. E.S. Gualberto, R.T. De Sousa, T.P. De Brito Vieira, J.P. Carvalho Lustosa Da Costa, C.G. Duque, The answer is in the text: Multi-stage methods for phishing detection based on feature engineering. IEEE Access **8**, 223529–223547 (2020)
20. M. Abutaha, M. Ababneh, K. Mahmoud, S. Al-Haj Baddar, URL phishing detection using machine learning techniques based on URLs lexical analysis, in *Proceedings of the 12th International Conference on Information and Communication Systems (ICICS)*, (IEEE, 2021), pp. 147–152
21. E.S. Aung, H. Yamana, URL-based phishing detection using the entropy of non-alphanumeric characters, in *Proceedings of the 21st International Conference on Information Integration and Web-based Applications & Services (iiWAS)* (iiWAS, 2019), pp. 385–392
22. J. Rashid, T. Mahmood, M.W. Nisar, T. Nazir, Phishing detection using machine learning technique, in *Proceedings of the 1st International Conference of Smart Systems and Emerging Technologies (SMARTTECH)* (IEEE, 2020), pp. 43–46
23. A.S.S.V. Lakshmi Pooja, M. Sridhar, Analysis of phishing website detection using CNN and bidirectional LSTM, in *Proceedings of the 4th International Conference on Electronics, Communication and Aerospace Technology (ICECA)* (IEEE, 2020), pp. 1620–1629
24. A. AlEroud, G. Karabatis, Bypassing detection of URL-based phishing attacks using generative adversarial deep neural networks, in *Proceedings of the 6th International Workshop on Security and Privacy Analytics (IWSPA)* (ACM, 2020), pp. 53–60
25. T. Chin, K. Xiong, H. Chengbin, Phishlimiter: A phishing detection and mitigation approach using software-defined networking. IEEE Access **6**, 42516–42531 (2018)
26. K. Gajera, M. Jangid, P. Mehta, J. Mittal, A novel approach to detect phishing attack using artificial neural networks combined with pharming detection, in *Proceedings of the 3rd International conference on Electronics, Communication and Aerospace Technology (ICECA)* (IEEE, 2019), pp. 196–200
27. E. Zhu, Y. Chen, C. Ye, X. Li, F. Liu, OFS-NN: an effective phishing websites detection model based on optimal feature selection and neural network. IEEE Access **7**, 73271–73284 (2019)
28. N. Megha, K.R. Remesh Babu, E. Sherly, An intelligent system for phishing attack detection and prevention, in *Proceedings of the International Conference on Communication and Electronics Systems (ICCES)* (IEEE, 2019), pp. 1577–1582
29. S. Sindhu, S.P. Patil, A. Sreevalsan, F. Rahman, M. Saritha, A. N. Phishing detection using random forest, SVM and neural network with backpropagation, in *Proceedings of the International Conference on Smart Technologies in Computing, Electrical and Electronics (ICSTCEE)* (IEEE, 2020), pp. 391–394
30. R. Almeida, C. Westphall, Heuristic phishing detection and URL checking methodology based on scraping and web crawling, in *Proceedings of the IEEE International Conference on Intelligence and Security Informatics (ISI)* (IEEE, 2020), pp. 1–6

31. Q. Cui, G.-V. Jourdan, G.V. Bochmann, I.-V. Onut, SemanticPhish: a semantic-based scanning system for early detection of phishing attacks, in *Proceedings of the APWG Symposium on Electronic Crime Research (eCrime)* (IEEE, 2020), pp. 1–12
32. H. Shirazi, B. Bezawada, I. Ray, Know thy domain name: Unbiased phishing detection using domain name based features, in *Proceedings of the 23rd ACM on Symposium on Access Control Models and Technologies (SACMAT)* (ACM, 2018), pp. 69–75
33. A.F. Nugraha, L. Rahman, Meta-algorithms for improving classification performance in the web-phishing detection process, in *Proceedings of the 4th International Conference on Information Technology, Information Systems and Electrical Engineering (ICITISEE)* (IEEE, 2019), pp. 271–275
34. S.-J. Bu, S.-B. Cho, Integrating deep learning with first-order logic programmed constraints for zero-day phishing attack detection, in *Proceedings of the IEEE International Conference on Acoustics, Speech and Signal Processing (ICASSP)* (IEEE, 2021), pp. 2685–2689
35. A. Cuzzocrea, F. Martinelli, F. Mercaldo, Applying machine learning techniques to detect and analyze web phishing attacks, in *Proceedings of the 20th International Conference on Information Integration and Web-based Applications & Services (iiWAS)* (ACM, 2018), pp. 355–359
36. J. Kumar, A. Santhanavijayan, B. Janet, B. Rajendran, B.S. Bindhumadhava, Phishing website classification and detection using machine learning, in *Proceedings of the International Conference on Computer Communication and Informatics (ICCCI)* (IEEE, 2020), pp. 1–6
37. S. Abdelnabi, K. Krombholz, M. Fritz, VisualPhishNet: zero-day phishing website detection by visual similarity, in *Proceedings of the ACM SIGSAC Conference on Computer and Communications Security (CCS)* (ACM, 2020), pp. 1681–1698
38. F.C. Dalgic, A.S. Bozkir, M. Aydos, Phish-IRIS: A new approach for vision based brand prediction of phishing web pages via compact visual descriptors, in *Proceedings of the 2nd International Symposium on Multidisciplinary Studies and Innovative Technologies (ISMSIT)* (IEEE, 2018), pp. 1–8
39. T. Phoka, P. Suthaphan, Image based phishing detection using transfer learning, in *Proceedings of the 11th International Conference on Knowledge and Smart Technology (KST)* (IEEE, 2019), pp. 232–237
40. Y. Lin, R. Liu, D.M. Divakaran, J.Y. Ng, Q.Z. Chan, Y. Lu, Y. Si, F. Zhang, J.S. Dong, Phishpedia: a hybrid deep learning based approach to visually identify phishing webpages, in *Proceedings of the 30th USENIX Security Symposium (USENIX Security 21)* (USENIX, 2021), pp. 3793–3810
41. B. van Dooremaal, P. Burda, L. Allodi, N. Zannone, Combining text and visual features to improve the identification of cloned webpages for early phishing detection, in *Proceedings of the 16th International Conference on Availability, Reliability and Security (ARES)* (ACM, 2021), pp. 1–10

Mallika Boyapati received a B.Tech degree in Electronics and Computer engineering from KL University, Vijayawada, India in 2016. pursued an M.Sc. degree in Applied computer science at Columbus State University, Columbus, Georgia, in 2018. From 2018 to 2021. Worked as a Graduate Research Assistant at Columbus state university as a data analyst. worked as a Senior Data Analyst at T-Mobile. Joined as a Ph.D. student at Kennesaw State University in Analytics and Data Science department. Working as a Graduate research assistant in the Information Systems Security Lab. My research interests include Data Analytics, Data Science, Machine Learning, and Natural Language Processing in the era of cybersecurity.

Bala Chaitanya Gutta received her Bachelor's degree in Computer Science from the Jawaharlal Nehru Technological University—Kakinada, India in 2020 and currently pursuing Master's degree in Computer Science at Kennesaw State University. She had worked as a software engineer at Wipro, Ltd for over a year. Her research interests include Machine Learning, Artificial Intelligence and Cyber Security. She is now working as a research assistant in the Computer Science department and serving as a member of Information and Intelligent Security Laboratory at KSU.

Md. Zakirul Alam Bhuiyan Ph.D., is currently an Assistant Professor of the Department of Computer and Information Sciences at Fordham University, NY, USA. Earlier, he worked as an Assistant Professor at Temple University. His research focuses on dependability, cybersecurity, and big data in emerging IoT/CPS applications. Dr. Bhuiyan authored/co-authored over 150 publications (including 55 SCI Q1), which appeared in many prestigious journals/conferences. Several research works of Dr. Bhuiyan have been recognized as the ESI Highly Cited Papers. He has been recognized as the 'highly-cited researcher' of the world for this work distinctions. He is a senior member of IEEE and a member of ACM.

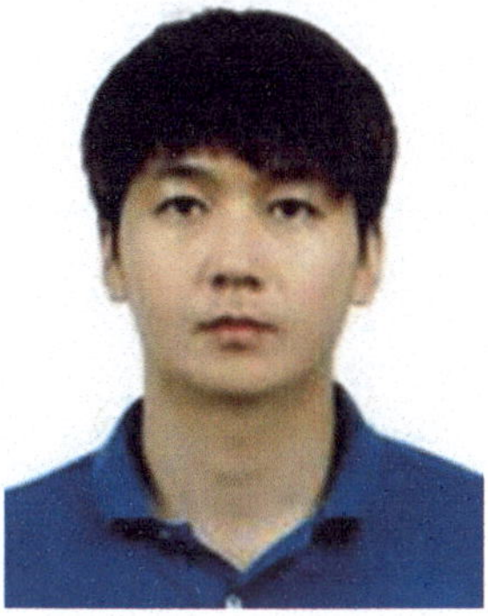

Junggab Son received the BSE degree in computer science and engineering from Hanyang University, Ansan, South Korea (2009), and the Ph.D. degree in computer science and engineering from Hanyang University, Seoul, South Korea (2014). From 2014 to 2016, he was a Post-doctoral Research Associate with the Department of Math and Physics, North Carolina Central University. From 2016 to 2018, he was a Research Fellow and a Limited-term Assistant Professor at Kennesaw State University. Since 2018, he has been an Assistant Professor of Computer Science and a Director of Information and Intelligent Security (IIS) Laboratory at Kennesaw State University. His research interests include applied cryptography, privacy preservation, blockchain and smart contract, malware detection, and security/privacy issues in artificial intelligent algorithms. He is a senior member of IEEE and a member of ACM.

Fault Tolerance and Security Management in IoMT

Rachida Hireche, Houssem Mansouri, and Al-Sakib Khan Pathan

Abstract In recent years, there has been a growing interest in collecting and storing healthcare data which eventually led to a revolution in this field. In fact, the development of IoT-enabled (Internet of Things-enabled) wearable devices like healthcare management software and smart medical sensors has effectively contributed to the rise of this technological revolution. Recently, we have witnessed a trend of increased use of IT (Information Technology) facilities and cyberspace. Cloud computing, which is one of the most significant technologies nowadays, plays a vital role in some mobile healthcare systems. As a result, it is highly expected that this trend would develop fast in the coming days and contribute to the field of IoMT (Internet of Medical Things) as a whole. In fact, the necessity of IoMT for remote healthcare services has significantly been realized during the recent outbreak of COVID-19 pandemic. Due to the prominent role and importance of IoMT, it is quite evident that such systems should be well protected and supported through efficient fault tolerant mechanisms and security mechanisms. In this chapter, we would explore the fault tolerance issues in such complex healthcare setting alongside the security assurance issues.

Keywords Fault tolerance · Healthcare · IoMT · IoT · Security management

---

R. Hireche (✉) · H. Mansouri
Laboratory of Networks & Distributed Systems, Computer Science Depart-Ment, Faculty of Sciences, Ferhat Abbas Setif University 1, Setif, Algeria
e-mail: hireche.rachida@univ-setif.dz

H. Mansouri
e-mail: mansouri_houssem@univ-setif.dz

A.-S. K. Pathan
Department of Computer Science and Engineering, United International University, Dhaka, Bangladesh

A. K. Pathan (ed.), *Towards a Wireless Connected World: Achievements and New Technologies*, https://doi.org/10.1007/978-3-031-04321-5_4

1 Introduction

Nowadays, Internet of Medical Things (IoMT) has become a building block for modern healthcare as it is able to operate with significantly stringent resources. Over the course of last two decades, it has been greatly enhanced to be used by healthcare providers for different purposes within this field: improving quality of treatments, managing diseases, reducing errors, improving patient experience, managing drugs, and even lowering costs. However, these applications are often prone to serious security issues which is a major impediment to the evolution and rapid deployment of this sophisticated technology. Issues related to this include mainly: identity theft, information theft, and data modification. In fact, these security problems represent real danger for the IoMT environment as medical data are often considered personal and sensitive.

One of the prominent cases of DDoS (Distributed Denial-of-Service) attacks took place in October 2016, which was launched on DNS (Domain Name System) service provider through an IoT botnet. The botnet used a malware named *Mirai*. The latter led to shutdown of huge portions of the Internet including Twitter, the Guardian, Netflix, Reddit, and CNN [1].

With the fact that such dangerous threats could be active at any point of time, the need arises for strong security mechanisms to protect the IoMT infrastructure. As we know, the first step to ensure security, which is a critical factor, is the complete understanding and appropriate categorization of existing and potential threats to the IoMT environment. It has been shown through several on-going research works that the implementation of secure IoMT applications is achievable by incorporating security measures with each involved technology. Moreover, the development of new IoMT technologies combined with Artificial Intelligence (AI), Big Data and Blockchain offers a variety of possible solutions [2]. The aim of this chapter is to study the existing literature and identify the factors and obstacles affecting the expected development of IoMT and its wide-spread use.

Following the Introduction, the rest of the chapter includes the following:

- In Sect. 2, we present the context of IoMT systems and their architecture, we specify the security requirements of IoMT systems, and also consider the current security techniques and their robustness against various existing attacks.
- In Sect. 3, we discuss different attacks against the IoMT system and classify the security techniques discussed to prevent or mitigate these attacks.
- In Sect. 4, we present for each layer of the IoMT system, the communication protocols and mechanisms used in different medical devices within the healthcare ecosystem. We also discuss the level of security for each mechanism studied as well as possible mitigation solutions.
- We conclude the chapter in Sect. 5 with some future research directions.

2 Internet of Medical Things (IoMT)

In order to understand the later sections, this section presents a general overview of IoMT systems, their architecture, the different security requirements as well as the available security techniques.

2.1 IoT and IoMT

The term, Internet of Things (IoT) refers to a wide range of interrelated objects and devices which use embedded systems like processors and sensors to collect information from the environment. After harvesting data, these devices analyze that. Then, through actuators, they act back and take action on the physical world [3]. By integrating every object for interaction through embedded systems, IoT enhances the ubiquity of the Internet. This leads to a highly distributed network of devices that can communicate with other devices and human beings [4].

Nowadays, the field of healthcare is witnessing a remarkable development thanks to the Internet of Things (IoT). With the ongoing development of different IoT technologies such as smart sensors and advanced lightweight communication protocols, it has been possible to interconnect many medical "*things*" to monitor and examine biomedical signals. Moreover, these IoT devices can even diagnose different diseases without any human intervention and thus they are called Internet of Medical Things (IoMT) [5]. Therefore, we can conclude that IoMT is mainly a network of devices which is connected to the Internet that uses sensors and electronic circuits to collect data in the form of biomedical signals from a patient [6]. Then, a processing unit processes these biomedical signals, a network device transmits the collected data over a network, a permanent or temporary storage unit is used to store data, and finally, a visualization platform is used with artificial intelligence schemes, so that it is capable of making decisions at the convenience of the physician.

2.2 Types of IoMT Devices

IoMT systems provide either needed or enhanced assistance for many medical conditions. Consequently, they can be classified into two main categories: Implantable Medical Devices (IMDs) which are necessary devices for specific medical conditions like pacemakers, and the Internet of Wearable Devices (IoWD) which are assistive devices to enhance the healthcare experience like smart watches.

Implantable Medical Devices (IMDs). As the name suggests, an Implantable Medical Device (IMD) is a device which is implanted to replace a missing biological structure or to support a damaged biological structure. Moreover, an IMD can even be used to enhance an existing biological structure. The main purpose of such implantable devices is monitoring signals from the patient's body and to send them to other medical systems [7]. They are mainly made up of tiny wireless modules and health sensors that collect like temperature, motion blood glucose and blood pressure. An example of such IMDs is the pacemaker which can be very useful for controlling abnormal heart rhythms. If the heart ever beats too fast or too slow from its normal rate, the pacemaker will work in an effective way to bring back the heart to its normal rate [8]. To keep such kind of devices in the human body for a long time, there are certain requirements for the IMD. Some of these requirements include low power consumption and small batteries that last a long time. The typical lifetime of a pacemaker, for example, is determined by how frequently we need to use it. Consequently, this can range from 6 to 10 years. And, it all depends on how frequently the device needs to pace the heart [9].

Infusion pumps, such as enteral, Patient-Controlled Analgesia (PCA), and insulin infusion pumps can be used in a variety of treatments [10]. Infusion pumps have been linked to a number of patient safety issues. As a result, the development of authentication mechanisms is critical. In real-world applications, remote pump control is a common requirement. This is why many authors concentrate on it. For example, to avoid the implementation of encryption, the authors in [11] have developed a new protocol that can be used in the communication of remote implantable devices (such as Medtronic insulin pump), and it will rely on plain text.

A glycemia (i.e., the presence, or the level, of glucose in one's blood) alarm system is presented in [12]. This system has the ability of calculating the amount of insulin dynamically to be administered to diabetes patients. Although the wireless communication scheme may increase the security threats on these electronic devices, it remains the best desired communication scheme for the implementation of these devices. Examples of this include cable breakage and infection [13]. Figure 1 shows some of the most used IMDs and their positions in the human body.

Internet of Wearable Devices (IoWDs). Individuals wear such devices to monitor their biometrics, which may help improve their overall health. This category contains a wide range of IoMT systems. Examples of IoWDs include [14, 15]:

- EEG (electroencephalography) and ECG (electrocardiography), which are used to monitor the heart and brain respectively.
- Fall detection band, blood pressure monitors and electrocardiogram (ECG) monitors [16].
- Smart watches that are quite famous currently for monitoring biometrics like heart rate and movement. When the individual is not active, the monitoring can detect slow and fast heartbeats. The new watches can also be used for fall detection and ECG readings to detect medical conditions such as atrial fibrillation (irregular heartbeat). They are now commonly used for non-critical patient monitoring [17].
- Activity sensors which can be used to monitor actions like running and sleep.
- Accelerating sensors which are capable of tracking the patient's rehabilitation.
- Respiratory rate sensors monitoring the patient's breathing and muscle activity.
- Sensors and fitness trackers.

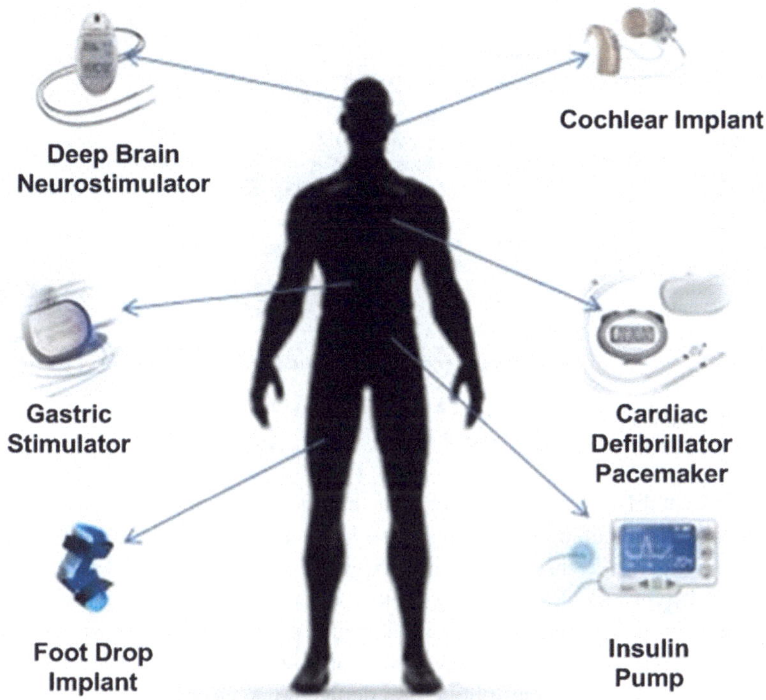

Fig. 1 Most used IMDs and their positions in the human body

However, due to battery life limitations and sensor accuracy, these devices are unlikely to be used to replace IMDs in critical situations [18].

2.3 IoMT Systems Architecture

The existing IoMT systems [19] usually have four main stages: *Sensor Layer, Gateway Layer, Cloud Layer and Visualization/Action Layer,* as shown in Fig. 2. These layers include all the steps that data passes through, from the collection of patient biometric signals via wearable sensors/devices to the final step of storage and visualization by the patient or analysis with a physician in a healthcare application.

Sensor Layer. The major function of the Sensor Layer is to establish an effective and accurate sensing technology to collect various types of health-related data [20]. The system uses implanted or worn sensors (like a pacemaker or a smart watch) to collect the patient's biometric data. These data are transmitted through wireless protocols such as WI-FI, Bluetooth or over MedRadio frequency spectrum reserved for IMDs to the second layer [21].

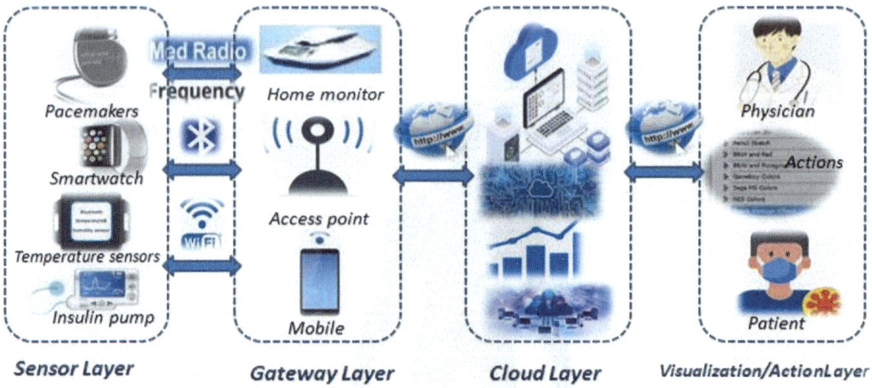

Fig. 2 IoMT system architecture

The attacks at this layer can be against the hardware or software. The system must be appropriately protected against these attacks so as to ensure the right functioning of the system and not to threaten the life of people using the IoMT.

Gateway Layer. As shown in Fig. 2, this layer acts as a bridge between IoMT sensors with low processing and storage capacity and the Cloud layer. The data is transferred to this layer without any processing. Devices that can be used in this layer include the patient's smartphone or a dedicated Access Point (AP), which can be typically more powerful than IoMT sensors. Some of their functions include performing some pre-processing operations as well as forwarding sensor data to the cloud through the Internet [22].

Cloud Layer. The retrieval and execution of the information obtained from the other layers, i.e., the sensor and gateway layer is performed at this level. Cloud servers control the systematic computing capacity. In addition to storage capacity, cloud servers also have the ability to make decisions based on the information obtained. In some critical heterogeneous IoMT applications, cloud servers can take action quickly based on emergency event detection mechanisms [23]. The analysis performed at the cloud layer includes processing data to find any changes in the patient's health. After being detected, the changes are presented to the physicians for any emergency response or patients for further actions. This layer provides a means of remote access to manage and control the various sensors.

The data in the cloud and visualization layer is mostly at rest - it is just as vulnerable as any other stage. Therefore, it is essential to protect it from unauthorized access. Attacks in this layer range from stealing account credentials to DoS/DDoS attacks [24].

Visualization/Action Layer. Data is displayed to the physician and the patient in this layer to allow for ongoing monitoring and control of the patient's condition. This layer also contains the procedures indicated by the physician in the event of a change in the patient's health; these processes can include quantity, indication, prescription or change of dosage of different medications.

2.4 IoMT Security Requirements

One of the major concerns of internet-accessible medical devices and healthcare network infrastructures is the security. In this section, we present the security requirements of future healthcare network infrastructures for IoMTs. This is based on CIANA (Confidentiality, Integrity, Availability, Non-Repudiation, and Authentication) considerations and includes the 11 security requirements listed below [22, 25, 26]:

(1) **Confidentiality/Privacy**. For the IoMT operations to be confidential, it is required to ensure that confidential information is not disclosed or made available to unauthorized parties [27, 28]. Confidentiality in the context of the IoMT refers to the protection of the medical information that the patient shares with his/her therapist, physician, or medical staff from any intrusion which can harm the patient (or a rogue entity can use the medical information against the individual) [29]. There are certainly rules for collecting and storing the patient's health data like adhering to legal and ethical privacy regulations such as GDPR (General Data Protection Regulation) and HIPAA (Health Insurance Portability and Accountability Act). The latter requires that only authorized individuals have access to the data. To protect the privacy of the patients' health data, adequate safeguards must be adopted so as to prevent any data breaches. Such measures should be handled seriously because cyber criminals do not only violate the patients' privacy but can also cause financial and reputational harm if they decide to sell that data in the illegal markets [29]. Fortunately, a range of approaches that can be used to ensure confidentiality are available. These approaches can make the patients' data unintelligible [28]. Currently, cryptography and access control lists are the techniques that best meet this requirement [22].

(2) **Integrity**. The data integrity requirement for IoMT health systems is to make sure that the data arriving at its intended destination has not been altered in any way during wireless transmission [30]. Integrity for IoMT data ensures that the patient's information, such as personal medical data and test results are accurate [28]. Nowadays, healthcare organizations are more aware than ever before about the importance of data integrity. The ability to detect possible unauthorized distortion or manipulation of data is critical to ensure that data has not been compromised. Therefore, appropriate data integrity mechanisms must be adopted to prevent the malicious attacks from altering transferred data. The legal and ethical GDPR state that medical providers must take the necessary steps to ensure that patient data is not altered i.e., it is accurate and up-to-date. Moreover, it insists that any altered personal data should be deleted or rectified as soon as possible [31]. The GDPR also emphasizes "*accuracy*" of data. It states that data owners should be able to request service providers to correct inaccurate information, and that service providers must respond to these requests within one calendar month. Similarly, HIPAA requires medical

providers to adopt measures to ensure that PHI (Patient Health Information) stored in systems can only be changed by legal authorization [31].

(3) **Availability**. Availability refers to the accessibility of services and data, provided by servers and medical equipment, to the affected users whenever they need them. Most importantly, these services and data will become unreachable in the event of DoS attacks. Any inaccessibility of data or services could result in life-threatening incidents for the patient, like the inability to provide early warning of a heart attack. Therefore, so as to ensure data availability to users and emergency services, any healthcare application must be *always-on*. By adopting preventive security measures and countermeasures to DoS attacks, healthcare providers can restore availability and access to personal data in a timely manner [32]. Therefore, to ensure availability, the system should be always updated to monitor any performance changes, provide suspicious data storage or transmission routes in case of DoS/DDoS attacks, and increase the performance of the systems to be able to solve any problem quickly.

(4) **Non-Repudiation**. It refers to the ability of holding any authorized user accountable for his/her actions. Simply put, non-repudiation ensures that no operation in the system can be denied [22]. This requirement prevents the authorized users from disclaiming previous commitments or actions in the system [28]. A patient might deny that some data belongs to him, when in fact the extracted data was sent from his sensors. Another case could be updating a few sensors firmware by an authorized developer, but the latter refuses to admit its validity. In many cases, if an authorized entity denies previous commitment or action, a specific procedure involving a trusted third party is usually required to resolve the situation [28]. Using digital signature techniques is the best way to meet this requirement [22].

(5) **Authentication**. This requirement refers to the ability to validate a user's identity when the user accesses the system. On the other hand, the process by which a user is verified as the original source of given data at some point in the past is known as message authentication. The most secure form of authentication is mutual authentication. In this authentication, the client and the server authenticate each other before exchanging secure key or data. Because of the lack of memory storage in several IoMT devices or insufficient CPU (Central Processing Unit) power to perform the cryptographic operations required by traditional authentication protocols, lightweight authentication protocols are becoming more popular [33].

(6) **Authorization**. It refers to confirming that authenticated users only execute commands that they are authorized to execute [34]. More specifically, authorization makes sure that only authorized entities can access to specific network services or resources, like patient's collected medical data. Permission to perform a given action, like issuing commands to medical IoMT devices or updating the medical IoMT device software is granted only for trusted expertise parties.

(7) **Anonymity**. This requirement ensures that the identity of the patient or physician remains hidden from unauthorized users when they interact with the

system, i.e., both the patient and the physician should remain anonymous. The identity of the patient/physician should not be exposed when they are in communication [35]. Passive attacks can see what you do but not who you are. This anonymity can be achieved (for instance) by using smart card like mechanisms.

(8) **Forward and Backward Secrecy**. Forward secrecy has been identified as a critical property of a variety of cryptographic primitives. It keeps the future transmitted data secure even if previous data have been compromised. However, even if the current data have been successfully attacked, backward secrecy makes sure that old data are safe. To achieve forward/backward secrecy, time-based authentication parameters must be used. The authors in [36] proposed a method that provides the secret both in front and behind the group's members. Furthermore, it provides a formal analysis of the new method's correction based on BAN (Burrows–Abadi–Needham) authentication logic.

(9) **Secure Key Exchange**. This is the requirement which means the ability to securely distribute keys among system nodes. One of the most efficient algorithms for data security is the Elliptic Curve-Diffie Hellman (ECDH) using key exchange [37].

(10) **Key Escrow resilience**. This requirement ensures that the system administrator is not allowed to impersonate any user authorized to use the system. This helps protect the system against internal threats. To meet this requirement, the Key Generation Server (KGS) only has half of the key and will be unable to compute the entire private key for both entities [38]. This requirement can be met by combining a cryptographic hash function (CHF) and asymmetric keys.

(11) **Session Key Agreement**. Following the authentication process, Session keys must be used by every node in the system. The work in [39] proposed a system in which each sensor node agrees on the generation of session keys. This scheme improves performance so that the authenticated device can calculate session key ahead of time.

2.5 *IoMT Security Techniques*

For securing IoMT systems, several techniques are available by this time. Based on [22] (see Fig. 3), these techniques are classified into three types (mainly): symmetric, asymmetric, and keyless. Cryptographic algorithms are used in both symmetric and asymmetric techniques, whereas keyless techniques are non-cryptographic.

(1) **Symmetric Cryptography**. Symmetrical key Cryptographic algorithms are the fundamental building blocks of any secure system that requires confidentiality. They are typically used to encrypt bulk messages transmitted between two systems. The keys used for encryption and decryption in these cryptographic algorithms are the same for both communicating entities, and this is

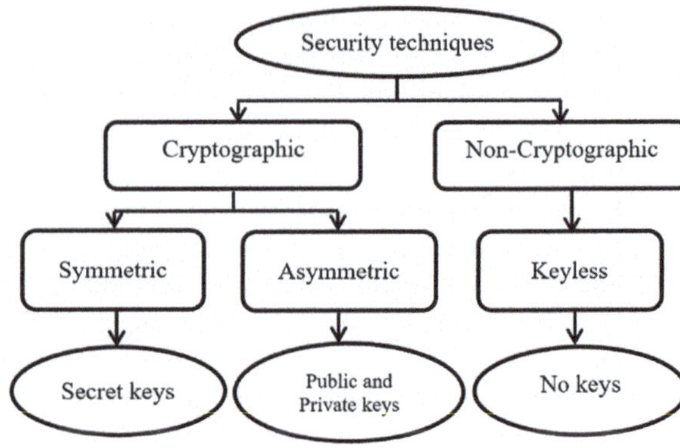

Fig. 3 Security techniques

shown in Fig. 4 [40]. This key must be generated and distributed prior to any communication.

In this subsection, we will look into how symmetric cryptographic algorithms can be integrated into IoMT systems.

Continuous Facial Recognition. It is the technology that allows IoMT systems to authenticate users by scanning their faces. Identity hashing and continuous facial recognition are the two steps in this technique. The ID is hashed only once, at the start of the session. After passing the identification hash test, continuous facial recognition is performed throughout the session [41]. Biometric authentication is performed in this step. Each authorized person has a set of images taken and saved with their respective roles. This technique can effectively secure the system in a medical environment due to its continuous scanning of the user's face while using the system.

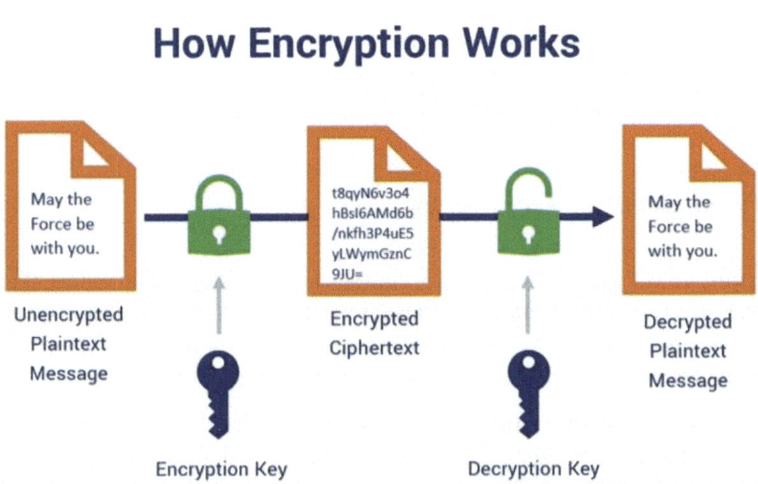

Fig. 4 Symmetric cryptography operation flow

Hierarchical Access. This technique enables patients' data stored in the cloud layer to be accessed in a hierarchical manner. One approach makes use of a hierarchical role based model and gives permission based on the role of the user [26]. All authenticated nurses, for example, can dispense medications; however, in order to prescribe a new medication, a doctor is required. To support this hierarchical access, the work in [26] used the Chinese Remainder properties. It is a technique in which any patient's data can be accessed by a user with a higher privilege. The user with a lower privilege, on the other hand, can access a portion according to his role. Additionally, the work in [41] proposed a hierarchical key allocation scheme that supports dynamic updates, in particular, the concept of security against key indistinguishability. As a foundation, the authors employed a symmetric encryption scheme.

Gait-Based Technique. Gait recognition refers to the task of identifying people based on how they walk. To generate unique symmetric keys, this method employs the human walking pattern. The work in [42] demonstrates that depending on the gait, additional tasks such as gender recognition or age estimation can be processed. When more than one walk-based task is jointly trained, the identification task converges faster than when trained independently, and multi-task pattern recognition performance is equal to or better than more complex single-task pattern recognition.

CHF with XOR. Converting data of arbitrary size to data of fixed size through a one-way mathematical function is known as CHF (Cryptographic Hash Function) [43]. In order to determine whether one of its operands is different, exclusive-OR (XOR) can be used. Within the healthcare field, a sensor ID or a shared key (or any other initial parameters) can be XORed and then hashed. Then, the hashed parameters are distributed from the key generation server to the sensor and gateway nodes. These nodes are enabled by the parameters to generate keys [44]. Experimental results and theoretical analysis indicate that when combining CHF, a symmetric key, and XOR operator, the scheme significantly reduces the computational cost compared to schemes using asymmetric encryption and presents a lower security risk compared to lightweight schemes, as demonstrated in [45] and [46]. The hash function is also used in this technique to support unique identification parameters. However, initial parameters must be added manually to all nodes by the system administrators during the system's initialization step.

(2) **Asymmetric Cryptography**. Asymmetric cryptography, also known as Public Key Cryptography (PKC), refers to cryptographic algorithms that encrypt and decrypt data using a pair of related keys, the public key and the private key, to prevent unauthorized access. Everyone has access to the public key, but only the owner has access to the private key. Two popular algorithms in this technique are Rivest–Shamir–Adleman (RSA) and Elliptic-Curve Cryptography (ECC) [47, 48]. However, due to its subtle characteristics, ECC is the most widely used cryptographic technique for securing IoMT systems. A 160-bit ECC key is as good as a 1024-bit RSA key and is 15 times faster [49]. Figure 5 [40] illustrates asymmetric encryption which uses two keys, mathematically linked but distinct to encrypt (public key) and decrypt data (private key).

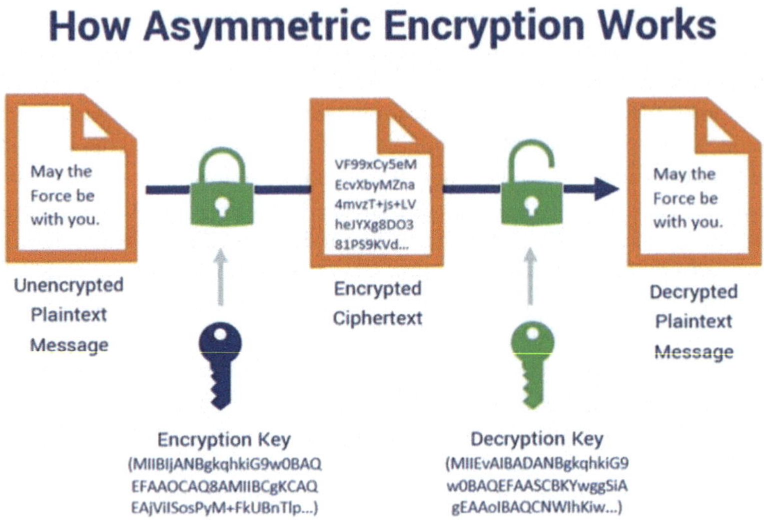

Fig. 5 Asymmetric cryptography operation flow

CHF with ECC. When used in conjunction with ECC keys, the CHF feature allows the establishment of a secure, *certificateless* channel between patients and their physicians [25]. The two techniques are combined to provide a secure method for sharing keys between different layers of IoMT. After the nodes receive the hashed values, they can be used to generate their asymmetric keys. This technique can also reduce the overhead associated with certificate management for cloud data storage and sharing [50].

Homomorphic Encryption (HE). Homomorphic encryption allows for the secure transmission and storage of confidential information across and within a computer system [51]. HE attempts to help in the encryption process by allowing certain types of computations to be performed on ciphertext. This process ends up with an encrypted result that is also in ciphertext. Its output is the result of operations on the plaintext. However, this technique is different from others because it does not allow the medical staff to see the patient data. Only the patients can have access to their data, except in emergencies. This is helpful for some IoMT sensors, like smartwatches.

There are three types of HE schemes: partial HE (PHE), which can perform a single mathematical operation an infinite number of times; somewhat HE (SHE), which can only perform a limited number of operations; and fully HE (FHE), which supports an infinite number of operations. Thus, among the three schemes, the FHE is the most suitable for fast data aggregation without compromising data confidentiality [49]. Optimal HE (OHE) is an FHE variant that is best suited for hospital healthcare monitoring systems. The key is authenticated during encryption, and the best key is chosen using the Step Size Fire Fly (SFF) optimization algorithm. This strategy can be used to generate the encrypted key while achieving maximum key breaking time and minimal computational time while maintaining high security [52].

Digital Signatures. These techniques are frequently used to validate the authenticity of data/commands by signing and verifying them with the sender's private and public keys, respectively [53]. Digital signatures can be embedded into sensor firmware in IoMT systems using an add-on software shim, allowing it to validate and intercept sensor wireless communications [54]. The sensor's firmware must store a list of authorized users' public keys in order to validate these techniques. The work in [55] propose a scheme for authenticating a device that includes multi-factor authentication, digital signatures, and device capability. The proposed scheme not only efficiently authenticates the device via multi-factor authentication, but also it authenticates the authentication server via digital signatures.

Smart Cards. Smart cards in healthcare systems are thought to have enormous potential for improving healthcare delivery as well as lowering healthcare costs. Because of its reliance on physical keys, this technique is different from the previous techniques [56]. With ECC keys serving as the first factors, the physical keys serve as the second for authentication. To gain access to a system IoMT, the user must first enter an access key before using their smart card. Apparently, this technique helps the system resist cyber break-ins if one of the two factors is compromised. This is why smart cards are quite common these days.

(3) **Keyless Techniques**. In this subsection, we explain the keyless techniques that provide security without using pre-shared keys.

Biometric Technique. Owing to its simplicity, this technique has become the most used technique to ensure IoMT systems. This technique uses biometric sensors to identify users' physical characteristics such as, fingerprint sensors, which can read the fingerprint image, and ECG-based sensors that record heartbeat activities in order to encrypt data. There are different fingerprint authentication algorithms such as: Delaunay triangulations, polar coordinates and Minutia Cylinder-Code (MCC) [57]. The performance and complexity of the applied algorithm determines the performance of the device used. The Finger to Heart (F2H) IMD fingerprint authentication algorithm based on Minutia Cylinder-Code (MCC) is proposed to ensure the safety of IMDs such as pacemakers and defibrillators. This improved algorithm significantly reduces both message size in transmission and device computational overhead, while conserving IMD's limited resources [58].

Token-Based Security. The use of passwords or predefined keys presents many problems that limit their applicability for various IoMT applications. Whether software or hardware, tokens can be used for user authentication. The use of lightweight token-based user authentication (TBL UA) for IoMT devices, based on the token technique, improves the robustness of authentication [59]. Radio Frequency Identification (RFID) can also be used as a hardware token in a hospital information system (HIS) for secure sensor logistic management [60]. The work in [61] proposes an implementation of MQTT (Message Queue Telemetry Transport) protocol token authentication in constrained devices. According to the results of the usability and performance tests, the system can perform valid and expired token authentication in a reasonable amount of time.

Blockchain Technology and AI. Due to their impact with their advanced distributed security and remarkable role in securing other fields like finance, Blockchain and Artificial Intelligence (AI) have become the key technology for the requirements of IoMT systems, mainly to bring transaction and data processing at the cloud layer [62]. In IoMT systems, the blockchain technology is used as a security management to share information between the patient and other parties like the doctors. AI systems, on the other hand, can detect intrusions or anomalous behavior in patient data and network flows. Nevertheless, these techniques still face some challenges that allow them to be implemented in the IoMT systems that are discussed in [63, 64].

3 Risks and Attacks in IoMT

In this section, we will discuss the possible physical and network attacks that threaten the IoMT systems and how to avoid or mitigate them.

3.1 Physical Attacks

In this type of attack, the attacker must be physically close to the network or devices of the system in order to launch the attack wirelessly [65]. To extract security keys or patient data, the attacker targets the physical components of the IoMT systems. Some of the common types of physical attacks are the following:

Physical Security Token Loss. It is when the attacker steals a physical security token, like a smart card or proximity card, from an authorized user in order to have access to the system. The security requirements violated in this case are authentication, authorization, anonymity, and forward secrecy. As the smart card or proximity card alone is insufficient to hijack the system, authentication based on ECC combined with smart cards can be used to protect the system against this type of attack [56].

Impersonation attacks. The attacker pretends to be a legitimate entity or an authorized user to access resources to which he is not authorized. Bluetooth Impersonation Attacks (BIAs) are effective against any Bluetooth device, and they are undetectable because the Bluetooth standard does not require notifying end users of the outcome of an authentication procedure or the lack of mutual authentication [66]. To avoid such attacks, cryptographic techniques such as, CHF and biometrics should be employed.

Tampering. It is an attack in which the attacker physically modifies the data of the IoMT systems [67]. Any modification in a device like RFID or communication link is considered a tampering attack. Altering the IoMT data by attaching external devices and attacking sensors is also considered a tampering in an emergency. However, this attack can be mitigated if symmetric keys are combined with facial recognition or if keyless methods are employed [41, 57].

Side Channel. These attacks rely on information achieved from the encryption device's side channels. In addition to plaintext and ciphertext messages, they are used to recover the secret key using electromagnetic analysis, power consumption or, differential power consumption during encryption/decryption of various messages and during computation of various security protocols [68]. In addition to cryptography techniques, the Datagram Transport Layer Security (DTLS) protocol can be used to avoid such attacks as the work presented in [69] recommended. On the other hand, Blockchain technology and AI were demonstrated as additional detection and mitigation strategies in [62].

Radio Frequency (RF) Jamming/Desynchronization. This is another serious type of attack on the IoMT systems. Because IoMT sensors are limited in energy by the battery, they may cause battery discharge. Blockchain and AI technologies have the potential to mitigate the effects of these intrusions by finding alternate routes or cutting off the canal's connection to the attacker [70].

Fake Node Injection. In this intrusion technique, to control data flow between two legitimate nodes of the network, the attacker drops a fake node between them [65].

Permanent Denial of Service (PDoS). Also known as Phlashing, PDoS is a type of DoS attack in which hardware sabotage completely destroys an IoMT device. The attacker launches the attack using a malware to destroy firmware or to upload corrupted BIOS (Basic Input Output System) [45].

Sleep Denial Attack. In this attack, the battery powered devices are kept awake by the attacker who feeds them with wrong inputs. The batteries eventually get exhausted and thus cause the devices to shutdown [65].

Malicious Code Injection. In this intrusion technique, a malicious code is injected onto a physical device by the attacker. By compromising this device, the attacker may be able to launch other attacks as well [65].

The physical attacks, their effects, and the solutions proposed are summarized in Table1.

3.2 Network Attacks

Bluetooth and Internet connections (wireless) can be targets of various types of attacks at different layers of the IoMT system. Stealing or fabricating patients' data, creating congestion, jamming, or connection blocking can affect normal operations or result in a total communication failure, which is usually the primary objective of these kinds of attacks.

Man-In-The-Middle (MITM). It is an attack that targets the communication between two IoMT devices and gives access to their private data. In this attack, the attacker is able to eavesdrop or monitor the communication between the two devices [67]. The intercepted data can be modified by the attacker before it is sent to

Table 1 List of physical attacks, effects and proposed solutions

Physical attack	Effects	Proposed solution	Solution references
Physical security token loss	Authentication; Authorization; Anonymity; Forward secrecy	Asymmetric (two-factor)	[69, 71]
Impersonation/Presentation		Asymmetric; Keyless	[25, 57, 58, 69],
Tampering/Malicious code injection	Data confidentiality; Data Integrity	PUF (Physically Unclonable Function) based Authentication; Symmetric (two-factor); Keyless	[41, 50, 57, 69] [57]
Side channel attack	Collect Encryption Keys; Data confidentiality; Data Integrity	Masking technique; Authentication using Physically Unclonable Function (PUF); Keyless	[50, 62, 69, 72]
Radio frequency (RF) jamming/Desynchronization	Battery discharge; Availability	CUTE Mote; Keyless	[70, 73]
Permanent denial of service (PDoS)	Hardware sabotage completely destroyed	NetwOrked Smart object (NOS) Middleware	[74]
Fake node injection	Control data flow and drops a fake node	Pervasive Authentication Protocol (PAuthKey)	[75]
Sleep denial	Node put on awake or shutdown	CUTE Mote; Support Vector Machine (SVM)	[73, 76]

its original destination. For instance, a patient biometric data, which is transmitted between any two layers of the IoMT system, may be altered or modified. As explained in [77], this is possible with the use of Unmanned Aerial Vehicles (UAVs) that result in a Drone-in-the-Middle (DitM) attack. MITM can be made even more powerful if the UAV is linked to a cloud, allowing it to perform more intensive computation in a relatively shorter amount of time.

DoS/Distributed DoS (DDoS). Unlike DoS attacks, which were perpetrated by a single node, a DDoS attack involves multiple sources attacking a specific target by flooding it with messages or connection requests with the goal of making service unavailable, preventing legitimate users of a service (i.e., from using it) [78]. Network fragmentation can also occur because of such attacks. Typically, the cloud layer is the main target for these attacks so as to make the system unavailable to users [79].

Consequently, availability is the violated requirement in this type of attacks. Similar to Radio Frequency (RF) Jamming attacks, Blockchain technology and AI can find alternative paths or terminate the connection to the channel controlled by the attacker, and thus can mitigate these attacks [70].

Clock Synchronization. IoMT systems, like all real-time systems, require a clock synchronization protocol. The latter is the target of this type of attack. The secure key exchange is the violated requirement in this attack. This attack is considered serious because the attacker can make other attacks (such as relay, replay, and MITM) difficult to detect [22]. However, the combination of ECC with smart cards can be used to mitigate this kind of intrusion [56].

Sniffing. Sniffing attacks passively intercept data sent between two nodes. This attack results in a breach of patient data confidentiality as the attacker can see the data transmitted between the system's layers [77]. Thus, the data confidentiality is the violated requirement in this attack. To mitigate this type of attack, any encryption algorithm, whether symmetric, asymmetric, or keyless can be used.

Relay. The intercepted data, after a successful sniffing attack, can be relayed to a third node without modifying it by the attacker. For instance, the intercepted patient data can be redirected to the attacker's device before being sent to its final destination [70]. The authorization requirement is violated by this attack. Techniques such as hierarchical access and secure session keys can be used to mitigate this.

Replay. In this case, a signed packet may be captured by the attacker who would resend the packet several times to the destination [52]. As a result, a DoS/DDoS attack is possible. The authorization requirement is violated with this attack. To mitigate these attacks, a *timestamp*, which is part of some cryptography techniques, can be used [62].

Brute Force. Typically, in this type of intrusion, the attackers use automated software that generates different password combinations until it succeeds. The strength of these attacks stems from the fact that the passwords chosen by the user are inherently weak, or it employs default generated passwords or username as password [42]. An example, which is a significant problem for IoMT devices, is the dictionary attack. The latter relies on passwords or known words in dictionaries. After capturing the encrypted/decrypted data with machines or more powerful tools, these attacks can also be carried out offline. A dictionary attack is considered a dangerous attack for IoMTs, because the password selection criteria can be guessed with a simple python script [80]. Security requirements for authentication and authorization are violated through such attacks; however; they can be mitigated with the use of keyless methods like biometrics.

Selective Forwarding. In this attack, some messages may be simply altered, dropped, or selectively forwarded to other nodes in the network by a malicious node [52]. As a result, the destination receives incomplete information.

RFID Spoofing. To gain access to the information printed on the RFID tag, the attacker first forges an RFID signal [65]. Then, he/she can send his/her data as valid using the original tag identifier [81].

RFID Unauthorized Access. An attacker can update (i.e. read, modify, or delete) data on RFID nodes because of the lack of proper authentication mechanisms, [82].

Table 2 List of network attacks, effects and proposed solutions

Network attack	Effects	Proposed solution	Solution references
MITM	Data confidentiality; Authorization	Symmetric/Asymmetric (two-factor); Keyless	[25, 71]
DoS/DDoS	Availability	Keyless	[70]
Sniffing	Data confidentiality	Symmetric/Asymmetric (two-factor); Keyless	[62, 77]
Relay	Authorization		[70, 71]
Replay			[41, 46, 56, 62]
Clock synchronization	Secure Key; Exchange	Asymmetric (two-factor)	[56]
Brute force	Authentication; Authorization	Keyless	[42]
Selective Forwarding	Data confidentiality; Data Integrity; Authentication; Authorization	Hash Chain Authentication technique with Rank Threshold; Monitor based approach (CMD)	[71, 83]
RFID spoofing/RFID unauthorized access		SRAM based PUF	[84]

Table 2 summarizes the network attacks, their effects and the corresponding solutions proposed.

4 Security in IoMT Communication Protocols

In this section, we explore the communication protocols of IoMT. According to [85], the IoMT system can be divided into three main layers: the perception, network, and application layers. There are two more sub-layers between these three main layers: the adaptation layer, which includes the protocols that communicate between the perception layer and the network layer [86], and the transport layer, which also includes the protocols that transport information between the network and application layers [87]. We also present for each layer the most documented security measures, mitigation and implementation for each protocol to secure modern healthcare infrastructures and networks.

Figure 6 shows the different layers of IoMT systems in relation to the OSI (Open Systems Interconnection) reference model. This classification is based on the protocols and functions that each layer requires. The perception layer is primarily used for hardware functions. The network layer is responsible for network functions, while the application layer is designated for user functions.

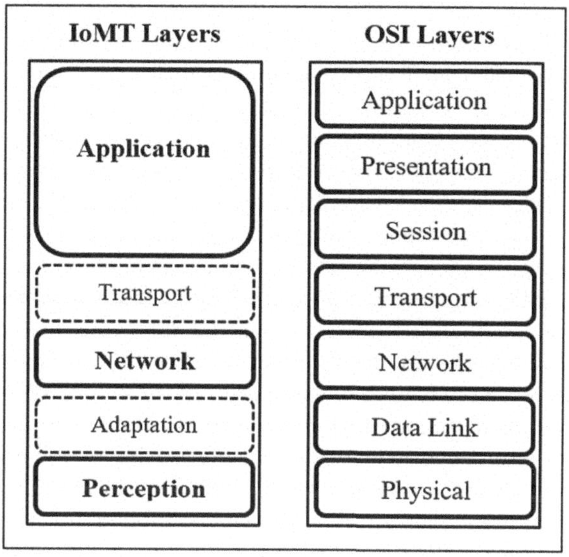

Fig. 6 IoMT versus OSI layers

4.1 Perception Layer

The majority of the perception layer protocols are based on or implement the IEEE 802.15.4 standard [88, 89]. To collect information about the patient's health status from sensors, health systems have used the following perception layer protocols and mechanisms:

RFID. Radio Frequency Identification (RFID) is a wireless object identification technology which uses radio frequency signals for very short range communications [90]. Autonomous RFID tag technology that is placed in or near the patient's body plays an essential role in the development of body health systems [91]. Moreover, passive RFID tags can be used in several situations such as; patient environment monitoring, physical access control [90, 91], and storage temperature monitoring for each type of drug [92, 93].

RFID is a technology that is used in devices with very low-power features, making common security mechanisms difficult to implement. However, researchers have proposed several noteworthy custom authentication mechanisms. An RFID tag authentication protocol is proposed in [94] that requires less storage and computation on the tag side. This protocol protects against replay, DoS, forward and backward tracing, and server impersonation, as well as provides privacy and security features. On the other hand, a hash-based RFID security protocol with forward privacy is presented in [95]. Its main aim is to protect the RF tag from tracking attacks by observing previous unsuccessful tag sessions. Furthermore, partial solutions to various limitations are identified and proposed in [96]. Examples include: dynamic password, synchronized secrets and custom system authentication systems.

NFC. NFC, or Near Field Communication, is a protocol that is used to connect IoT devices in a simple and low-cost manner [93, 97]. However, when NFC is used in IoT devices in the medical field, a number of biocompatibility issues arise. This infrastructure has the potential to provide convenient and low-cost power distribution and communication channels for a variety of medical devices. In addition, a battery or external electrical connection is not necessarily required in NFC-enabled medical devices for their custom operations [98]. An NFC device embedded in a cell phone, for example, can transmit pacemaker measurements to a monitoring doctor, control an insulin pump remotely, or activate an implanted neural simulator [92].

NFC implementations can be theoretically attacked by MITM attacks; however, it is extremely complicated to launch these attacks in real-world executions because of the NFC's architecture and distance limitations [99] (even if tried wirelessly). Moreover, a list of known security issues with the NFC protocol is presented in the existing literature like for instance, in [100], where some practical countermeasures are also suggested for each of the attacks mentioned. Furthermore, a single and multiple antenna design for the NFC controller component is suggested in [101], in order to mitigate attacks like, data corruption, low battery, and tag cloning.

Bluetooth/BLE. Bluetooth is a wireless technology that is based on the IEEE 802.15.1 standard. It is a low-power, low-cost wireless communication technology that can transmit data between mobile devices over a short distance (8–10 m with 2.4 GHz band). Bluetooth Low Energy is the ultra-low power, low-cost version of this standard (BLE or Bluetooth Smart) [90]. In addition, these features make Bluetooth/BLE more suitable for IoMT devices such as IoWDs and human interface (HID) devices [102].

Different attacks may threaten devices which are connected through BLE, and according to published research works, these threats are across all communication layers. Nevertheless, a variety of security controls to mitigate such attacks are provided by BLE implementation [103]. To achieve confidentiality and integrity, some solutions employ AES-CCM encryption. To authenticate data channel packet data units (PDUs), a 4-byte MIC module can also be used [104]. Furthermore, in order to protect Bluetooth Low Energy (BLE) technology from attacks, the authors in [105] propose a set of techniques and countermeasures that can be used to secure Bluetooth communications.

Z-Wave. Z-wave is a low-power wireless MAC protocol developed by Zensys. It is used for remote control applications and small commercial domains [90]. This protocol supports two types of devices: control devices and slave devices [106]. Z-wave can also support short messaging between IoMT devices for light, energy, and healthcare control [87].

Z-Wave provides confidentiality, source integrity, and data integrity services through AES (mostly 128) encryption, policy-driven and behavior detection mechanisms. The security command class included in the Z-wave allows application frames to be encapsulated in an encrypted and signed security frame. Symmetric encryption protects the frame by using AES with three shared keys known by every network node that needs the security service [107]. Furthermore, techniques like hiding the WLAN SSID (Service Set IDentifier), using WPA2 (Wireless Protected Access 2) instead of

WEP (Wired Equivalent Privacy), and Reverse Proxy Server can also provide extra protection for IoMT devices using Z-wave [108].

UWB. UWB (Ultra-wideband) technology is based on the IEEE 802.15.3 standard, which has recently gained popularity as a method of high-speed, short-distance indoor wireless communication [109]. One of the most intriguing aspects of UWB is its bandwidth of more than 110 Mbps, which is sufficient for most multimedia applications and is applicable for hospitals. UWB for medical systems is suggested in [110] because when communicating with implanted sensors, high signal attenuation requires a protocol that transcends channel limitations. It works by transmitting signals from sensors to a microcontroller [93]. For instance, a short distance communication technology is required by the electrocardiogram procedure and this is the aim of using UWB (among other protocols) [97, 111, 112].

Being a distance protocol, UWB is threatened by attacks that differentiate the distances between nodes. UWB adopts the Advanced Encryption Standard (AES) block cipher with counter mode (CTR) and cipher block chaining message authentication code (CBC-MAC) [113]. In [114], a Verifiable Multilateration (VM) algorithm that uses verification triangles to detect a distance enlargement attack is suggested. A location-based secure authentication scheme is proposed by other works like [115, 116] to prevent external attacks. In addition, [117] suggests the first modulation technique to prevent ED/LC (Early Detect/Late Commit) attacks regardless of communication range in the UWB with pulse reordering (UWB-PR).

Table 3 summarizes these issues discussed so far.

4.2 Network Layer

The network layer is responsible for the transmission and reception of the collected medical data. As a result, this layer serves as the foundational infrastructure layer for the healthcare platform. As such network devices transfer sensitive data, network security is a major concern in the field of healthcare [131]. The IEEE 802.15 standard is the foundation for the majority of the protocols in this layer [132]. The following protocols are the most commonly used for IoMT at this layer:

WiFi. Wireless Fidelity (Wi-Fi) is a middle-range (up to 100 m) protocol based on the IEEE 802.11 family of standards [133, 134]. A number of authors have proposed using Wi-Fi to communicate with monitoring devices in an IoMT system. For instance, the authors in [135] use this protocol on a network of 45 critical medical care devices, demonstrating that communication between these devices is effective and secure via Wi-Fi. Moreover, this protocol is used in a system for remote patient health monitoring in conjunction with Global System for Mobile communication (GSM) to simulate the transfer of medical data between two different geographical locations [136].

Wi-Fi Protected Access (WPA), Wi-Fi Protected Access 2 (WPA2), and Wi-Fi Protected Access 3 (WPA3) are the mechanisms used to secure Wi-Fi 802.11

× communications. WPA technology is characterized by providing more powerful encryption mechanisms [137].

ZigBee. ZigBee is a wireless communication protocol that conforms to the IEEE 802.15.4 standard and is intended for low-power, low-cost, low-speed wireless personal area networks that connect devices primarily for personal use [138]. This protocol is used by health zones to connect sensors to the coordinator, as well as

Table 3 Perception Layer protocols—security level, attacks and countermeasures

Protocol	Security level	Attacks	Countermeasures proposed	References
RFID	Several weaknesses in the active and passive RFID systems; Requires the integration of special security mechanisms into the system to ensure the fundamental security requirements	Side channel attacks backward/forward traceability	Encrypted RFID implementations; hash-based RFID security protocol;	[118–120]
NFC	Data exchange in close proximity; Several threats in transactions or contact processes (requires data encryption before any communication or transaction)	Eavesdropping; MITM; Data Modification; Data Insertion and Data Corruption attacks	Communication distance limitation; integrate standard cryptographic practices to protect its communication channel and data	[121, 122]
Bluetooth/BLE	The link keys may be stored incorrectly; The length of the encryption keys may be small or only 1 byte; No user authentication	Sniffing, DoS, MITM, PIN Cracking Attacks and Brute-Force Attacks	AES-CCM; AES-128 bits; Use link encryption and combination keys	[123, 124], [125, 126]
Z-Wave	Not enforcing a standard key exchange protocol; The source and destination fields of the MPDU (MAC protocol data unit) aggregation frame are implicitly trusted by Z-Wave devices	Key Reset, Black Hole, impersonation and node spoofing attacks	AES-128 encryption using three shared keys	[127]

(continued)

Table 3 (continued)

Protocol	Security level	Attacks	Countermeasures proposed	References
UWB	Incorrect access control configuration, Symbols with a long size	Same-Nonce, ED/LC attacks	AES block cipher with counter mode (CTR) and cipher block chaining message authentication code (CBC-MAC); The distance between nodes is secured by location and distancing protocols	[128–130]

between the coordinators themselves [139]. Implementing a fully working application layer protocol for healthcare environments, the ZigBee Health Care Profile is based on ZigBee Pro [140]. To enforce MAC layer security, ZigBee uses The IEEE 802.15.4 standard to employ higher layers. AES is used for symmetric key cryptography in implemented security mechanisms. Several other security modes are defined in [141, 142]. The authors in [143] propose a framework capable of predicting and protecting against various potential malicious attacks in the ZigBee network and responding appropriately by notifying the system administrator. It can also make instantaneous automated decisions based on real-time data defined by the system administrator.

WIA-PA. WIA-PA is a Chinese industrial wireless communication standard for process automation [144]. Despite being an industrial protocol, the work in [145] proposes WIA-PA as a transmission protocol in the internal networks of wireless sensor network, in medical remote monitoring system. The WIA-PA network's MAC layer security is based on IEEE STD 802.15.4–2006. Above the MAC layer, it provides two levels of security services: end-to-end security in the application sub layer and point-to-point security in the data link sub layer (DLSL). Furthermore, WIA-PA provides a secure access authentication mechanism for the entire network [146]. WIA-PA architecture was proposed by Wang et al. for device authentication [147]. Access is authorized through WIA-PA by using a join key shared by a device and a security manager. A security mechanism for WIA-PA and its protocol stack is also suggested and implemented in [148].

6LoWPAN. 6LoWPAN is an IPv6 adaptation layer that defines mechanisms for enabling IP connectivity for tightly resource constrained devices communicating over low power, lossy links such as IEEE 802.15.4 [93]. In the healthcare sector, IoMT sensors and local devices can be linked to IP networks via 6LoWPAN [149]. Moreover, the interconnection of sensors with middleware devices or Internet-connected routers is allowed by 6LowPAN [150]. Security protocols for different layers of the 6LoWPAN stack have been developed. The MAC security sub layer of IEEE

802.15.4 provides hop-to-hop security for the wireless medium, while the upper layer security is defined to provide end-to-end security between two remote peers [151]. The 6LowPAN security measures are classified into two taxonomies in [152]. The first is about communication outside of the 6LowPAN network (use DTLS (Datagram Transport Layer Security), HIP (Host Identity protocol) and IKE (Internet Key Exchange) technology). The second is about "*protocols inside communication*" (use IDS tool).

LoRaWAN. Originally developed by Semtech, LoRa (Long Range) is a physical layer protocol made to support low-power and wide area networks [153]. LoRaWAN, on the other hand, defines the network's communication protocol as well as the underlying system architecture [154]. An IoT-based health monitoring system is presented in [155]. In this system, the medical data collected by sensors is sent to an analysis module via secure, low-cost and low-power communication links, provided by an infrastructure LoRaWAN network. Moreover, an IoMT biofluid analyzer which uses LoRa and Bluetooth is presented in [156] in order to support long-range data transmission.

LoRaWAN uses the 128-bit Advanced Encryption Standard (AES128) to ensure complete network security, including mutual end-point authentication, data origin authentication, replay and integrity protection, and privacy. A 128-bit AES key (called AppKey) and a globally unique identifier based on EUI-64 are used to uniquely identify each LoRaWAN device [157, 158].

Table 4 summarizes the Network Layer protocols' security level, attacks and countermeasures proposed.

4.3 Application Layer

The application layer is responsible for managing the smart medical platform, which includes custom interfaces and role-based control panels for diagnostic decision making. The most commonly used IoMT protocols in the application layer are listed below:

HL7. HL7 is a set of standards that enable the exchange, integration, sharing, and retrieval of electronic health information between various health entities, allowing for the development of flexible and effective processes [167]. For its great importance, it is recognized as the most widely used application layer protocol in the healthcare systems [168]. The transparency of the information flow between health care systems is ensured by this protocol. In addition to clinical practice, HL7 supports the delivery, management and evaluation of health services [169].

Protecting data is the major aim from the security scope, because HL7 transmits data that may have a high impact. Many institutions rely on SSL VPNs (Secure Sockets Layer Virtual Private Networks) and similar solutions to protect the entire network. Deidentification/anonymization is helpful in protecting patient data [170].

CoAP. The Constrained Application Protocol (CoAP) protocol was originally designed for web transfer in the IoT with limited nodes and networks. The initial

Table 4 Network Layer protocols' security level, attacks and countermeasures

Protocol	Security level	Attacks	Countermeasures proposed	References
Wi-Fi	The devices' lack of granular authentication; Weakness and limited protection against DoS attacks and service integrity	Replay, Channel, DoS, Sniffing, MAC Spoofing, and packet analysis attacks	WPA and WPA2 security technology, 128-bits WEP authentication	[137]
ZigBee	Using unsecured key transport for pre-shared keys; PAN IDs do not have verification; There are no integrity checks in ACKs, and network keys are not properly registered	Key Sniffing, Association Flooding, Device Spoofing, DoS, jamming, Replay and Energy-consuming attacks	Symmetric cryptography, AES-CTR, AES-CCM, AES-CBC-MAC, 128-bit AES-based encryption system	[159–161]
WIA-PA	There is no public key or encryption algorithm, no intrusion prevention, and the first request is not encrypted	selective forwarding, Interference, Jamming, tampering, Traffic analysis attacks	Adaptive frequency switch (AFS), Adaptive Frequency Hopping (AFH), Timeslot hopping (TH) and message integrity (MIC)	[162]
6LoWPAN	The IP network and the radio signal are the targets of 6LowPAN attacks, Vulnerabilities at its fragmentation mechanism, Node's IP address remains unchanged	Signal jamming, Replay, Flooding and Traffic analysis attackers	DTLS cryptographic techniques; Internet Key Exchange technology (IKE); HIP host identification technology	[163–165]
LoRaWAN	Using a post-message dictionary, Resetting frame counters without recoding, Caching and replaying ACK packets, and Waking up sensors using forged gateway beacon transmissions	Replay, Plain text recovery, Denial of packet delivery, The battery exhaustion, Selective DoS and MITM attacks	AES -CMAC, MIC, AAES-CTR	[166]

motivation for developing this protocol was to meet the high requirements of the IoT as well as the need for a lightweight, low-rate protocol. This protocol is specifically suited to IoMT constrained nodes with limited memory and processing power [171]. CoAP, along with the MQTT protocol, is used in a proposed system in [172] for securing real-time health monitoring systems, to protect sensor data from security breaches during its continuous transmission over the layers. To avoid breaches such as data theft and DoS attacks, strong authentication techniques should be used. It is recommended to use an intrusion detection system to detect any malicious activity in the system [173]. DTLS can also be used to protect data [174].

MQTT. Message Queue Telemetry Transport (MQTT) is a standardized publish/subscribe Push protocol developed by IBM in 1999. This protocol is used by IoMT developers due to its low memory consumption and low bandwidth requirements; MQTT was designed to send data accurately even with long network delays and limited bandwidth [171]. A Blockchain-based medical application that connects various devices to an IoMT platform via MQTT is created in the work presented in [175]. In addition, the work in [136] proposed a system to connect a remote healthcare unit as it is inside the hospital, which uses the MQTT protocol to transfer measured data from the healthcare unit to the hospital's gateway.

Unfortunately, the MQTT protocol only supports authentication for the security mechanism, which does not encrypt data in transit by default. As a result, implementing this protocol raises concerns about confidentiality, authentication, and data integrity. MQTT brokers may require username/password authentication to ensure security, which is handled by the TLS/SSL (Secure Sockets Layer/Secure Socket Layer) protocol [176].

HTTP. There are different uses of this protocol in the IoMT. For example, it is used in a system that also includes a portable medical module with a pulse oximeter and an accelerometer that communicates with the microcontroller via a custom display to which a ZigBee module is connected. The goal of this system is to track the speed and direction of movement as well as the pulse and oxygen saturation of the blood [177]. Furthermore, it is used by the work presented in [12] to provide a system capable of dynamically assessing the amount of insulin needed to be administered to diabetic patients.

In order to make this protocol more secure, it is implemented on top of an encryption layer like SSL or TLS, to form its secure version https; with an '*s*' at the end to indicate that the data is exchanged securely via an encrypted tunnel using the SSL or TLS protocol. HTTPS client authentication is done below the protocol level (at the transport level). Only the server side of an SSL connection must use a certified public key from a server certification. This method is appropriate when the client wants to ensure that it is communicating with the intended server, but if the server needs to authenticate the client, it can use a traditional authentication mechanism (basic HTTP authentication or form authentication). On the other hand, Mutual SSL authentication, also known as two-way SSL authentication, necessitates the use of certified public keys by both the server and the client of the SSL connection. The server identifies the client in this case based on the client certificate used to establish the SSL connection [178].

Table 5 Application Layer protocols' security level, attacks and countermeasures proposed

Protocol	Security level	Attacks	Countermeasures proposed	References
HL7	No security integration; the size of HL7 and sources of messages are not validated by default	Spoofing attacks; DoS and flooding attacks	SSL VPNs; Deidentification/anonymization	[170]
CoAP	Type of DTLS implementation at the proxies level (multicast or unicast)	Parsing attacks; Caching attacks Amplification attacks; Cross-Protocol and Spoofing attacks	DTLS; TLS; CoAPs	[180, 181]
MQTT	No specific security mechanism	Eavesdropping; DoS; Timing attacks; Access, modify or redirect accessible data to an untrusted server	SSL/TLS	[182, 183]
HTTP	Insecure by default; Default HTTP implementations are not encrypted	Eavesdropping; injection and manipulation attacks	SSL/TLS (HTTPS version)	[184]

Tables 5 summarizes the Application Layer protocols' security level, attacks and countermeasures proposed.

5 Conclusions and Future Research Directions

The use of IoMT has recently grown in popularity. The majority of current studies are concerned about how medical and health-monitoring devices can help reduce healthcare spending and improve patient health. As a result, securing this technology has become extremely important since this IoMT is vulnerable to different attacks mainly because of its heavy reliance on wireless connectivity. These attacks can breach the system and invade the privacy of patients and affect the medical services' confidentiality, integrity and availability. Throughout this chapter, we have shown and explained the major security problems, challenges and drawbacks facing IoMT. In addition, we have discussed the way to secure the IoMT domains and their associated assets through varied suitable security measures to enhance IoMT services as well as the way to better the patients' health and experience via different techniques. Moreover, we have highlighted the importance of an effective security policy of different wireless communication protocols used by the IoMT system in order to keep it secured, private, trusted, and accurate.

In short, the purpose of this chapter is to highlight the relations between various technical and non-technical solutions to guarantee a secure and efficient system in all IoMT domains. Therefore, the chapter in hand gives some open research areas on security issues in IoMT both for traditional and novel-technology based solutions. To conclude, the need for developing robust security solutions using the latest technologies like Artificial Intelligence, Big Data and Blockchain is significantly growing as the IoMT are nowadays widely applicable.

References

1. I. T. Dunlap, The 5 Worst Examples of IoT Hacking and Vulnerabilities in Recorded History, https://www.iotforall.com/5-worst-iot-hacking-vulnerabilities/. Accessed: 30 Jan. 2022
2. A.J. Bamidele, R. Ogundokun, S. Misra, Cloud and IoMT-Based Big Data Analytics System During COVID-19 Pandemic, in *Efficient Data Handling for Massive Internet of Medical Things* (Springer, 2021), pp. 181–201
3. A-S.K. Pathan, Z.M. Fadlullah, S. Choudhury, M. Guerroumi, Internet of Things for smart living, Spec. Issue Wirel. Netw. Springer, 2019 **27**, 4293–4295 (2021), https://doi.org/10.1007/s11276-019-01970-3
4. J. Gubbi, R. Buyya, S. Marusic, M. Palaniswami, Internet of Things (IoT): a vision, architectural elements, and future directions. Futur. Gener. Comput. Syst. **29**(7), 1645–1660 (2013)
5. R.A. Khan, A.-S.K. Pathan, The state of the art Wireless body area sensor networks—a survey. Int. J. Distrib. Sens. Netw., SAGE publications **14**(4) (2018). https://doi.org/10.1177/1550147718768994.
6. S.S. Ahamad, A.-S.K. Pathan, A formally verified authentication protocol in secure framework for mobile healthcare during COVID-19-like pandemic. Connect. Sci., Taylor & Francis **33**(3) (2021). https://doi.org/10.1080/09540091.2020.1854180.
7. M. Haghi, K. Thurow, Habil, R. Stoll, M. Habil, Wearable devices in medical internet of things: scientific research and commercially available devices. Healthc. Inf. Res. **23**(1), 4–15 (2017)
8. R. Altawy, A.M. Youssef, Security tradeoffs in cyber physical systems: a case study survey on implantable medical devices. IEEE Access **4**, 959–979 (2016). https://doi.org/10.1109/ACCESS.2016.2521727
9. Pacemaker, Mayo Clinic, https://www.mayoclinic.org/testsprocedures/pacemaker/about/pac-20384689. Accessed 25 Dec. 2021
10. B.R. Larson, Y. Zhang, S.C. Barrett, J. Hatcliff, P.L. Jones, Enabling safe interoperation by medical device virtual integration. IEEE Design & Test **32**(5), 74–88 (2015). https://doi.org/10.1109/MDAT.2015.2464813
11. T. Belkhouja, X. Du, A. Mohamed, A.K. Al-Ali, M. Guizani, New plain-text authentication secure scheme for implantable medical devices with remote control, in *Proceedings of the GLOBECOM 2017 IEEE Global Communications Conference*, Singapore, 4–8 December 2017, pp. 1–5, https://doi.org/10.1109/GLOCOM.2017.8255015.
12. S. Sicari, A. Rizzardi, A. Coen-Porisini, How to evaluate an Internet of Things system: models, case studies, and real developments. Softw. Pract. Exp. **49**(11), 1663–1685 (2019)

13. J.E. Ferguson, A.D. Redish, Wireless communication with implanted medical devices using the conductive properties of the body. Expert Rev. Med. Devices **8**(4), 427–433 (2011). https://doi.org/10.1586/erd.11.16
14. N. Scarpato, A. Pieroni, L. Di Nunzio, F. Fallucchi, E-health-IoT universe: a review. Int. J. Adv. Sci. Eng. Inf. Technol. **7**(6), 2328–2336 (2017)
15. S. Neethirajan, Recent advances in wearable sensors for animal health management. Sens. Bio-Sens. Res **12**(44), 15–29 (2017)
16. A. Phaneuf, Latest trends in medical monitoring devices and wearable health technology. (Business Insider, 2019)
17. Heart health notifications on your Apple Watch, Apple, https://support.apple.com/en-us/HT208931. Accessed 25 Dec. 2021
18. A. Kos, V. Milutinović, A. Umek, Challenges in wireless communication for connected sensors and wearable devices used in sport biofeedback applications. Futur. Gener. Comput. Syst. **92**, 582–592 (2019)
19. H. Jahankhani, J. Ibarra, Digital forensic investigation for the Internet of Medical Things (IoMT). J. Forensic Leg. Investig. Sci. **5** (029) (2019)
20. O. AlShorman, B. AlShorman, M. Al-khassaweneh, F. Alkahtani, A review of internet of medical things (IoMT)—based remote health monitoring through wearable sensors: a case study for diabetic patients. Indones. J. Electr. Eng. Comput. Sci. **20**(1), 414–422 (2020)
21. Medical Device Radiocommunications Service (MedRadio), Federal Communications Commission, https://www.fcc.gov/medical-device-radiocommunications-service-medradio. Accessed 25 Dec. 2021
22. A. Ghubaish, T. Salman, M. Zolanvari, D. Unal, A. Al-Ali, R. Jain, Recent advances in the Internet of Medical Things (IoMT) systems security. IEEE Internet Things J. **8**(11), 8707–8718 (2020)
23. I.U. Din, M. Guizani, S. Hassan, B.-S. Kim, M.K. Khan, M. Atiquzzaman, S.H. Ahmed, The Internet of Things: a review of enabled technologies and future challenges. IEEE Access **7**, 7606–7640 (2018)
24. F.S.D. Lima Filho, F.A. Silveira, A. de Medeiros Brito Junior, G. Vargas-Solar, L.F. Silveira, Smart detection: an online approach for DoS/DDoS attack detection using machine learning. Secur. Commun. Netw. (2019)
25. P. Kasyoka, M. Kimwele, S. MbanduAngolo, Certificateless pairing-free authentication scheme for wireless body area network in healthcare management system. J. Med. Eng. Technol. **44**(1), 12–19 (2020)
26. T. Belkhouja, S. Sorour, M.S. Hefeida, Role-based hierarchical medical data encryption for implantable medical devices, in *2019 IEEE Global Communications Conference (GLOBECOM)*, 9–13 Dec. 2019, (2019), pp. 1–6
27. M. Papaioannou, M. Karageorgou, G. Mantas,V. Sucasas, I. Essop, J. Rodriguez, D. Lymberopoulos, A survey on security threats and countermeasures in Internet of Medical Things (IoMT), in *2019 IEEE Global Communications Conference (GLOBECOM)*, 9–13 Dec. (2019), pp. 1–6
28. A.J. Menezes, P.C. Van Oorschot, S.A. Vanstone, *Handbook of Applied Cryptography*. (CRC Press, 2018)
29. J. Hash, P. Bowen, A. Johnson, C.D. Smith, D.I. Steinberg, *An Introductory Resource Guide for Implementing the Health Insurance Portability and Accountability Act (HIPAA) Security Rule*, Technology Administration (American Health Information Management Association, Illinois, 2008)
30. Talend, What is Data Integrity and Why Is It Important?, https://www.talend.com/resources/what-is-data-integrity/. Accessed 25 Dec. 2021
31. Y. Sun, F.P.-W. Lo, B. Lo, Security and privacy for the Internet of medical things enabled healthcare systems: a survey. IEEE Access **7**, 183339–183355 (2019)
32. T. Bienkowski, GDPR is explicit about protecting availability, https://www.netscout.com/blog/gdpravailability-protection (cit. on p. 18) (2018)

33. A. Alrawais, A. Alhothaily, C. Hu, X. Cheng, Fog computing for the Internet of Things: security and privacy issues. IEEE Internet Comput. **21**(2), 34–42 (2017)
34. R. Kumar, R. Tripathi, Towards design and implementation of security and privacy framework for internet of medical things (iomt) by leveraging blockchain and ipfs technology. J. Supercomput. **77**, 7916–7955 (2021)
35. J.P.A. Yaacoub, M. Noura, H.N. Noura, O. Salman, E. Yaacoub, R. Couturier, A. Chehab, Securing internet of medical things systems: limitations, issues and recommendations. Futur. Gener. Comput. Syst. **105**, 581–606 (2020)
36. J.S. Lee, C.C. Chang, K.J. Wei, Provably secure conference key distribution mechanism preserving the forward and backward secrecy. Int. J. Netw. Secur. **15**(5), 405–410 (2013)
37. A. Abusukhon, N.M. Anwar, Z. Mohammad, B. Alghannam, A hybrid network security algorithm based on Diffie Hellman and Text to Image Encryption algorithm. J. Discret. Math. Sci. Cryptogr. **22**(1), 65–81 (2019)
38. J. Batamuliza, D. Hanyurwimfura, A secure and efficient anonymous certificateless signcryption for key distribution scheme for smart grid, in *2020 21st International Arab Conference on Information Technology (ACIT)* (IEEE, 2020), pp. 1–7
39. N. Park., M. Kim, H.C. Bang, Symmetric key-based authentication and the session key agreement scheme in IoT environment, in *Computer Science and its Applications*. Springer, Berlin, Heidelberg, (2015) pp. 379–384.
40. Casey Crane, Asymmetric versus symmetric encryption: definitions & differences, https://www.thesslstore.com/blog/asymmetric-vs-symmetric-encryption/. Accessed 25 Dec. 2021
41. V.H. Tutari, B. Das, D.R. Chowdhury, A continuous role-based authentication scheme and data transmission protocol for implantable medical devices. In *2019 Second International Conference on Advanced Computational and Communication Paradigms (ICACCP)*, 25–28 Feb. 2019, (2019), pp. 1–6.
42. M.J. Marin-Jiménez, F.M. Castro, N. Guil, F. De la Torre, R. Medina-Carnicer, Deep multitask learning for gait-based biometrics, in *2017 IEEE International Conference on Image Processing (ICIP)* (IEEE, 2017), pp. 106–110.
43. S. Azad, A.-S.K. Pathan, *Practical Cryptography: Algorithms and Implementations Using C++*. ISBN: 978-1-48-222889-2, (CRC Press, Taylor & Francis Group, USA, 2014)
44. K. Juretus, I. Savidis, Reducing logic encryption overhead through gate level key insertion, in *2016 IEEE International Symposium on Circuits and Systems (ISCAS)*, 22–25 (2016). https://doi.org/10.1109/ISCAS.2016.7538898
45. B.A. Alzahrani, A. Irshad, A. Albeshri, K. Alsubhi, A provably secure and lightweight patient-healthcare authentication protocol in wireless body area networks. Wireless Pers. Commun. **177**(1), 47–69 (2020)
46. Z. Xu, C. Xu, W. Liang, J. Xu, H. Chen, A lightweight mutual authentication and key agreement scheme for medical Internet of Things. IEEE Access **7**, 53922–53931 (2019)
47. S.B. Sadkhan, F.H. Abdulraheem, A proposed ANFIS evaluator for RSA cryptosystem used in cloud networking, in *2017 International Conference on Current Research in Computer Science and Information Technology (ICCIT)*, 26–27 April 2017, https://doi.org/10.1109/CRCSIT.2017.7965561
48. A.A. Shaikh, N.S. Vani, An extended approach for securing the Short Messaging Services of GSM using multi-threading elliptical curve cryptography, in *2015 International Conference on Communication, Information & Computing Technology (ICCICT)*, 15–17 Jan. (2015), https://doi.org/10.1109/ICCICT.2015.7045733.
49. V.J. Jariwala, D.C. Jinwala, Chapter 4—adaptableSDA: secure data aggregation framework in wireless body area networks, in *Wearable and Implantable Medical Devices*, eds. by N. Dey, A. S. Ashour, S. James Fong, C. Bhatt, vol 7 (Academic Press, 2020), pp. 79–114
50. T. Bhatia, A.K. Verma, G. Sharma, Towards a secure incremental proxy re-encryption for e-healthcare data sharing in mobile cloud computing. Concurr. Comput.: Pract. Exp. **32**(5), 1–16 (2020)
51. K.E. Makkaoui, A. Beni-Hssane, A. Ezzati, Can hybrid homomorphic encryption schemes be practical?, in *2016 5th International Conference on Multimedia Computing and Systems (ICMCS)*, 29 Sept.–1 Oct. (2016), https://doi.org/10.1109/ICMCS.2016.7905580

52. G. Kalyani, S. Chaudhari, An efficient approach for enhancing security in Internet of Things using the optimum authentication key. Int. J. Comput. Appl. **42**(3), 306–314 (2020)
53. G.M. Abdullah, Q. Mehmood, C.B.A. Khan, Adoption of lamport signature scheme to implement digital signatures in IoT, in *2018 International Conference on Computing, Mathematics and Engineering Technologies (iCoMET)*, 3–4 March (2018), https://doi.org/10.1109/ICOMET.2018.8346359
54. C. Easttom, N. Mei, Mitigating implanted medical device cyber security risks, in *IEEE 10th Annual Ubiquitous Computing. Electronics & Mobile Communication Conference (UEMCON)*, vol. 2019 (2019), pp. 0145–0148
55. Z.A. Alizai, N.F. Tareen, I. Jadoon, Improved IoT device authentication scheme using device capability and digital signatures, in *2018 International Conference on Applied and Engineering Mathematics (ICAEM)* (IEEE, 2018) pp. 1–5
56. A. Kumari, S. Jangirala, M.Y. Abbasi, V. Kumar, M. Alam, ESEAP: ECC based secure and efficient mutual authentication protocol using smart card. J. Inf. Secur. Appl. **51**, 1–12 (2020)
57. G. Zheng, W. Yang, C. Valli, L. Qiao, R. Shankaran, M.A. Orgun, S.C. Mukhopadhyay, Finger-to-Heart (F2H): authentication for wireless implantable medical devices. IEEE J. Biomed. Health Inform. **23**(4), 1546–1557 (2019)
58. G. Zheng, W. Yang, M. Johnstone, R. Shankaran, C. Valli, Securing the elderly in cyberspace with fingerprints, in *Assistive Technology for the Elderly*, eds. by N.K. Suryadevara, S.C. Mukhopadhyay (Academic Press, 2020) pp. 59–79
59. M. Dammak, O.R.M. Boudia, M.A. Messous, S.M. Senouci, C. Gransart, Token-based lightweight authentication to secure IoT networks, in *16th IEEE Annual Consumer Communications & Networking Conference (CCNC)* (IEEE, 2019), pp. 1–4
60. W. Youssef, A.O. Zaid, M.S. Mourali, M.H. Kammoun, RFID-based system for secure logistic management of implantable medical devices in Tunisian health centers, in *2019 IEEE International Smart Cities Conference (ISC2)* (2019), pp. 83–86
61. A. Bhawiyuga, M. Data, A. Warda, Architectural design of token based authentication of MQTT protocol in constrained IoT device, in *2017 11th International Conference on Telecommunication Systems Services and Applications (TSSA)* (IEEE, 2017), pp. 1–4
62. S. Saif, S. Biswas, S. Chattopadhyay, Intelligent, secure big health data management using deep learning and blockchain technology: an overview, in *Deep Learning Techniques for Biomedical and Health Informatics*, eds. by S. Dash, B. Acharya, M. Mittal, A. Abraham, A. Kelemen, vol 68 (Springer International Publishing, Cham, 2020) pp. 187–209
63. A.A. Hady, A. Ghubaish, T. Salman, D. Unal, R. Jain, Intrusion detection system for healthcare systems using medical and network data: a comparison study. IEEE Access **8**, 106576–106584 (2020)
64. L. Gupta, T. Salman, M. Zolanvari, A. Erbad, R. Jain, Fault and performance management in multi-cloud virtual network services using AI: a tutorial and a case study. Comput. Netw. **165**, 106950 (2019)
65. M.M. Ahemd, M.A. Shah, A. Wahid, Iot security: a layered approach for attacks and defenses, in *2017 International Conference on Communication Technologies (ComTech)* (IEEE, 2017) pp. 104–110
66. D. Antonioli, N.O. Tippenhauer, K. Rasmussen, BIAS: bluetooth impersonation attacks, in *IEEE Symposium on Security and Privacy (SP)*. (IEEE, 2020), pp. 549–562
67. I. Andrea, C. Chrysostomou, G. Hadjichristofi, Internet of things: security vulnerabilities and challenges, in *IEEE Symposium on Computers and Communication (ISCC)*. (IEEE, 2015), pp. 180–187
68. F.-X. Standaert, Introduction to side-channel attacks, in *Secure Integrated Circuits and Systems* (Springer, Boston, MA, 2010), pp. 27–42
69. S. Maji, U. Banerjee, S.H. Fuller, M.R. Abdelhamid, P.M. Nadeau, R.T. Yazicigil, A.P. Chandrakasan, A low-power dual-factor authentication unit for secure implantable devices, in *2020 IEEE Custom Integrated Circuits Conference (CICC)* (2020), pp. 1–4
70. X. Chen, H. Zhu, D. Geng, W. Liu, R. Yang, S. Li, Merging RFID and blockchain technologies to accelerate big data medical research based on physiological signals. J. Healthc. Eng. 1–17 (2020)

71. D. Koutras, G. Stergiopoulos, T. Dasaklis, P. Kotzanikolaou, D. Glynos, C. Douligeris, Security in iomt communications: a survey. Sensors **20**(17), 4828 (2020)
72. M.N. Aman, K.C. Chua, B. Sikdar, A light-weight mutual authentication protocol for iot systems, in *GLOBECOM 2017–2017 IEEE Global Communications Conference* (2017), pp. 1–6
73. T. Gomes, F. Salgado, A. Tavares, J. Cabral, Cute mote, a customizable and trustable end-device for the internet of things. IEEE Sens. J. **17**(20), 6816–6824 (2017)
74. S. Sicari, A. Rizzardi, D. Miorandi, A. Coen-Porisini, Reato: reacting to denial of service attacks in the internet of things. Comput. Netw. **137**, 37–48 (2018)
75. P. Porambage, C. Schmitt, P. Kumar, A. Gurtov, M. Ylianttila, Pauthkey: a pervasive authentication protocol and key establishment scheme for wireless sensor networks in distributed iot applications. Int. J. Distrib. Sens. Netw. **10**(7), 1–15 (2014)
76. X. Hei, X. Du, J. Wu, F. Hu, Defending resource depletion attacks on implantable medical devices, in *IEEE Global Telecommunications Conference GLOBECOM 2010*. (IEEE, 2010), pp. 1–5
77. S.C. Sethuraman, V. Vijayakumar, S. Walczak, Cyber Attacks on Healthcare Devices Using Unmanned Aerial Vehicles. J. Med. Syst. **44**(1), 1–10 (2020)
78. Rambus, Industrial iot: Threats and countermeasures, https://www.rambus.com/iot/industrial-iot/. Accessed 25 Dec. 2021
79. P. Kukielka, Z. Kotulski, New Unknown Attack Detection with the Neural Network–Based IDS, in *Chapter 11, The State of the Art in Intrusion Prevention and Detection*, ed. by A.-S.K. Pathan. ISBN 9781482203516 (CRC Press, Taylor & Francis Group, USA, 2014)
80. O. Shwartz, Y. Mathov, M. Bohadana, Y. Elovici, Y. Oren, Opening Pandora's Box: Effective Techniques for Reverse Engineering IoT Devices, in *International Conference on Smart Card Research and Advanced Applications* (Springer, Cham, 2017), pp. 1–21
81. F. I. Khan, S. Hameed, Understanding security requirements and challenges in internet of things (iots): a review, arXiv:1808.10529 (2018)
82. H. Ding, J. Han, Y. Zhang, F. Xiao, W. Xi, G. Wang, Z. Jiang, Preventing unauthorized access on passive tags, in *IEEE INFOCOM 2018-IEEE Conference on Computer Communications*, Honolulu, HI, USA, IEEE, 16–19 April 2018, pp. 1115–1123
83. D. Zheng, A. Wu, Y. Zhang, Q. Zhao, Efficient and privacy-preserving medical data sharing in internet of things with limited computing power. IEEE Access **6**, 28019–28027 (2018)
84. U. Guin, A. Singh, M. Alam, J. Canedo, A. Skjellum, A secure low-cost edge device authentication scheme for the internet of things, in *2018 31st International Conference on VLSI Design and 2018 17th International Conference on Embedded Systems (VLSID)* (IEEE, 2018) pp. 85–90
85. A. Al-Fuqaha, M. Guizani, M. Mohammadi, M. Aledhari, M. Ayyash, Internet of things: a survey on enabling technologies, protocols, and applications. IEEE Commun. Surveys Tutor. **17**(4), 2347–2376 (2015)
86. S. Deshmukh, S.S. Sonavane, Security protocols for Internet of Things: a survey, in *Proceedings of the 2017 International Conference on Nextgen Electronic Technologies: Silicon to Software (ICNETS2)*, Chennai, India, 23–25 March 2017, pp. 71–74
87. M. Bagga, P. Thakral, T. Bagga, A Study on IoT: Model, Communication Protocols, Security Hazards & Countermeasures, in *2018 Fifth International Conference on Parallel, Distributed and Grid Computing (PDGC)* (IEEE, 2018), pp. 591–598
88. K. Devadiga, IEEE 802.15.4 and the Internet of Things, Technical Report; Aalto University School of Science: Espoo, Finland, 2011. https://wiki.aalto.fi/download/attachments/59704179/devadiga-802-15-4-and-the-iot.pdf?version=1
89. R. Jain, *Wireless Protocols for IoT Part I: Bluetooth and Bluetooth Smart*. (Washington University in Saint Louis, Saint Louis, MO, USA, 2016). https://www.cse.wustl.edu/~jain/cse574-16/ftp/j_11ble.pdf
90. H. Javdani, H. Kashanian, Internet of things in medical applications with a service-oriented and security approach: a survey. Health Technol. **8**(1), 39–50 (2018)

91. S. Amendola, R. Lodato, S. Manzari, C. Occhiuzzi, G. Marrocco, RFID technology for IoT-based personal healthcare in smart spaces. IEEE Internet Things J. **1**(2), 144–152 (2014)
92. GS1, EPC™ Radio-Frequency Identity Protocols Generation-2 UHF RFID Specification for RFID Air Interface, GS1: Brussels, Belgium, (2015) pp. 1–152, https://www.gs1.org/sites/default/files/docs/epc/Gen2_Protocol_Standard.pdf. Accessed 25 Dec. 2021
93. L.M. Dang, M.J. Piran, D. Han, K. Min, H. Moon, A survey on internet of things and cloud computing for healthcare. Electronics **8**(7), 768 (2019)
94. Proceedings of the First ACM Conference on Wireless Network Security, WiSec '08, Alexandria, VA, USA, 31 March–2 April 2008; Association for Computing Machinery, New York, NY, USA (2008) pp. 140–147
95. D.Z. Sun, J.D. Zhong, hash-based RFID security protocol for strong privacy protection. IEEE Trans. Consum. Electron. **58**(4), 1246–1252 (2012)
96. I. Cvitic, M. Vujic, S. Husnjak, Classification of security risks in the IoT environment, in *Proceedings of the Annals of DAAAM and Proceedings of the International DAAAM Symposium*, Vienna, Austria, 21–24 October 2015 (2015) pp. 731–740
97. V. Hassija, V. Chamola, V. Saxena, D. Jain, P. Goyal, B. Sikdar, A survey on IoT security: application areas, security threats, and solution architectures. IEEE Access **7**, 82721–82743 (2019)
98. ECMA International, Near Field Communication-Interface and Protocol (NFCIP-1), ECMA International: Geneva, Switzerland, (2013) pp. 52, https://www.ecma-international.org/wp-content/uploads/ECMA-340_3rd_edition_june_2013.pdf. Accessed 25 Dec. 2021
99. H. Eun, H. Lee, H. Oh, Conditional privacy preserving security protocol for NFC applications. IEEE Trans. Consum. Electron. **59**(1), 153–160 (2013)
100. G. Madlmayr, J. Langer, C. Kantner, J. Scharinger, NFC Devices: Security and Privacy, in *Proceedings of the 2008 Third International Conference on Availability, Reliability and Security*, Barcelona, Spain, 4–7 March 2008, pp. 642–647.
101. N.E. Tabet, M.A. Ayu, Analysing the security of NFC based payment systems, in *Proceedings of the 2016 International Conference on Informatics and Computing (ICIC)*, Mataram, Indonesia, 28–29 October (2016) pp. 169–174
102. Cypress. PSoC® Creator Component Datasheet-Bluetooth Low Energy (BLE) 3.10, Description SIG adopted Profiles and Services Comprehensive APIs, (2015) pp. 408–943, https://www.cypress.com/file/232821/download. Acessed 25 Dec. 2021
103. J.B. SIG, Bluetooth Specification, v. 3.0. EEE Spectr. (2009)
104. M. Frustaci, P. Pace, G. Aloi, G. Fortino, Evaluating critical security issues of the IoT world: present and future challenges. IEEE Internet Things J. **5**(4), 2483–2495 (2018)
105. A.M. Lonzetta, P. Cope, J. Campbell, B.J. Mohd, T. Hayajneh, Security vulnerabilities in Bluetooth technology as used in IoT. J. Sens. Actuator Netw. **7**(3), 28 (2018)
106. G. Choudhary, A.K. Jain, Internet of Things: A survey on architecture, technologies, protocols and challenges, in *Proceedings of the 2016 International Conference on Recent Advances and Innovations in Engineering (ICRAIE)*, Jaipur, India, 23–25 December 2016, pp. 1–8
107. C.W. Badenhop, S.R. Graham, B.W. Ramsey, B.E. Mullins, L.O. Mailloux, The Z-Wave routing protocol and its security implications. Comput. Secur. **68**, 112–129 (2017)
108. M.B. Yassein, W. Mardini, T. Almasri, Evaluation of security regarding Z-Wave wireless protocol, in *Proceedings of the Fourth International Conference on Engineering & MIS 2018*, Istanbul, Turkey, 9–11 April 2018; Association for Computing Machinery: New York, NY, USA, (2018) pp. 1–8
109. S.R. Ramson, D.J. Moni, A case study on different wireless networking technologies for remote health care. Intell. Decis. Technol. **10**(4), 353–364 (2016)
110. H. Fotouhi, A. Causevic, K. Lundqvist, M. Björkman, Communication and security in health monitoring systems–a review, in *IEEE 40th Annual Computer Software and Applications Conference (COMPSAC)*. vol. 1 (IEEE, 2016), pp. 545–554
111. A. Chehri, H.T. Mouftah, Internet of Things-integrated IR-UWB technology for healthcare applications. Concurr. Comput.: Pract. Exp. **32**(2) (2020), e5454

112. W. Yin, X. Yang, L. Zhang, E. Oki, ECG monitoring system integrated with IR-UWB radar based on CNN. IEEE Access **4**, 6344–6351 (2016)
113. X. Zhang, M. Wei, P. Wang, Y. Kim, Research and implementation of security mechanism in ISA100.11a networks, in *Proceedings of the 2009 9th International Conference on Electronic Measurement Instruments*, Beijing, China, IEEE, 16–19 August 2009, pp. 4–716–4–721
114. Y. Zeng, J. Cao, J. Hong, S. Zhang, L. Xie, Secure localization and location verification in wireless sensor networks: a survey. J. Supercomput. **64**, 685–701 (2013)
115. Y. Wang, X. Ma, G. Leus, An UWB ranging-based localization strategy with internal attack immunity, in *Proceedings of the 2010 IEEE International Conference on Ultra-Wideband*, Nanjing, China, IEEE, 2(2010) pp. 1–4
116. M. Flury, M. Poturalski, P. Papadimitratos, J.P. Hubaux, J.Y. Le Boudec, Effectiveness of Distance-Decreasing Attacks Against Impulse Radio Ranging, in *Proceedings of the 3rd ACM Conference on Wireless Network Security, WiSec'10*, Hoboken, NJ, USA, 22–24 March 2010, pp. 117–128
117. M. Singh, P. Leu, S. Capkun, UWB with pulse reordering: securing ranging against relay and physical-layer attacks, in *Proceedings of the 26th Annual Network and Distributed System Security Symposium (NDSS 2019)*, San Diego, CA, USA, 24–27 February 2019
118. S. Sundaresan, R. Doss, W. Zhou, RFID in Healthcare–Current Trends and the Future, in *Mobile Health* (Springer, Berlin/Heidelberg, Germany, 2015), pp. 839–870
119. S.A. Weis, S.E. Sarma, R.L. Rivest, D.W. Engels, Security and privacy aspects of low-cost radio frequency identification systems, in *Security in Pervasive Computing* (Springer, Berlin/Heidelberg, Germany, 2004), pp. 201–212
120. S. Upson, RFID Systems May Disrupt the Function of Medical Devices, IEEE Spectr, 2008, https://spectrum.ieee.org/computing/embedded-systems/rfid-systems-may-disrupt-the-function-of-medical-devices. Accessed 25 Dec. 2021
121. M. Roland, J. Langer, J. Scharinger, Applying relay attacks to Google Wallet, in *Proceedings of the 2013 5th International Workshop on Near Field Communication (NFC)*, Zurich, Switzerland, 5 February 2013, pp. 1–6
122. E. Haselsteiner, K. Breitfuß, Security in Near Field Communication (NFC) Strengths and Weaknesses (2006)
123. J. Dunning, Taming the blue beast: a survey of bluetooth based threats. IEEE Secur. Priv. **8**(2), 20–27 (2010)
124. LeCroy, CATC Merlin II-Bluetooth V1.2 Protocol Analyzer, LeCroy, Santa Clara, CA, USA, 2003, http://cdn.teledynelecroy.com/files/pdf/lecroy_merlinii_datasheet.pdf. Accessed 25 Dec. 2021
125. K.M. Haataja, K. Hypponen, Man-in-the-middle attacks on bluetooth: a comparative analysis, a novel attack, and countermeasures, in *Proceedings of the 2008 3rd International Symposium on Communications, Control and Signal Processing*, St Julians, Malta, 12–14 March 2008, pp. 1096–1102
126. N.B.N.I. Minar, M. Tarique, Bluetooth security threats and solutions: a survey. Int. J. Distrib. Parallel Syst. **3**(1) (2012), 127
127. B. Fouladi, S. Ghanoun, SensePost UK Honey, i'm home!!, hacking zwave home automation systems, Black Hat USA, Las Vegas, Nevada, 2013, https://code.google.com/archive/p/z-force/. Accessed 25 Dec. 2021
128. P. Sarigiannidis, E. Karapistoli, A.A. Economides, Detecting Sybil attacks in wireless sensor networks using UWB ranging-based information. Expert Syst. Appl. **42**(21), 7560–7572 (2015)
129. A. Compagno, M. Conti, A.A. D'Amico, G. Dini, P. Perazzo, L. Taponecco, Modeling enlargement attacks against UWB distance bounding protocols. IEEE Trans. Inf. Forensics Secur. **11**(7), 1565–1577 (2016)
130. N. Vidgren, K. Haataja, J.L. Patiño-Andres, J.J. Ramírez-Sanchis, P. Toivanen, Security threats in ZigBee-enabled systems: vulnerability evaluation, practical experiments, countermeasures, and lessons learned, in *Proceedings of the Annual Hawaii International Conference on System Sciences*, Wailea, HI, USA, vol. 7–10 (2013), pp. 5132–5138

131. M.R. Fuentes, N. Huq, Securing connected hospitals, by Trend Micro, 2018, https://documents.trendmicro.com/assets/rpt/rpt-securing-connected-hospitals.pdf. Accessed 25 Dec. 2021
132. T. Poongodi, A. Rathee, R. Indrakumari, P. Suresh, IoT Sensing Capabilities: Sensor Deployment and Node Discovery, Wearable Sensors, Wireless Body Area Network (WBAN), data acquisition, in *Principles of Internet of Things (IoT) Ecosystem: Insight Paradigm*. ed. by S.L. Peng, S. Pal, L. Huang (Springer International Publishing, Cham, Switzerland, 2020), pp. 127–151
133. M. Sain, Y.J. Kang, H.J. Lee, Survey on security in Internet of Things: state of the art and challenges, in *Proceedings of the International Conference on Advanced Communication Technology, ICACT*, vol. 19–22 (IEEE, Bongpyeong, Korea, 2017), pp. 699–704
134. T. Salman, R. Jain, A Survey of Protocols and Standards for Internet of Things, arXiv 2019, arXiv:1903.11549
135. G. Calcagnini, E. Mattei, F. Censi, M. Triventi, R. Lo Sterzo, E. Marchetta, P. Bartolini, Electromagnetic compatibility of WiFi technology with life-supporting medical devices, in *Proceedings of the World Congress on Medical Physics and Biomedical Engineering*, Munich, Germany, 7–12 September 2009 (Springer, Berlin/Heidelberg, Germany, 2009), pp. 616–619
136. B.A. Mubdir, H.M.A. Bayram, Adopting MQTT for a multi protocols IoMT system. Int. J. Electr. Comput. Eng. *12*(1), 2088–8708 (2022)
137. H. Peng, WIFI network information security analysis research, in *Proceedings of the 2012 2nd International Conference on Consumer Electronics, Communications and Networks (CECNet)*, Yichang, China, IEEE, 21–23 April 2012, pp. 2243–2245
138. M.A. Uddin, A. Stranieri, I. Gondal, V. Balasubramanian, Continuous patient monitoring with a patient centric agent: a block architecture. IEEE Access **6**, 32700–32726 (2018)
139. D.C. Yacchirema, C.E. Palau, M. Esteve, Enable IoT interoperability in ambient assisted living: active and healthy aging scenarios, in *Proceedings of the 2017 14th IEEE Annual Consumer Communications & Networking Conference (CCNC)*, Las Vegas, NV, USA, 8–11 January 2017, pp. 53–58
140. Zigbee Alliance Inc, ZigBee Specification, Zigbee Alliance Inc. (2015), pp. 1–378, https://zigbeealliance.org/wp-content/uploads/2019/11/docs-05-3474-21-0csg-zigbee-specification.pdf. Accessed on 25 Dec. 2021
141. S. Plosz, A. Farshad, M. Tauber, C. Lesjak, T. Ruprechter, N. Pereira, Security vulnerabilities and risks in industrial usage of wireless communication, in *Proceedings of the 2014 IEEE Emerging Technology and Factory Automation (ETFA)*, Barcelona, Spain, IEEE, 16–19 September 2014.
142. O. Olawumi, K. Haataja, M. Asikainen, N. Vidgren, P. Toivanen, Three practical attacks against ZigBee security: attack scenario definitions, practical experiments, countermeasures, and lessons learned, in *Proceedings of the 2014 14th International Conference on Hybrid Intelligent Systems*, Hawally, Kuwait, 14–16 December 2014, pp. 199–206
143. S.M. Rana, M.A., Halim, M.H. Kabir, Design and implementation of a security improvement framework of Zigbee network for intelligent monitoring in IoT platform. Appl. Sci. **8**(11), 2305 (2018)
144. T. Zhong, C. Mengjin, Z. Peng, W. Hong, Real-time communication in WIA-PA industrial wireless networks, in *Proceedings of the 2010 3rd International Conference on Computer Science and Information Technology*, Chengdu, China, 9–11 July 2010, Vol 2, pp. 600–605
145. C.R. Su, J. Hajiyev, C.J. Fu, K.C. Kao, C.H. Chang, C.T. Chang, A novel framework for a remote patient monitoring (RPM) system with abnormality detection. Health Policy Technol. **8**(2), 157–170 (2019). https://doi.org/10.1016/j.hlpt.2019.05.008
146. X. Wang, L. Cui, Z. Guo, Advanced technologies in ad hoc and sensor networks, in *Proceedings of the 7th China Conference on Wireless Sensor Networks*, vol. 295 (2014), pp. 288
147. W. Liang, X. Zhang, Y. Xiao, F. Wang, P. Zeng, H. Yu, Survey and experiments of WIA-PA specification of industrial wireless network. Wirel. Commun. Mob. Comput. **11**(8), 1197–1212 (2011)

148. W. Min, X. Zhang, W. Ping, K. Kim, Y. Kim, Research and implementation of the security method based on WIA-PA standard, in *Proceedings of the 2010 International Conference on Electrical and Control Engineering*, Wuhan, China, 25–27 June 2010, pp. 1580–1585
149. H. Fotouhi, A. Caušević, M. Vahabi, M. Björkman, Interoperability in heterogeneous low-power wireless networks for health monitoring systems, in *Proceedings of the 2016 IEEE International Conference on Communications Workshops (ICC)*, Kuala Lumpur, Malaysia, 23–27 May 2016, pp. 393–398
150. J. Olsson, 6LoWPAN Demystified, Texas Instruments: Dallas, TX, USA, 13 (2014), https://www.ti.com/lit/wp/swry013/swry013.pdf?ts=1645202797751&ref_url=https%253A%252F%252Fwww.google.com%252F. Accessed 25 Dec. 2021
151. P. Chen, Yokogawa electric corporation, using ISA100.11a wireless technology to monitor pressure and temperature in a refinery (2011)
152. Y. Benslimane, K. Benahmed, H. Benslimane, Security mechanisms for 6LoWPAN network in context of Internet of Things: a survey, in *Renewable Energy for Smart and Sustainable Cities*, ed. by M. Hatti (Springer International Publishing, Cham, Switzerland, 2019), pp. 49–69
153. i-Scoop, LoRa and LoRaWAN: the technologies, ecosystems, use cases and market by i-Scoop, https://www.i-scoop.eu/internet-of-things-guide/lpwan/iot-network-lora-lorawan/. Accessed 25 Dec. 2021
154. J. Haxhibeqiri, E. De Poorter, I. Moerman, J. Hoebeke, A survey of LoRaWAN for IoT: from technology to application. Sensors **18**(11), 3995 (2018)
155. A. Mdhaffar, T. Chaari, K. Larbi, M. Jmaiel, B. Freisleben, IoT-based health monitoring via LoRaWAN, in *Proceedings of the IEEE EUROCON 2017–17th International Conference on Smart Technologies*, Ohrid, Macedonia, 6–8 July 2017, pp. 519–524
156. P.A. Catherwood, D. Steele, M. Little, S. Mccomb, J. Mclaughlin, A community-based IoT personalized wireless healthcare solution trial. IEEE J. Transl. Eng. Health Med. **6**, 1–13 (2018)
157. A. Gemalto, Semtech, LoRaWAN™ security a white paper prepared for the LoRa alliance, https://lora-alliance.org/sites/default/files/2019-05/lorawan_security_whitepaper.pdf. Accessed 25 Dec. 2021
158. A. Yegin, T. Kramp, P. Dufour, R. Gupta, R. Soss, O. Hersent, D. Hunt, N. Sornin, LoRaWAN protocol: specifications, security, and capabilities, in *LPWAN Technologies for IoT and M2M Applications*. ed. by B.S. Chaudhari, M. Zennaro (Academic Press, Cambridge, MA, USA, 2020), pp. 37–63
159. L.N. Whitehurst, T.R. Andel, J.T. McDonald, Exploring security in ZigBee networks, in *Proceedings of the 9th Annual Cyber and Information Security Research Conference*, Oak Ridge, TN, USA, 8–10 April 2014, pp. 25–28
160. E. Ronen, A. Shamir, A.O. Weingarten, C. O'Flynn, IoT goes nuclear: creating a ZigBee chain reaction, in *Proceedings of the 2017 IEEE Symposium on Security and Privacy (SP)*, CA, USA, 22–26 May 2017, pp.195–212
161. X. Cao, D.M. Shila, Y. Cheng, Z. Yang, Y. Zhou, J. Chen, Ghost-in-zigbee: energy depletion attack on zigbee-based wireless networks. IEEE Internet Things J **3**(5), 816–829 (2016)
162. Y. Qi, W. Li, X. Luo, Q. Wang, Security analysis of WIA-PA protocol, in *Advanced Technologies in Ad Hoc and Sensor Networks*, vol. 295 (Springer, Berlin/Heidelberg, Germany), pp. 287–298 (2014)
163. G. Glissa, A. Meddeb, 6LoWPAN multi-layered security protocol based on IEEE 802.15. 4 security features, in *Proceedings of the 2017 13th International Wireless Communications and Mobile Computing Conference (IWCMC)*, Valencia, Spain, 26–30 June 2017, pp 264–269
164. J.-S. Lee., Y.-W. Su, C.-C. Shen, A comparative study of wireless protocols: Bluetooth, UWB, ZigBee, and Wi-Fi, in *IECON 2007–33rd Annual Conference of the IEEE Industrial Electronics Society*, (IEEE, 2007), pp. 46–51
165. R. Hummen, J. Hiller, H. Wirtz, M. Henze, H. Shafagh, K. Wehrle, 6LoWPAN fragmentation attacks and mitigation mechanisms, in *Proceedings of the 6th ACM Conference on Security and Privacy in Wireless and Mobile Networks*, WiSec April 2013, Budapest, Hungary, 2013, pp. 55–66

166. X. Yang, E. Karampatzakis, C. Doerr, F. Kuipers, Security vulnerabilities in LoRaWAN, in *Proceedings of the 2018 IEEE/ACM Third International Conference on Internet-of-Things Design and Implementation (IoTDI)*, Orlando, FL, USA, 17–20 April 2018, IEEE, 2018, pp. 129–140
167. S.S. Arrieta, O.J.S. Parra, R.M.P. Chaves, Design of PHD solution based on HL7 and IoT, in *Future Data and Security Engineering*. ed. by T.K. Dang, J. Küng, R. Wagner, N. Thoai, M. Takizawa (Springer International Publishing, Cham, Switzerland, 2018), pp. 405–409
168. A.F. Santamaria, F. De Rango, A. Serianni, P. Raimondo, A real IoT device deployment for e-Health applications under lightweight communication protocols, activity classifier and edge data filtering. Comput. Commun. **128**, 60–73 (2018)
169. J. Hong, P. Morris, J. Seo, Interconnected personal health record ecosystem using IoT cloud platform and HL7 FHIR, in *Proceedings of the 2017 IEEE International Conference on Healthcare Informatics (ICHI)*, Park City, UT, USA, 23–26 August 2017, pp. 362–367
170. A. Duggal, HL7 2. x security, in *Proceedings of the The 8th Annual HITB Security Conference*, Amsterdam, The Nertherlands, 10–14 April 2017
171. C. Gündoğan, P. Kietzmann, M. Lenders, H. Petersen, T.C. Schmidt, M. Wählisch, NDN, CoAP, and MQTT: a comparative measurement study in the IoT, in *Proceedings of the 5th ACM Conference on Information-Centric Networking* (2018), pp. 159–171
172. A. Hussain, T. Ali, F. Althobiani, U. Draz, M. Irfan, S. Yasin, S. Shafiq, Z. Safdar, A. Glowacz, G. Nowakowski, M.S. Khan, S. Alqhtani, Security framework for IoT based real-time health applications. Electronics **10**(6), 719 (2021)
173. S.N. Swamy, D. Jadhav, N. Kulkarni, Security threats in the application layer in IOT applications, in *Proceedings of the 2017 International Conference on I-SMAC (IoT in Social, Mobile, Analytics and Cloud) (I-SMAC)*, Palladam, India, 10–11 February 2017, pp. 477–480
174. M. Brachmann, O. Garcia-Morchon, M. Kirsche, Security for practical coap applications: Issues and solution approaches, in *GI/ITG KuVS Fachgesprch Sensornetze (FGSN), Paderborn* (University Stuttgart, Stuttgart, Germany, 2011)
175. T. Dey, S. Jaiswal, S. Sunderkrishnan, N. Katre, HealthSense: a medical use case of Internet of Things and Blockchain, in *Proceedings of the 2017 International Conference on Intelligent Sustainable Systems (ICISS)*, Palladam, India, 7–8 December 2017, pp. 486–491
176. V. Karagiannis, P. Chatzimisios, F. Vazquez-gallego, J. Alonso-zarate, Sensus: smart water network, rans. IoT Cloud Comput. **3**, 1–10 (2016)
177. G. Suciu, V. Suciu, A. Martian, R. Craciunescu, A. Vulpe, I. Marcu, S. Halunga, O. Fratu, Big data, internet of things and cloud convergence—an architecture for secure e-health applications. J. Med. Syst. **39**(11), 1–8 (2015)
178. Y. Liu, G. Zhang, W. Chen, X. Wang, An efficient privacy protection solution for smart home application platform, in *2016 2nd IEEE International Conference on Computer and Communications (ICCC)*, Chengdu, China, IEEE, (2016), pp. 2281–2285
179. J.F. Reschke, The 'basic' http authentication scheme, in *Internet Engineering Task Force (IETF) Internet Engineering Steering Group (IESG) 2015*. https://httpwg.org/specs/rfc7617.html, Accessed 25 Dec. 2021
180. R.A. Rahman, B. Shah, Security analysis of IoT protocols: a focus in CoAP, in *Proceedings of the 2016 3rd MEC International Conference on Big Data and Smart City (ICBDSC)*, Muscat, Oman, 15–16 March 2016, pp. 1–7
181. S. Arvind, V.A. Narayanan, An overview of security in CoAP: attack and analysis, in *Proceedings of the 2019 5th International Conference on Advanced Computing and Communication Systems (ICACCS)*, Coimbatore, India, 2019, pp. 655–660
182. D. Dinculeana, X. Cheng, Vulnerabilities and limitations of MQTT protocol used between IoT devices. Appl. Sci. **9**(5), 848 (2019)

183. S. Andy, B. Rahardjo, B. Hanindhito, Attack scenarios and security analysis of MQTT communication protocol in IoT system, in *Proceedings of the 2017 4th International Conference on Electrical Engineering, Computer Science and Informatics (EECSI)*, Yogyakarta, Indonesia, 2017, pp. 1–6
184. D. Moustis, P. Kotzanikolaou, Evaluating security controls against HTTP-based DDoS attacks, in *Proceedings of the IISA*, 10–12 July 2013 (IEEE, Piraeus, Greece, 2013), pp. 1–6

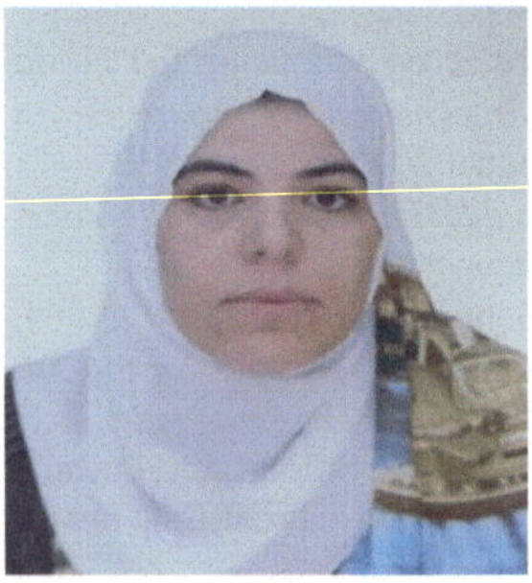

Rachida Hireche has a computer engineer diploma in 2006 on parallel and distributed systems specialty from the University of Mentouri Brothers Constantine, Algeria. And Master diploma in 2019 on information and communications science and technology specialty from Abdelhafid Boussouf University Mila, Algeria. She is the manager of development service in the University of Abdelhafid Boussouf Mila since 2010. She is currently a Ph.D. candidate in computer sciences department, faculty of sciences, University of Farhat Abbes Setif 1, Algeria. Her research interests focus on Fault Tolerance and Security Management in Medical Internet of Things (IoMT).

Houssem Mansouri obtained the engineering degree in computer science from University of Farhat Abbes Setif 1, Algeria in 2004 and his master's degree in computer science from the University of Abderrahmane Mira Bejaia, Algeria in 2007. He received his Ph.D. on fault tolerance in mobile environment from the doctoral school in computer sciences of the University of Abderrahmane Mira Bejaia in 2016. He also obtained the diploma of "enabling to supervise research works" in 2019 from University of Farhat Abbes Setif 1. He is actually associate professor in computer science. He is working since 2008 until now as part-time lecturer at the computer science department in faculty of sciences, Farhat Abbes Setif 1, where he held the position of academic deputy chairman of computer science department from 2010 to 2014 and the position of Head of E-learning Unit between 2015 and 2020. He is the head of the laboratory of networks and distributed systems since 2021. His research interests are fault tolerance and security in networks and distributed systems.

Al-Sakib Khan Pathan is a Professor at Computer Science and Engineering department, United International University (UIU), Bangladesh. He is also serving as a Ph.D. Co-supervisor (external) at Computer Sciences Department, University Ferhat Abbas Setif 1, Algeria. He received Ph.D. degree in Computer Engineering in 2009 from Kyung Hee University, South Korea and B.Sc. degree in Computer Science and Information Technology from Islamic University of Technology (IUT), Bangladesh in 2003. In his academic career so far, he worked as a faculty member in various capacities in various institutions like at the CSE Department of Independent University, Bangladesh (IUB) during 2020–2021, Southeast University,

Bangladesh during 2015–2020, Computer Science department, International Islamic University Malaysia (IIUM), Malaysia during 2010–2015; at BRACU, Bangladesh during 2009–2010, and at NSU, Bangladesh during 2004–2005. He served as a Guest Professor at the Department of Technical and Vocational Education, Islamic University of Technology, Bangladesh in 2018. He also worked as a Researcher at Networking Lab, Kyung Hee University, South Korea from September 2005 to August 2009 where he completed his MS leading to PhD. His research interests include wireless sensor networks, network security, cloud computing, and e-services technologies. Currently he is also working on some multidisciplinary issues. He is a recipient of several awards/best paper awards and has several notable publications in these areas. So far, he has delivered 32 Keynotes and Invited speeches at various international conferences and events. He was named on the List of Top 2% Scientists of the World, 2019 and Top 2% Scientists on the World, 2020 by Stanford University, USA in 2020 and 2021. He has served as a General Chair, Organizing Committee Member, and Technical Program Committee (TPC) member in numerous top-ranked international conferences/workshops like INFOCOM, GLOBECOM, ICC, LCN, GreenCom, AINA, WCNC, HPCS, ICA3PP, IWCMC, VTC, HPCC, SGIoT, etc. He was awarded the IEEE Outstanding Leadership Award for his role in IEEE GreenCom'13 conference and IEEE Outstanding Service Awards twice in recognition and appreciation of the service and outstanding contributions to the IEEE IRI'20 and IRI'21. He is currently serving as the Editor-in-Chief of International Journal of Computers and Applications and Journal of Cyber Security Technology, Taylor & Francis, UK; Editor of Ad Hoc and Sensor Wireless Networks, Old City Publishing, International Journal of Sensor Networks, Inderscience Publishers, and Malaysian Journal of Computer Science, Associate Editor of Connection Science, Taylor & Francis, UK, International Journal of Computational Science and Engineering, Inderscience, Area Editor of International Journal of Communication Networks and Information Security, Guest Editor of many special issues of top-ranked journals, and Editor/Author of 30 books. One of his books has been included twice in Intel Corporation's Recommended Reading List for Developers, 2nd half 2013 and 1st half of 2014; 3 books were included in IEEE Communications Society's (IEEE ComSoc) Best Readings in Communications and Information Systems Security, 2013, several other books were indexed with all the titles (chapters) in Elsevier's acclaimed abstract and citation database, Scopus and in Web of Science (WoS), Book Citation Index, Clarivate Analytics, at least one has been approved as a textbook at NJCU, USA in 2020, one is among the Top Used resources on SpringerLink in 2020 for UN's Sustainable Development Goal 7 (SDG7)—Affordable and Clean Energy and one book has been translated to simplified Chinese language from English version. Also, 2 of his journal papers and 1 conference paper were included under different categories in IEEE Communications Society's (IEEE ComSoc) Best Readings Topics on Communications and Information Systems Security, 2013. He also serves as a referee of many prestigious journals. He received some awards for his reviewing activities like: one of the most active reviewers of IAJIT several times; Elsevier Outstanding Reviewer for Computer Networks, Ad Hoc Networks, FGCS, and JNCA in multiple years. He is a Senior Member of the Institute of Electrical and Electronics Engineers (IEEE), USA.

A Comprehensive Study on Cybersecurity Challenges and Solutions in an IoT Framework

Reenie Tanya and Balika J. Chelliah

Abstract Internet of Things has changed the networking arena where we are moving towards increasing the range of devices that can be connected to the wireless world. However, it has also opened up fronts for security needs on another level. Intruders have also upped their game and utilize this idea of connecting daily devices to the Internet to dig into the lives of the users without their consent. This chapter provides an extensive view on the various security issues faced under four classifications: Vulnerabilities of the connected devices, Software applications that run on these devices, data transition stage in wireless mode and the final part where the received data is stored for further analysis. Emphasis has been made on the different features that are desired on these above mentioned levels and the well-known and practiced solutions available today. The objective is to give a solid platform for Researchers looking to enrich the Security front of the IoT framework, of the various current challenges and fixes.

1 Introduction

1.1 Overview

The Internet of Things (IoT) is spreading like a wildfire consuming a multitude of different domains including monitoring of the environment, healthcare, smart homes, education, smart cities, security fronts, mobility and is instigating the rise of Industry 4.0. As a consequence, the number of IoT devices is constantly on the rise in both public and private environments, progressively adding in daily objects. It is therefore true that the security aspect of these devices occupies a major position in order to avoid threats like leakage of private data, Denial of Service Attacks (DoS),

R. Tanya (✉) · B. J. Chelliah
SRM Institute of Science and Technology, Ramapuram, Chennai, India
e-mail: reeniet@srmist.edu.in

B. J. Chelliah
e-mail: ballikaj@srmist.edu.in

A. K. Pathan (ed.), *Towards a Wireless Connected World: Achievements and New Technologies*, https://doi.org/10.1007/978-3-031-04321-5_5

unauthorized access and so on. Unfortunately, many low-end IoT devices do not give due importance to security mechanisms and hence turns into easy targets for intruders to access information for illegitimate purposes.

In the recent years, the attention has shifted to the risks related to the utilization of such small scale IoT devices in service since it has access to sensitive information or critical data such as multimedia and real-time health monitoring data. Furthermore, some security attacks against commercial IoT devices have been posted in the social media, contributing to public awareness of the security aspect of the IoT world. Addressing the security challenges for an IoT framework requires a deep understanding of the landscape in which the devices are deployed. This includes analysis of devices present and their vulnerable points so that the critical security flaws might be met. In the excitement to take advantage of IoT, industries adopting this technology must take steps to assess security breaches and control loss of the operational systems. However, the intruders are more aware of these loopholes and careless gateways provided for them to attack the system.

Attackers look to deviate the device's features from their original functionality. The attack ideas detected so far can be classified into four broad types [1] based on their behavior: (1) Attacker ignores the intended physical capabilities and considers into only as a standard computing device connected to the LAN or Internet. This compromised device will be used as an entry point into the network to exploit the other connected devices (2) The attacker tries to delimit or sabotage the working of the IoT device in order to disrupt the daily activity of an organization or household (3) Altering the basic functionality of a device and making it to function in a different manner like using a temperature control device to heat up the room when cooling function is needed (4) The most dangerous class of attack is where the attacker uses the connected devices for hazardous purposes like using a smart air conditioner to initiate a fire or using the smart home security system to lock the owners out. Such attacks on IoT devices cause irritation and perils for the users and with the IoT devices increasing exponentially day-by-day, it becomes mandatory to extend security measures for an IoT skeleton.

1.2 History of IoT Threats

The major aspect of IoT is to connect device to gather and share information. The conception of these connected devices at home, healthcare, businesses and other organizations creates a vulnerable network for cyber criminals to exploit. This booms as an imminent challenge as we move towards the idea of connecting more day-to-day devices to the Internet. Devices in a smart home are the most targeted ones since they are a powerhouse of data. Studies show that 75% of the IoT devices are defenseless and hence they are susceptible of getting compromised. Given below are some serious cybersecurity-scares reported around the world.

(1) **Surveillance cameras**

Dio-V, who owns few Xiaomi Mijia surveillance cameras [2] tapped on accidentally into other home camera's feeds while trying to upload content from his camera to a Google Nest hub. The stream from other neighbours included stills of infant sleeping or people on the porch. The owner reported the incident to Reddit first along with 8 or so examples of such images. This incident led to Google disabling of all Mi Home products temporarily.

However, this isn't the first time such a breach of surveillance cameras is reported. Wyze, manufacturer of smart security cameras also committed a [3] blunder storing unsecured user data in a public access manner and requested all customers to pair/set up devices again. Ring, a home security products concern owned by Amazon was pulled for a lawsuit in the U.S, where more than one hacking incidents were reported. Amazon-owned Blink XT2 surveillance cameras were also found vulnerable by cybersecurity researchers from Tenable. They published a report with as much as seven detected loopholes which will allow hackers to view a remote feed of the camera and launch a Distributed Denial of Service (DDoS) attacks. This lead to Amazon releasing a firmware version 2.13.11 to solve these vulnerabilities. Such violation of privacy poses a great challenge as we move towards Internet of Video Things (IoVT) and smart city framework loaded with security cameras.

(2) **Smart Home devices**

Smart home structure provides with large amount of one of the most valuable assets of today- personal data. Those who recognize the fundamental value of data might look to extract and use it for illegitimate purposes. A family from Milwaukee suffered an unpleasant experience when their smart home setup was taken over by hackers. They resorted to playing loud music, speaking through cameras and meddling with the temperature of the room. The family had to change their network ID to flush out the intrusion. Comcast, in their Xfnity Cyber Health report 2020, stated that hackers are hitting smart homes at an average of 104 security threats impacting around 19 million homes a month [4].

Multiple reports were filed on intrusion in smart bulbs. Murtuza Jadliwala, a Research expert at the University of Texas, San Antonio had been working for about two years on the idea that even the smallest technology can be used for nefarious purposes when connected to the Internet [5]. He proved that once an attacker has access to the infrared spectrum, a covert-channel can be created which can be used as a gateway to exfiltrate user's information which can be decoded and the received text or image can be reconstructed.

From the smallest device like a smart light to larger ones like TVs and Speakers, anything can be used by attackers once introduced to the Internet. The FBI issued a warning in 2019 for smart TV owners [6] with tips to protect their device against cyber criminals, who can gain access to volume and channel controls and can also monitor the home activities after successful infiltration.

DEFCON, the annual hackers conference, popped up a security vulnerability in smart speakers like the Amazon Echo. A live demonstration was given at the conference by two presenters, Tencent's WuWu HuiYu and Qian Wenxiang, [DEFCON] as they hacked into a modified Echo [7] and turned it into a spying device. They offered a presentation called Breaking Smart Speakers: We Are Listening to You, which gave a clear report on how they modified an Echo device, used it to turn their attention to an out of the box Echo and then broke in by connecting the two devices on the same LAN [8]. Intruders can take advantage of the Complete Home Audio Daemon, a software component of the Echo. The bug was able to relay any audio received from the other speaker with no visible signs of transmission whatsoever. Though it is not an easy hack, since it requires reconstruction of one Echo device, still it raises concerns owing to the growing popularity of such devices.

A comprehensive evaluation of these attacks in various real-life settings confirms their feasibility and affirms the need [9] for new privacy protection mechanisms.

(3) **Smart Airports**

With Internet of Things spreading its wings into every sector possible, Services and systems in the Aviation sector also are increasingly being powered by the same. Smart sensors are employed to control the environmental conditions inside the airport, automate passenger related actions and enhance the airport security [10]. The systems implemented in airports are the fruit of extensive research work done to prevent potential terrorist attacks. In 2018, Singapore has included civil aviation as a crucial element of the "Critical Infocomm Infrastructure Protection" program, intended to the Cyber Security Plan (CSP). The global view is that the aviation sector for civil utilisation requires improvement of the technologies and means of enhancing the security front of the airports. This lead to the signing of Civil Aviation Cyber Security Action (CACSA) by several associations such as ICAO, CANSO, ICCAIA and IATA [11]. This collaborative effort laid the foundation for the promotion of the development of a more vigilant culture with respect to cyber threats and how to be guarded against them. The most convenient systems in the airports that are at risk are: international airport computer systems, in-flight control systems and air traffic management systems.

Internet is used for every minor operation such as informational exchange and messaging. Due to this reason, the airport security system will be under danger with side-effects on airport operations. The impact of potential threats booming because of the introduction of Wi-Fi was demonstrated by various research experts [12] over the years. Hugo Teso, a Spanish researcher performed a live demonstration [13] at the EASA (European Aviation Safety Agency) conference on how to make an airplane crash by controlling the aircraft control system with a simple mobile application that he had designed. Ruben Santamarta also staged a demonstration in which [14] he was able to access the navigation system of the flight by interfering with the satellite communications simply by using a WiFi accessible flight entertainment system. In 2015, according to an FBI report, Chris Roberts, was able to pull off the same feat with just a command CLB (climb) and the plane's engine responded to the prompt [15]. This drives the point that there is a need to adopt and develop Cyber-defence techniques to address the challenges of building a smart airport framework in order to avoid the tremendous consequences such as sensitive information leak, network intrusions, passenger reservation disruption.

(4) **Smart phones**

Smartphones have become the third arm for mankind these days. They serve as a companion, an assistant and is capable of performing a plethora of activities for the owner. However, they also serve as an ocean of sensitive personal data of the user as well which opens up opportunities for the intruders. Research scholars from England and Sweden came up with a malware that can exploit a smartphone's microphone to pull out critical information like the device's passwords and sensitive codes [16]. They claimed that they've uncovered the first acoustic side-channel attack which will give information that is being typed on their touch-screen devices. When the user taps the screen, a sound wave is propagated on the surface and in the air which can be captured by the smartphone's microphone, which can be in turn used by a malicious app to infer the text entered by the user. Analysis by cybersecurity experts have revealed that there's been an increase of 37% in smartphone phishing attacks worldwide between the end of 2019 and beginning of 2020 [17].

Smartphones can be injected with malware attacks which proves to be challenging for the users and manufacturers. Kaspersky, in their annual report stated that around 78% of malicious files [18] detected in smartphones were malware programs that targeted mobile devices. Other concerns detected in a smartphone were included in the same report and it deals with information leakage, using Unsecured Wi-Fi and phishing attacks. Public networks are subjected to network spoofing where hackers are able to use these access points as traps for naïve users. Since smartphones are the most common device in the world of regular people, it makes human beings the weakest link in the cybersecurity chain and hence requires special attention to the security detail.

(5) **Everyday Devices**

IoT is all about connecting day-to-day devices to the Internet and utilizing them for making human job easier. These machines are also prone to cyber-attacks since they are in close contact with their users. As an instance, a coffee machine connected to the Internet proves to be very useful as we can send a command from our smartphones to brew a coffee and can collect it when the 'complete' notification is received from the machine. However, such common machines used by more than one user in a network does not give much priority to its protection measures against intruders. Martin Hron of Avast, the security giant, in his blog explained about how they were able to turn a coffee maker into a dangerous machine demanding ransom from its users by modifying the firmware. The quest was to prove that not only weak routers or internet exposure is a threat, but the IoT device in itself is vulnerable to attacks and can be easily gained control of even without owning the network or the router [19]. The process of hacking into the coffee machine was simple: Download the firmware available in the Internet, Monitor the network traffic between the communicating parties and reverse engineer the companion app.

If we shift our focus from the coffee machine to other machines for public use in an organization, like a smart printer or a fax machine, the story remains the same. Security research firm, Quocirca, in their annual report "Global Print Security

Landscape, 2019" have proved that mart printers are the gateway for cyber criminals into an organization's network. Threats relating to smart printers can be classified into two broad areas: documents that the printer produces and the printer itself [20]. The data that is being printed might be confidential and it can be stolen once the attacker gains control of the printer. There are three ways in which an attacker can use a smart printer: (1) It can be used as an entry point to the network (2) It can be sabotaged causing disruption in the daily activities and in turn the business workflow (3) It can be recruited to botnets which are then employed to launch distributed denial-of-service attacks that can provide the attacker a lot of free processing power without the knowledge of the owner.

Another vulnerability was deduced in a connected fax machine that could allow an intruder to steal delicate data through a company's network using just a telephone link and a fax number. Check Point researchers Eyal Itkin, Yannay Livneh and Yaniv Balmas demonstrated how a fax machine can be a weak point in a network and termed the act as "Faxploit" at DEFCON 26 [21]. They started off by reverse engineering the firmware and detecting the environment in which the loaded firmware was being executed. The next step was to build a serial debugging interface and connecting it to the machine. After identifying the open sources used by the firmware, vulnerable points can be spotted out to gain control of the device.

These incidents serve as an eye-opener as the exposure of connected devices have enormously increased and has brought a challenging threat to the cybersecurity world. A research study has revealed that by the end of 2020, there will be 492 million motion and signal GPS beacons to support the infrastructure of the existing Internet of Things network [1]. The lack of embedded security and programming walls to protect these devices is raising a red flag that this technology is holding a greater risk than any other technology.

1.3 Need of Cybersecurity in IoT

Internet of things refers to multiple IP-based devices connected and interacting with each other and the physical world. With industries and government sectors also jumping into the field of IoT, there is a striving force to connect things and employ the data that is generated by these connected devices daily. The procreation of connected objects, system, services and devices has opened up huge opportunities and ways to benefit the society. Data generated by these devices prove to be immensely valuable when subjected to analysis to provide insights and intelligence. In fact, the risks of IoT setup being hacked [9] are immense in comparison with any predeceasing technology that has existed in three aspects: (1) the huge number of connected devices to be monitored increases the risk multi-fold (2) The devices/ sensors being geographically distributed and (3) IoT, being a newbie in the IT world employs heterogeneous platforms and hence opens up vulnerabilities due to the multi-dimensionality. These risks are further complicated by other limitations like energy efficiency, communication mode, computational capability and storage capacity of the IoT devices.

Such factors limit the adoption of traditional security models and hence call for new answers [22] for security challenges.

There is also a need in developing a cybersecurity culture among the stakeholders like the manufacturers and the final consumers. Customers look forward to low cost sensors and actuators. Such devices were initially designed to exist in isolated systems for which the security threats are limited. This leads to the fact that these manufacturers do not hold a strong proficiency in cybersecurity and are hence ignorant of the issues that arise while connecting these devices to global network. Therefore, it is best practice to start cyber security counter measures from the design phase of the devices [23]. It is also necessary to provide security functionalities at every level of the IoT network. This is the major reason as to why the security front of IoT is still under research and more mechanisms are being suggested.

2 Assortment of Cyber-Threats and Solutions

Cyber-threats are many in number and various surveys have attempted to classify them according to the layers pertaining to it or the genre of solutions it requires or the features of an IoT framework. An extensive survey is done on these threats and is provided below based on how and where data is processed. The classification is as follows (Fig. 1):

- Hardware devices

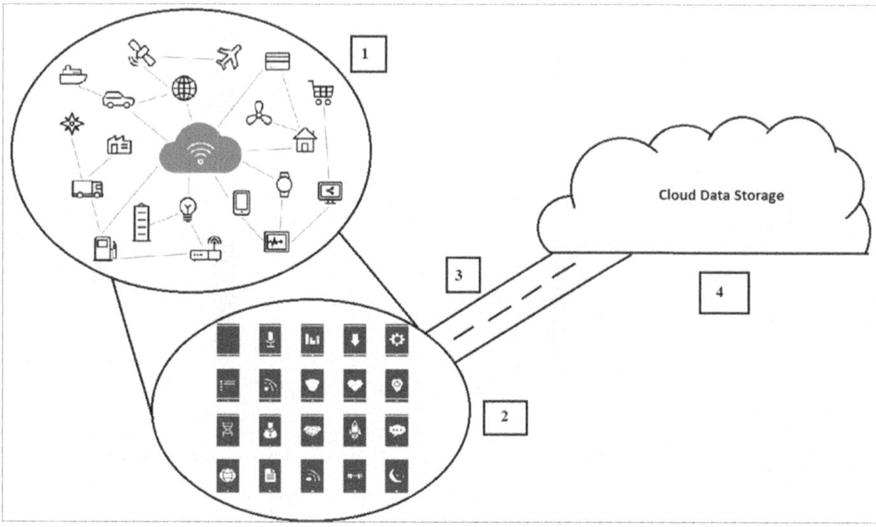

Fig. 1 IoT Framework-attack prone areas. 1. IoT devices, 2. Smart applications, 3. Transmission of data, 4. Cloud storage

- Software resources
- Data Transit
- Data Storage.

Furthermore, we discuss on the layer of the IoT protocol stack, the level of security and the corresponding desirable features of the security framework.

The IoT protocol stack can be depicted as an augmentation to the traditional TCP/IP protocol structure and holds three major layers [24]:

- Physical layer—related to the physical IoT devices and sensors to gather real time data.
- Network Layer—Layer that provides connection of the physical layer to the environment to transmit information.
- Application Layer—Layer that is in interaction with the user to provide the service requested.

Additional Layer:

- Support layer—A logical layer that handles the database of the collected data for analysis.

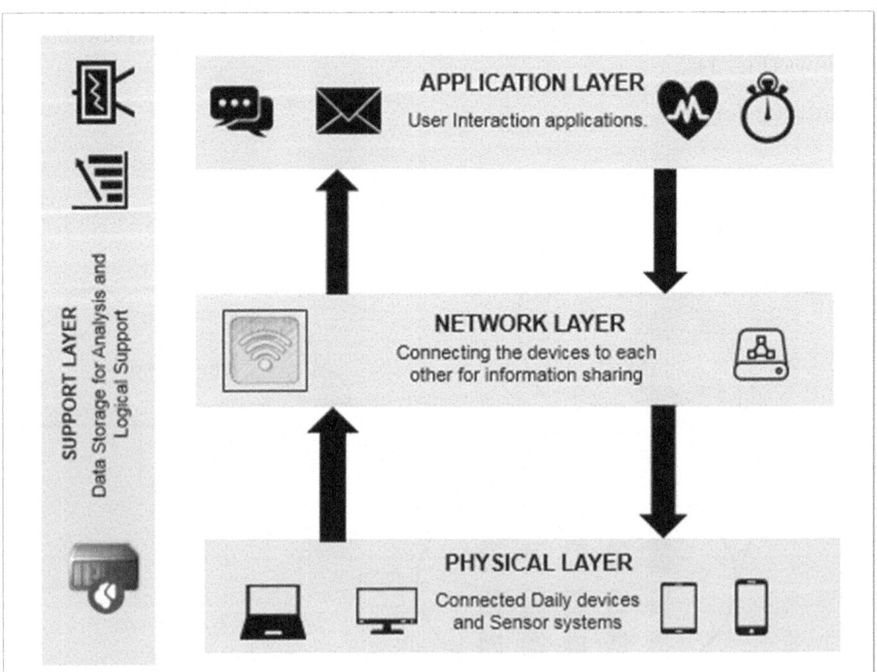

Fig. 2 IoT 4-layered architecture

The layered architecture (Fig. 2) has been widely accepted and each of the layer has its own specific technologies and protocols along with its weakness and challenges. In fact, the protocol stack is studied layer by layer to analyse the security drawbacks and respective robust solutions.

The rest of the chapter will introduce various threats classified as mentioned above and will be illustrated based on four aspects—Threats, layer involved, IoT features involved and corresponding challenges according to operational level involved.

2.1 Hardware Devices Related Threats

The basic flow of data is initiated from the devices that are connected to the Internet. These devices are designed to recognize, sense and collect information from the environment and hence are also termed as the sensors. Such devices are placed in surroundings where they are exposed to physical attacks and to attacks concerning the conditions they are existing. Confidentiality is another major issue in this category since it can be sabotaged by replay attacks without much trouble. Device can also be physically stolen or can be replaced by another malicious node which threatens the integrity and the chain of information can be severely altered. Hence, the attacks related to hardware devices can be broadly [25] classified into: (1) Node tampering issues and (2) Malicious Node injection. Node tampering elates to causing intentional damage to the sensor nodes to disrupt the flow of work, while Malicious node injection is to include a compromised node to indulge into the network.

Layer: The layer that holds these devices is the first layer of the protocol stack- the physical layer. It contains the PHY and MAC functionalities for base-level communications. It is the first step in the workflow where data is gathered from the surroundings. It comprises of the physical devices, controller, sensors and the smart nodes that are the entry points of an IoT network. Its capabilities include data collection and data control. Security challenges of this layer comes from the different configurations of the devices, energy requirements of the devices and other frameworks involved.

Features:

(1) *Inter-relationship*: IoT devices are no longer stand-alone devices that simply gather information but are in need of collaborating with other devices as well to provide well-formed data. In fact, many of the devices can also be implicitly controlled by other devices or other environmental conditions or other protocols to be followed when communicating with a remote cloud server. For e.g., agricultural devices like automatic watering system takes into account the external temperature and humidity to decide whether to initiate the watering system or not. The scholars investigating the challenges in IoT structure are unaware of the effects of interdependence of these devices on the security feature. They focus on the protection of a single device but the objective should be to plan a clear safety boundary of these IoT devices or to implement an access control

method or user privilege management of these devices [26] due to their interdependence. A permission model that dictates fine-grained rules must be framed since the behaviours of these devices might change according to the stimulus from the environment or other devices.

(2) *Computing constraints*: IoT sensor devices face many limitations in terms of cost and physical conditions and hence roots for less storage and computing resources to compensate. These devices must be lightweight and small in size, because of which these devices face restrictions in deploying proper defences for the system. Additionally, sensors embedded in industrial, military devices must work for a longer time without any charging facilities and should meet stringent power consumption requirements. It also induces a long latency delay by utilizing too much computing resources in order to perform complex authentication procedures and encryption techniques. This poses the difficulty of scheming a system protection mechanism that will require minimal software and hardware resources and will satisfy the application's security requirements.

(3) *Human interference*: IoT devices are becoming indispensable in one's life. They are destined to increase exponentially in the future as well. No matter how many security algorithms come up to ensure safety of these devices, human interference in handling them plays a major role in altering the protection scheme. An IoT device is constructed by a manufacturer, utilized by an end-user, administered and monitored by an operator and studied upon by a researcher. Most of the manufacturers consider adding security measures takes a toll on their profit margin and therefore most of the devices are sold in the market with nil or mere minimal security instructions or in-built counter measures. Manufacturers fail to provide support for users to install firmware against ever evolving malware infections which in turn profits the intruders. Customers, on the other hand, lack the awareness on the consequences of privacy compromise and data control and management. For eg, in 2016, Mirai virus simply used the default username and password set to exploit the connected devices. [27]. It is also difficult for the naïve users to realise if their devices are compromised or not. When these devices are employed in a large scale, an administrator is instated for management purposes. An administrator first looks into the hardware malfunctions if a device behaves strangely. They must also be aware of attackers using specialised devices to intrude into the network. Finally, a researcher should take up the responsibility of digging in deeper into these vulnerabilities and must have a better insight than the other groups of people involved for the betterment of the technology.

Challenges-Operational Levels: The main focus with reference to hardware related attacks is on the access level of the network. The objective is to provide access to the right devices and the right amount of access level to each device. The major features under consideration are:

(1) *User Authorization*: The users requesting service from the IoT applications must be given the correct level of access to the Physical layer devices. Users

needing access for administrative tasks for the network must succeed an authorization procedure.

(2) *Device Authorization*: Just like how users have to be authorized before they can access the devices, another objective is to authorize any device that enters the network. Any IoT network must be flexible to accommodate new devices, heterogeneous devices or devices from different manufacturers. Each device has its own security algorithms running and it must be compatible with the network's requirements. When all devices are from the same manufacturer, they will possess the same authentication credentials and when one is compromised, all other are prone to the same fate. In case of heterogeneous devices, the challenge is to frame an authentication procedure that is compatible with all types of devices.

Threats:

(1) *Side channel attacks*: The objective of this type of attack is to emanate information without the knowledge of the user by analysing the side signals such as power consumption, electromagnetic emissions or encryption computations [25]. Increasing efficiency in terms of energy increases the chance of power-based side channel attack. In PSCA, the power dissipation of the hardware devices is measured while computing key for encryption and then analysed statistically to recover the encryption key which will lead to compromising the device. Hence, there is a need to attain a trade-off between side-channel security of encryption, performance, power and area.

(2) *Denial of Service attack*: The intruder makes the device to deny users of its proposed service by tampering with the internal circuit of the IC. The devices can be forced to exit sleep mode in order to increase the power consumption and to drain the batteries or by disrupting the radio communications. The intruder can also utilize the resources for his own gain stealthily without the legitimate user's consent and thus leading to no resources left for the latter. DoS also leads to reduction in the network capacity thus hampering the system's performance. DoS attacks are of two categories: Distributed Denial of service and Simple DoS. It can result in disabling the network and also detaches its access from a larger network.

(3) *Faulty Nodes*: Sensor nodes are prone for physical drubbing with the aim of infecting the network with malicious nodes. The legitimate nodes can either be disabled, destroyed or compromised by the attackers to diminish the quality of service provided by the network. These devices are termed as faulty nodes [28] and calls for solutions that will be able to detect such nodes and carry forth necessary action to safeguard the network. Such faulty nodes will be used as an ingress point to the entire network which in turn poses danger to sensitive information such as routing path tables or secret cryptographic keys.

(4) *Outage attack*: Outage attack is when a node stops working or stops performing its normal operation. Reasons behind such attacks may be unwitting mistakes done in manufacturing process, unauthorized access, draining of the battery's power or even malicious code injection. One of the major outage incidents

in history is the outage that resulted from injecting Stuxnet [29] into Iran's nuclear project. The injected malware influences the process control system so that it goes blind to the abnormal performance induced. Hence this causes the system to stay awake even if the users try to shut down due to an emergency.

(5) *Counterfeiting*: The attacker tracks the identity given to an IoT device and revises it through distortion of the tag within. Tags are included in an IoT device to help with the identification of individual machines or objects connected to the automatically and remotely as and when required. Much less information is needed for carrying out such attacks since it simply involves getting control of the tag to be manipulated. The intruder can extract the identifier information through a proper channel. Digital Object Architecture (DOA) is a name-attribute binding service that helps in uniquely identifying digital devices distributed all over the world [30] is a widely acclaimed solution for the counterfeit attack.

Security Measures:

(1) *Protection against Hardware Trojans*: With Internet of Things taking over the world, System-on-Chip (SoC) Security is under the radar. HT infusion can cause leakages of data, Denial of Service or abnormal functioning of the device. Numerous Classification algorithms have been suggested for the detection of such HT attacks, but since it increases the computational load a lighter solution is needed. Initial circuit and IP cores attributes [31] can be verified for malicious infections. However, these measures fail to explicitly detect the Trojans, rather it provides a platform for improving the security strategies. Going a step further, using these circuit features and a suitable machine learning model, the efficiency of detecting dubious lines and nets increased. The features used differed [32] from various suggested models like gate-level features, side channel information, consumption of power or delay in transmission.

(2) *Anomaly detection in Traffic*: An IoT device has to interact with various other devices and networks to transmit measured information. Based on any discrepancies in the amount of data transmitted over a network in a period of time, various attacks can be pointed out like botnet intrusion, denial-of-service, Man-in-the-Middle Attack, etc.., Researchers have suggested to use ML models or even Recurrent Neural Networks to detect these discrepancies [33]. The initial research had models that were trained on complete labelled data. But, in real-world, landing on such data is not easy and training causes high resource demand. Hence, the focus shifted to designing models that use device configuration information, device manufacturing information or rhythm of data transfer for classifying normal and abnormal behaviour or third party cloud analytic services for traffic analysis.

(3) *Wireless Channel Security Technologies*: The channels used for wireless communication were enriched to avoid noise, interference and fade. Since the dawn of IoT, the primary focus is to enhance its security [34]. Known as the wiretap coding, secure coding is done to avoid the imperfections in such channels from being used by intruders to achieve legit information in transit.

Wiretap coding ensures reliable transmission of data even in noisy channels and keeps the data away from eavesdroppers. Another method to keep the channels secure is to provide it with an authentication method using its Channel State Information (CSI) verified by means of channel probing. Every time a sink receives a message, the initial CSI can be compared with the current one to authorize the source.

Wireless Network resources can also be made use of to produce a higher rate of security. The resources that are dealt here are the ones in the spatial domain such as the multi-antennas and transmitting nodes. How these resources are allocated to legit user devices and their relaying technique also play a part against intruders. Following this perspective, signals sent forth must be processed securely as well. Beamforming [35] is the standard solution used and is the process of combining signals from multiple antennas with attention on coverage and leakage ratio. This technique can also be enhanced to increase the level of secrecy.

2.2 *Software Resources Related Attack*

The main target here is the software running on the devices. Such attacks manipulate the underlying algorithm thus inducing delinquency of the device executing it. Some devices are responsible for running complex software applications but may be similar to a general purpose computer with limited security measures. Such systems when linked to the Internet will pose imminent danger to the underlying node. The applications also get to decide on the subset of devices [36] that an IoT device gets to communicate with. This will also pose an opportunity of hacking into the network by influencing the node to communicate with other malicious devices.

Layer: The layer involved with software attacks is the application layer which is responsible for the execution of service providing applications. Functionalities are performed by APIs that enables interaction between back-end and front-end systems. These APIs do not possess a special security front and this creates a hole for the intruders. Software vulnerabilities exist in all forms of network where the attacker may push the IoT device to abnormal activities that can prove to be harmful to the network.

Features:

(1) *Heterogeneity*: In order to accommodate varied cases in study, specific IoT devices are employed for specific tasks that interact with each other and the environment it is placed in. This heterogeneity poses a problem in the view of different protocols being put in order to ensure hassle-free communication. From the system security point of view, it is a Himalayan task to design a common system defence for the variety of devices. Hence, the researchers must be aware of all security vulnerabilities of the different devices [37] and the security issues of the protocols.

(2) *Solitary applications*: Certain implantable medical devices and sensors placed in agricultural fields or industrial sectors must engage for a long period of time without direct human access. They are controlled remotely through the applications and their underlying wireless adaptor systems. When these systems or applications become vulnerable, it poses a danger. And since the attacks are also made remotely, it increases the difficulty of being detected. Moreover, such devices are also constrained in terms of size and hence won't be upholding great security counter-measures which acts in favour of the intruders as well. Such devices also are in strong influence of the physical environment. All these factors taken into consideration makes designing a solitary trusted defence system that can be controlled from a remote environment. It is also a challenge when we consider the latency measure and computing resources available.

Challenges-Operational Level: The prey for the attackers in the application layer or the service layer are the underlying programs that are in execution. The objective is to verify the authenticity of the applications and since these applications have APIs as their focal points, these APIs must also be subjected to security countermeasures. Another loophole that an attacker can use are constraints or conditions that have been initialised for initiation of certain actions which can be manipulated for malicious reasons.

(1) *Application verification*: As mentioned earlier, the application layer hosts a variety of services that are running on a remote platform. Service include remote administration and control, monitoring and troubleshooting and log maintenance. These remote cloud platforms are found to have several potential vulnerabilities due to poor authentication schemes, insecure underlying connection protocol and access levels fixed. Most of the applications lack sensitive data protection, proper third-party application collaborations, unrestricted API control and an unsafe level of access permissions. Hence security measures must be in place to avoid the applications from gaining access to all capabilities of the device and to prevent sensitive user information and evet information leakage.
(2) *Security of the APIs*: All the applications that are providing services and interacting with the user need APIs to function. Right from data collection to data sharing and analysis, there are specific APIs functioning on these devices. However, these API's security front is being traded for its performance aspect and remains the most ignored threat in an IoT framework. This, in turn makes these API vulnerabilities an easy scoop for the intruders. The most widely employed APIs in an IoT environment are the Web APIs and backend cloud APIs and are implemented either through Simple Object Access Protocol (SOAP) or the Representational State Transfer (REST) protocols. Such protocols must employ mechanisms that prevents the users from accessing these API functions outside their prescribed level. For example, a normal user with a read-only permission must be prevented from gaining an administrative role of the API.

(3) *Programming constraints*: Many IoT frameworks in the view of data analysis are more focussed on the quality and performance improvement of the service. It is necessary to keep in account that the data generated by the devices are valuable sources for breaching. Another concern is that most of the actions carried out by the IoT devices are outcomes of an automated program that follows a certain behaviour model. When these devices are placed in a physical environment that is dynamic in nature, the static behaviours dictated by these programs misbehave. Based on these details, it is understood that the security analytics must be performed on a multi-dimensional aspect [38] that collaborates data from multiple domains and draws anomalies that could pose danger in the future. For e.g., a smart thermometer is provided with a programming constraint to open the windows when the indoor temperature rises above 26 degrees Celsius. This condition can be misused by any attacker in close proximity to the device. The attacker can simply increase the temperature to activate this condition. In this case, the application must be armed to correlate from information from other sources like temperature based on location or the CPU usage statistics and must be able to deduce the false activation done.

Threats:

(1) *Insecure software applications/OS*: The objective of building an IoT framework is to collect information from the environment which will be utilised by various third-party applications for analysis and results. When the applications that are in implementation are framed on weak code or is a victim of misconfiguration, it becomes a cakewalk for the intruders to cause mishap. The clients are allowed to access the services provided through web interfaces and these interfaces act as the point of entry to their hosting environment. The underlying code must ensure that these applications do not provide room for unauthorised users to peck the data illicitly.

(2) *Sybil attack*: One node will pose as multiple identities to other interconnected nodes in the network [39]. It is done so to manipulate the system and to reduce the effectiveness of fault-tolerant schemes. It also influences the system to generate incorrect reports spread advertisements and spam among the connected users. It can also lead to more harmful activities like disseminating malware and phishing websites to thieve user's private information. Since most of the Sybil attackers behave like legit users, it becomes a difficult task to weed out such pseudo-identities. They are highly prevalent in social IoT networks where fake identities are used by intruders to peek into the network.

(3) *Middle-ware Attacks*: The application layer comprises of numerous services and applications collaborating to access the generated data and to provide the intended services to the end user. These applications [40] communicate via interfaces for a better communication. The same applies for the variety of IoT devices that have to fraternize to gather data and hence must commune with each other in a secure way as well. Therefore, different types of IoT environments interact through a middle-ware [41]. When these middle-wares

are attacked, the attackers are in a position to gain control of the interacting applications or the connected devices.

Security measures:

(1) *Key Establishment*: Constrained Application Protocol (CoAP) has been the standard application layer protocol and it holds a security binding with the Datagram Transport Layer Security (DTLS) for protecting the communication end to end. However, such end-to-end security cannot be ensured because of the proxies [42] that are to be deployed in between in order to make the network more efficient and scalable. Hence, OSCORE (Object Security for Constrained RESTful Environments) suggested using key exchange protocol for ensuring security [43] using the Concise Binary Object Representation (COSE) for establishing object-based security. This was followed by a Lightweight Authenticated Key Exchange (LAKE) taking into consideration the resource constrained environment of IoT framework. In accordance to the above mentioned milestones, EDHOC (Ephemeral Diffie-Hellman Over COSE) was put forth with the objective of establishing of both end- to- end security and keeping in mind the lightweight framework. EDHOC is now being used by various other works with additional features for authorization or for signature endearing. There have been advancements in the key establishment strategy also in terms of reducing the overhead of the network [44] and utilization of computing resources.

(2) *Against Sybil attacks*: Sybil attack is when a node can don various logical identities and the network is unable to verify whether all its entities are legit. The defence schemes framed for Sybil attacks can be classified as three approaches [45]: Behavioural profiling based, Social Graph based Detection and Mobile Sybil Detection. Behavioural aspects of the Sybil nodes can be used to differentiate them from legit nodes. However, common attributes [46] like befriending, sharing and liking posts, frequency of posting are all very similar between them and since other factors like click patterns (Average clicks per session), average time spent online per login entry, average time spent online per day and the intermittent time can all be used by a Machine Learning Algorithm to predict the illegal nodes. The second approach is to identify Sybil nodes from the relationships between the existing nodes. The objective is to provide a node enough insights to term a connected node as Sybil node based on its relationship graph. The well renowned example would be the SybilGuard [47] and its predecessor SybilLimit [48] that etches the probability of a node being a Sybil node. The third approach is for networks where the nodes are mobile and a social graph is not possible. In such cases, mobile channel information authorization failures, community information can be studied to detect Sybil nodes or cryptographic signature methods can also be used to authorize a node.

(3) *Fighting DDoS attacks*: Distributed Denial of Service attack at the application layer is different from DDoS attacks carried out in other fronts like pulling down a device or causing network traffic at the network layer. Here, on the application perspective, the attack can be carried out on a particular resource like the

processing unit, storage unit, database itself or the logical sockets involved. The attackers may use the unauthorized software applications or the programmer's negligence to sneak into the system. For example, SQL injection can be used where the naïve SQL Queries can be used to inject malicious data or even delete an entire database. Though there are innumerable defence mechanisms [49] pertaining to the underlying protocols and devices used, simple cautions can also help like including puzzles like captchas for authorising a user and to keep away bots, using better secure models of programming and monitoring the traffic for any extraordinary hikes.

(4) *Protection against third-party apps*: Smart apps can invade the registered mobile device by either gaining more facilities than actually committed or by gaining more commands and attributes that are needed. Hence, the first step to protect the device against such attacks is to detect the extent of the smart apps permission using an analysis tool. Various tools have been suggested by authors to automate this process like the SmartApps [50], which exports a capability model that mentions the devices and its facilities that an App basically has the permission to interact and access, which is compared to the commands or attributes to detect the illegal privileges enjoyed. Another interesting perspective is to study the flow of information between source and sink of an IoT app. Tools analyses foul flows by examining device attributes, device specific information, geolocation information, user specific information and the app stored variables [51, 52].

(5) *User-Level Security*: Users can be educated about the importance of security and usage of anti-malware systems to protect their devices. They can also setup strong passwords and change them frequently to avoid illegal access. Unknown links must not be entertained as well since they can be a malware in disguise. Scholars believe that when users are aware and alert [53], hackers at the application level can be averted.

2.3 Data Transmission Related Attack

The most vulnerable spot in an IoT network is the process of transmitting data. The data gathered by the edge devices are to be sent to the remote cloud for further analysis. Intruders aim to attack the network's privacy and integrity. Attacks on account of the data transit is mainly due to the distant access and remote exchange of data.

Layer: Information forwarding or routing of data to various centres over the internet is being carried out by the Network layer. Internet gateways, switches, routers and cloud computing infrastructure are a part of this layer. The network congestion, data being transmitted, traffic management resources can be exploited by these intruders for malicious purposes. Another issue comes from the fact that devices in interaction do not follow standard protocols of security and sensitive sensory data is being transmitted between them [54]. The protection of such data and the devices that are involved is very crucial.

Features:

1. *Data correlation*: IoT devices are increasing in number because it makes life easier for their owners but in-turn holds critical information about them. This strong correlation between the devices and their users is a striking feature of the IoT technology and also is a hurdle to cross over. Our digital devices are given permissions to control intimate information like our heart rate, blood pressure, our home's temperature, places we have visited, miles walked and food eaten. The applications depend on the data collected by these devices and they are to be transmitted to a remote server for analysis during which it can be stolen or replaced. There must also be a regulation that will instruct the use of this data by the server providers since they can be used for illegal profits by selling it to third party advertising agencies.
2. *Cross-Domain Networking*: Devices such as wearable devices, smartphones and smart vehicles work in a mobile environment and hence they cannot stay connected to one stable network. They will have to hop from network to network to stay alive. This characteristic can be used by the intruders to inject malicious code through these networks to catalyse the spread. Hence, to tackle such threats, cross-domain network identification and trusted permission protocols must be in place. When data is carried from one network to another, the authentication of these devices, data privacy management and data source identity protection must be monitored.

Challenges-Operational Level: The prime challenges regarding security during data transit are in two aspects- ability of the network to ensure security in any case of failures and the capability of an IoT system to remain functional during unexpected shutdown of nodes.

(1) *Security at any cost*: As mentioned above, security is indispensable in the network layer both to the data being exchanged and the communicating devices. The network layer must have an overlaid security counter-attack as multiple devices communicate. This security measure must be resilient to attacks, meaning, it must not be compromised however the devices are subjected to attacks.
(2) *Fault-tolerant*: With the physical devices being placed in external conditions, they are prone to either physical attacks or malfunction which makes the data acquired unreliable. In that case, the network must be able to carry out its destined work even one or more devices are down. Such occasional misbehaviour or malfunctioning of these nodes must not hamper the functioning of the system.
(3) *Anomaly Detection*: Provision of an extra layer of security in case of device infection comprises the anomaly detection component. The large number of connected devices makes it almost impossible to provide such protective mechanisms. The analysis report published by Nokia [40] on the detection of botnet

activity and it evolution states that around 78% of the network activity was performed by the botnets and not the legit users. There can also be a false positive scenario where the security mechanism itself may pop up incorrect detection of malicious nodes due to the lack of background information. The restricted access that is implemented on the devices are not sufficient for determining its authenticity.

(4) *Traffic monitoring*: Some security mechanisms need information about the connected devices to perform malware detection during data transmission. The information needed is the identity of the source of the data to authenticate the packet before it is transmitted in order to avoid packets from unauthorised nodes. If such malicious packets enter the stream, the bandwidth can be taken over and DoS attack results. Hence, it becomes a challenge to perform security assessments before data is transmitted over the network and malware removal rules need to be framed.

(5) *Traffic traces*: Apthorpe et al. [55] has proved that a network traffic monitoring application can access sensitive data about the communicating devices. They claim that traffic monitors can derive the number of unique devices connected from the service payload of the destination or the IP addresses of the gateway to the cloud or third-party apps. Secondly, the identity of these devices can also be revealed based on the DNS queries of each individual packet. Another concern is that based on simple receiving and transmission rates of each stream, the possible user communications can also be revealed [56].

Threats:

(1) *Protocols-based attacks*: Network transmission of the packets from a source to a destination always involves standard communication protocols that enables the identity management of the connected devices and routing of packets. There are sub-protocols that take care of specific actions in the transmission process. One instance would be the Neighbour Discovery Protocol (NDP) used in IPv6 that is employed for identifying the nearby devices' physical addresses, address resolution maintenance and detection of node duplication [57]. When this operation is under execution, there must be a proper authentication mechanism to avoid DoS attack, since these identified nearby devices can be compromised.

(2) *Replay Attacks*: When the packets are being transmitted over the network, there arises a situation where the IPv6 packets have to be fragmented for mapping into a predefined frame size according to the IEEE802.15.x standard. These packets have to be reconstructed in order in the destination devices. This situation can pose a serious problem for the destination device in terms of buffer overflow or device rebooting. But a more serious issue will be that it opens up an opportunity for the intruders to inject malicious fragmented packets causing the Replay attack [58]. The packets in transit can be subjected to manipulation or can be replaced if the security front is weak.

(3) *DoS attack*: When the service providers are unable to serve the legit users due to exploitation of the available resources by intruders, we term it as Denial of Service attack. During packet transmission, there are various chinks in the

process that can be misused by attackers. The first scenario is when the destination device runs out of buffer space for incoming packets. Every device that is communicating possesses a buffer to receive packets that are being sent and when the attacker sends in incomplete packets [59], this buffer space runs out and hence results in a DoS attack.

Second scenario that can potentially cause a DoS is session hijacking. The communicating devices enter into a temporary established interaction time known as a session, during which more than one messages will be interacted. DoS attack can be launched by trying to keep the session open [60] by continuing forge message forwarding. Since the session is still on, the legit devices will try to re-send the packets thus causing other devices to wait for their opportunity.

(1) *Sinkhole Attack*: Every network employs a routing protocol that provides the optimum path for the sender to forward packets to the destined node. When a malicious device is provided the access to the network, it resorts to tampering the routing tables and provides different routes for the packets to the destination [61]. When this is accomplished, these devices will be able to alter the data [62] gathered by other devices in the network and can also modify even the packet payload or cause delay by dropping a fake message.

Security Measures:

Encryption: The major objective of encryption is to maintain confidentiality during transmission of data. When encryption is implemented, it can avoid any intrusion or possible eavesdropping during the transit of data. Owing to the attacks mentioned above related to data transmission, the final resort of the intruder is to capture the transmitted packets and utilize it, however this can be prevented by encrypting the data. It is the process of converting the normal message into a cipher text using hash function that can be reversed using a secret key. Encryption is broadly classified into symmetric and asymmetric. In symmetric encryption, the same secret key is used for both encryption and decryption, whereas in asymmetric mode, each receiver needs a private key that can be derived from the shared public key. The issue with the symmetric mode is that the key in itself must be transmitted in a secure manner to maintain the objective of secure transmission. If the intruder gets hold of the key, the transmitted file can be well decrypted into the intended message. Beyond these classifications, encryption algorithms can also be classified into stream and block ciphers. The variation between these two lies in the way the data is converted into ciphers. In stream cipher, the encryption is done one byte at a time or one bit at a time, whereas, in block ciphers, 128 bits (one block at a time) is encrypted at a time. Block ciphers prove to be better than stream ciphers if the size of the file is known. However, stream ciphers must be the choice if the plain text is sent in a continuous stream. The strength of an encryption algorithm lies in the size of the key. Any Encryption algorithm is subject to brute force attack, where the intruder resorts to trying out various combinations of codes to try and crack the shared key. The length of time that it can withstand such efforts, measured in millennia, decide how strong the algorithm is. Some of the widely used efficient encryption algorithms are compared in Table 1.

(1) *Protocol attacks countermeasures*: As mentioned above, standard protocols are used for data transmission which provides loopholes for the intruders to attack the system. There are also some customised protocols for resource and power constrained wireless networks like the RPL which is a routing protocol. MQTT (Message Queue Telemetry protocol) is a lightweight protocol that enables two devices to communicate with each other by means of a simple publish-subscribe messaging method. Though security for these protocols can be enhanced using a SSL/TLS and certificates, it becomes cumbersome due to the heterogeneity of the connected devices and the storage considerations. Thus, a more dynamic, robust security mechanism is required as a counterpart for these kind of attacks based on protocols.

(2) *Attack detection*: The first step would be to keep detection strategies in place [87]. The approaches undertaken for such detecting are classified into four categories:

(3) Signature based detection—the pattern of the attack when it occurs, is stored in the Intrusion detection database, so that when a similar attack occurs in the future, the attack is identified beforehand.

(4) Anomaly based detection—Any deviation from the normal behaviour of the network is introspected as an attack. At several intervals of time, the activity of the system is compared to the standard activities of the network and any discrepancies found will be investigated for a possible attack.

(5) Specification based detection: Any networking system will hold certain specifications or features like the components involved, routing tables, protocols followed and communicating nodes. Each of the component has its own features defined based on which, any divergences will be classified as a possible attack and an alert is given.

(6) Deep Learning Approach: Various Neural networks like RNN (Recurrent Neural Network) [88], SOM (Self Organizing Map) [89] can also be used for detecting attacks based on the knowledge base built. These detection schemes are dynamic and adaptable to the environment. These systems are capable of identifying varied types of attacks like sinkhole attack, DDoS attack, wormhole attacks and any other anomalies that may occur.

(7) *Countermeasures*: Routing protocols may give rise to various attacks [90] like blackhole attack (packets being dropped), grayhole attacks (packet being selectively forwarded), wormhole attacks (two malicious nodes conniving to receive packets from the network), sinkhole attacks (intruder advertises as the next best hop and receives packets) or node replication attacks (Clone attacks) [91]. Apart from these attacks, false routing information can be circulated to the neighbouring nodes, the network can be divided ending in cutting off communication between two legit nodes or messages can be forwarded in a loop resulting in depletion of energy and resources. Some of the defences built against such attacks prove to be worthy.

Table 1 Encryption algorithm-comparison

	Year developed	Designed by	Key length	Block Size	Cipher used	Network used	No. of rounds	Rounds designed	Attacks known	Performance speed	Avalanche effect	Advantages	Disadvantages	Commercial use
DES [63–67]	1977	IBM	56 bits	64 bits	Symmetric Block Cipher	Feistel network	16	Expansion, XOR operation with round key, S-Box (Substitution) and P-Box (Permutation)	Simple DES: Brute Force, known plaintext attack (Meet-in-the-middle attack), chosen plaintext attack	Fast in relation to its successors	Over 50%	Initially took 22 days to break the key using Brute Force approach	Vulnerable to all known attacks	Proved to be inefficient, hence not in use
2-DES [68]	1995	IBM	112 bits	64 bits	Symmetric Block Cipher	Feistel network	16 × 2	Simple DES Repeated Twice	Meet-in-the-middle attack, Brute force	Slower than Simple DES	Over 50%	Two times protection compared to Simple DES	Delays the time of brute force attack and is not resilient	Proved to be inefficient, hence not in use
3-DES [69–72]	1999	IBM	168 bits	64 bits	Symmetric Block Cipher	Feistel network	16 × 3	Simple DES Repeated Thice	Meet-in-the-middle attack	Slower than Simple DES	Over 50%	Easier to implement. Does increase the security level and solves issues faced by DES and 2-DES	Resource Efiiciency	To be deprecated for applications from 2023

(continued)

Table 1 (continued)

	Year developed	Designed by	Key length	Block Size	Cipher used	Network used	No. of rounds	Rounds designed	Attacks known	Performance speed	Avalanche effect	Advantages	Disadvantages	Commercial use
AES [73–77]	2001	Vincent Rijmen and Joan Daemen	128 bits, 192 bits and 256 bits	128 bits	Symmetric Block Cipher	Substitution-permutation network	10 for 128-bits, 12 for 192 bits, 14 for 256 bits	Byte-by-Byte Substitution, Shifting of Rows, Mixing of Column and Key Addition	Cache attacks (Theoretically proven)	Fastest among the peer algorithms	100%-one-bit change in the plaintext might cause changes in all the final 128bit ciphertext	Larger key implies greater security and yet faster	Repeated table lookups leading to Cache miss. This time lapse can be adversely used by intruders	Still in use
RSA [78–81]	1977	Rivest, Shamir & Adelman	1024 to 4096 bit	variable size	Asymmetric Block Cipher	Public key cryptosystem	Not Repeated	Generation of public and private key using Euler Totient function and two prime numbers. Public key alone transmitted. Decrypted using private key	Short message attack, Cycling attack, Unconcealed message attack, Chosen Cipher attack, Factorization attack	Key generation and encryption are faster Decryption is slow	50%	1. It employs public key encryption2. Embeds features of attaching a signature along with the message so that the receiver will be able to verify the sender as well	The algorithm falls short on the computation speed	Not used for wireless communication because of its large key size
BlowFish [82–85]	1993	Bruce Schneier	32 bits to 448 bits	64 bits	Symmetric Block Cipher	Feistel network	16	Division of 64 bits into two halves, XOR with subkey, BlowFish function, 4 S-Boxes	No known attacks	Much Faster than DES	Avalanche occurs between the key and the right half of the plain text at the conclusion of every round	Efficient speed of computation and to simplify the coding process	Extensive key generation phase amount to the complexity for a brute force attack	Domestic and Exportable use. TwoFish [86] a better alternative

Regarding hole attacks, a number of strategies have been developed by various researchers. The solutions have been identified on different levels like nodes level, topology of the network, routing algorithm and so on. Researchers suggested to equip the routing algorithm in itself to detect and avoid the possible holes in the network [92] or utilize a honeypot to lure the attackers/hole nodes or to study their behaviour pattern [93, 94]. Other possible front is to choose the right topology [95]. Authors have taken up a survey on detection of balckholes [96] in relation with the topology of the network used and have come up with a solution that mesh topology can withstand hole attacks better than other topologies. Another arena under research is to use an external node [97] purely for the detection of such holes and to report to a master node that will remove the suspected node from the network [98].

2.4 Data Storage Related Attack

Data gathered by the sensors and devices are sent to a remote server to be analyzed. This data must be stored in order to perform such statistical analysis and must be protected in this state from intruders. Data centers that hold such critical data must be given a layer of security in-order to keep away intruders.

Layer: The support layer of the IoT framework aids the various services offered by the application layer. The layer in focus performs two major functions—Storing data and Analysis of data. Data received from the IoT devices are stored for further analysis. This analysis includes data accumulation, data reporting, data mining, machine learning and data science. Identification of the owner of the data and being notified as to when the data loses its value is an important challenge to be addressed. Centralized data centres, remote cloud systems play a huge role in the effective working of an IoT network in providing the required services which calls for their inclusion in the security umbrella.

Features:

(1) *Massive data collection*: The amount of data generated, collected, transmitted and analysed is enormous. This serves as the primary target for any attacker that looks to exploit the system. Malware attacks few interconnected devices and engages as botnets to infect the network it is connected to. As public networks of IoT devices increase, the botnets would look to infest the infrastructure that will gravely impact the social security. Botnets are also capable of launching Distributed Denial of Service Attacks (DDoS) remotely in a large scale [37]. Detecting these botnet infections becomes extremely challenging because of the power constraints and the lack in system defence in low-cost devices. Since detection is an issue, stopping the spread becomes a greater challenge.

(2) *Access Model*: Data that is collected through the IoT physical layer must be kept secure from unauthorized access from other collaborating agencies. Data can be stored in a data lake in its native format till it is needed for analysis. Such critical data can be stolen or sold illegally to marketing agencies for making

profit. It can also be used against competing agencies or organizations that are stakeholders. With various governments also embracing smart applications that hold sensitive citizen data, proper access permission model becomes indispensable to ensure data privacy. The permission model must also be scalable as the data lake increases.

Challenges-Operational Level:

(1) *Intrusion-free*: As stated above, the data gathered by the IoT devices are stored in the remote servers for analysis. The data generated must be transmitted over a network and must be made sure is free of any illegal alteration. The integrity of the data must be maintained during transmission and verifying the purity of data is a major breaker.
(2) *Access control*: Data in storage will be used by different parties for different purposes. Access control continues to be a huge obstacle to be dealt with. The first level of users, to whom only the result is to be expressed must be kept away from the background data based on which the solution was derived [99]. Privacy of the data must be ensured both during the storage and during the transmission. Access to the client's critical information must be strictly monitored by the security structure in place.
(3) *Identity Concealment*: Several stakeholders collaborate to utilize the information available and it is necessary that the source of this data must not be revealed to other users. The identity of the source of the data must remain concealed to third parties to maintain the privacy of the contributing users and to avoid any intrusion into such users' personal information.

Threats:

(1) *Fabrication*: Data is stored in large amount in the cloud, transmitted from the lower end devices. Intruders willing to attack the network look to alter the data stored [28]. Information uploaded by the IoT devices must be stored intact even in a shared cloud environment. This again boils down to the access control of users sharing and contributing data. When the identity of a legit contributor is hacked or duplicated by an intruder, it can prove to be disastrous. The intruder can access and falsify the data stored.
(2) *Data Breach*: Large Scale data storage leads to a large consequence when data is breached. Data is said to be breached when an unauthorized or illegal location of data is done and the valuable data is stolen. This call for an upscale of security [100] on the user, owner and cloud service part.
(3) *Data deletion*: When multiple users share a piece of data in the cloud and some of the users request for deleting the same, data must be deleted safely. The data must not be deleted for all, but the access provided for the ones who have requested must be cut-off [101]. Intruders can take advantage of this policy to delete some data illegally. The prime solution is to delete the data completely only if the owner requests for deletion.

Security Measures:

(1) *Encryption*: Encryption has been proved effective in protecting the data stored in the cloud as well. The idea is to transfigure the original data into unreadable data using a key that is formed as a result of some pre-defined algorithm. There are innumerable algorithms [102] that vary in their core concept but the objective remains the same. Some encryption algorithms employ the user's identity information itself with the key so that it serves as user authentication as well. Others made use of an additional attribute set that only when matched will allow the users to access the data. Caution also has to be taken to safeguard the storage of this key and the transfigured data.

(2) *Blockchain*: Blockchain enables secure communication between two non-trustees without the aid of a third-party or a mediator. The data is identified in terms of a transactions in the form of a connected chain which when disturbed notifies malpractice [103]. Introduced for the security of cryptocurrency, it has now found its way to other domains as well where security is needed. It overcomes a lot of hurdles faced [104] in IoT Security like

  (i) Increasing the trust between the communicating parties since each of them can have an unaltered record of transactions.
  (ii) Data integrity since there is no way to alter the blocks without recalculating the ensuing block's hashes
  (iii) Brings in peer-to-peer communication because it eliminates the need of a centralized node and hence the speed of exchange increases
  (iv) Identity Access Management (IAM) of contributing devices since Blockchain employs authorized User Registration and management and Ownership Management
  (v) Data privacy using Smart contracts. Smart contracts are nothing but an access treaty stating the permissions provided for each of the parties.

(3) *Cloud Decentralization*: Cloud models have always followed a centralized structure with multi-tenant architecture. However, decentralization (i.e.) moving the processing entities closer to the data contributors have been studied and they seem to have a good impact [105] on the security considerations. Technologies like Edge computing employing IoT-Edge devices, Mobile Cloud computing known as Cloudlets and usage of Virtual Machines are different approaches with the objective of decentralization owing to improved latency, reduced traffic, reduced delays and better resource utilization. Even though, this reduces the attack plane of the cloud, it does increase the vulnerability of each individual device interacting. It can be reduced by instantiating peer-to-peer communications driven by certain policies in place.

(4) *Trust Assessment*: IoT networks seek the help of a cloud setup due to lack of computing and storage capabilities. The cloud in turn adopts the security and trust issues of the IoT network. There are innumerable Cloud Service Providers who offer various kind of services, but the trustworthiness of these services must be verified before interaction. The trust assessment models in general are

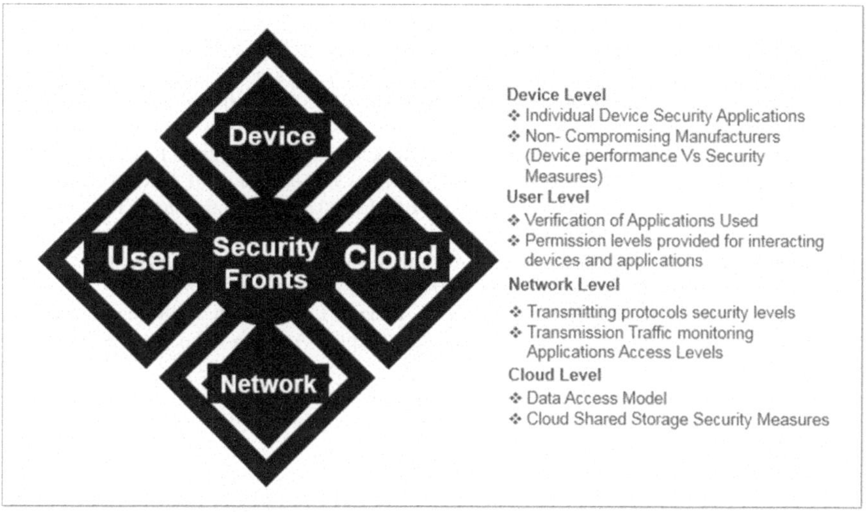

Fig. 3 Security fronts to be monitored

with respect to the service features like reliability, security, availability and scalability. Since it does effect the efficiency of the security front of an IoT framework, several factors like cost, time, QoS parameters, feedback ratings are displayed to the Consumers to help them choose the best candidate.

3 Summary

IoT framework has opened the networking front to any device that is used by mankind. However, when trying to upscale, security becomes a major concern since it has to handle sensitive data (Fig. 3). Owing to the architecture of an IoT framework, security concerns have been studies under four categories: Device-level, Applications level, During Data transmission and Data storage. The desirable security features with respect to that category along with major threats and some well-known solutions have been showcased.

References

1. E. Ronen, A. Shamir, Extended functionality attacks on IoT devices: the case of smart lights, in *2016 IEEE European Symposium on Security and Privacy (EuroS&P)* (IEEE, 2016), pp. 3–12
2. R. Hager, Google shuts down Xiaomi access to Assistant following Nest Hub picking up strangers' camera feeds (Update: Fully resolved), https://www.androidpolice.com/2020/01/17/uh-oh-xiaomi-camera-feed-showing-random-homes-on-a-google-nest-hub-including-still-images-of-sleeping-people/. Last Accessed 17 Jan 2020

3. I. Shakeel, How IoT is raising cybersecurity concerns, https://resources.infosecinstitute.com/topic/iot-raising-cybersecurity-concerns/. Last Accessed 30 Nov 2017
4. Hackers attack homes on average 104 times a month, says new Comcast report, https://www.gearbrain.com/are-smart-home-devices-secure-2649035325.html. Last Accessed 29 Nov 2020
5. A. Maiti, M. Jadliwala, Light ears: information leakage via smart lights, in *Proceedings of the ACM on Interactive, Mobile, Wearable and Ubiquitous Technologies*, vol. 3.3 (2019), pp. 1–27
6. A. Holmes, The FBI just issued a warning about the risks of owning a smart TV—here are its suggestions for protecting your privacy, https://www.businessinsider.in/tech/news/the-fbi-just-issued-a-warning-about-the-risks-of-owning-a-smart-tv-here-are-its-suggestions-for-protecting-your-privacy/articleshow/72387047.cms. Last Accessed 5 Dec 2019
7. Tencent Blade Team, "Shellcode, reports of Amazon Echo, which we have presented on Defcon26", https://github.com/tencentbladeteam/Exploit-Amazon-Echo. Last Accessed 13 Aug 2018
8. W. Huiyu, Q. Wenxiang, Security Researchers at Tencent Blade Team, "Breaking Smart Speakers: We are Listening to you", https://www.defcon.org/html/defcon-26/dc-26-speakers.html#HuiYu.
9. S. Thiagarajan, Global Head, Cyber Security Practice, Tata Consultancy Services, "Raising Your IoT Security Game", https://www.tcs.com/content/dam/tcs/pdf/perspectives/edition-10-raising-your-iot-cyber-security-game.pdf
10. N. Koroniotis, N. Moustafa, F. Schiliro, P. Gauravaram, H. Janicke, A Holistic review of cybersecurity and reliability perspectives in smart airports. IEEE Access **8**, 209802–209834 (2020). https://doi.org/10.1109/ACCESS.2020.3036728
11. IFALPA: cyber threats: who controls your aircraft? IFALPA Position Papers, No. 14POS03 (2013), http://www.ifalpa.org/store/14POS03%20-%20Cyber%20threats.pdf
12. G. Suciu, A. Scheianu, A. Vulpe, I. Petre, V. Suciu, Cyber-attacks—the impact over airports security and prevention modalities, in *Trends and Advances in Information Systems and Technologies*, ed. by Á. Rocha, H. Adeli, L. Reis, S. Costanzo (2018) 319-77700-9_16
13. R. Marcel, T. Gerald, Cyber-attack warning: could hackers bring down a plane? Spiegel Online International (2015), http://spon.de/aevsu
14. N. McAllister, No, you CAN'T hijack a plane with an Android app. The Register (2013), http://goo.gl/news/0PmU
15. Z. Kim, Feds say that banned researcher commandeered a plane. Wired (2015), http://www.wired.com/?p=1782748
16. I. Shumailov, L. Simon, J. Yan, R. Anderson, Hearing your touch: a new acoustic side channel on smartphones (2019). arXiv preprint arXiv:1903.11137.
17. "Phishing Spotlight Research Report", https://www.lookout.com/phishing-spotlight-report-lp
18. "Kaspersky- Top 7 Mobile Security Threats in 2020" (2020), https://www.kaspersky.co.in/resource-center/threats/top-seven-mobile-security-threats-smart-phones-tablets-and-mobile-internet-devices-what-the-future-has-in-store
19. M. Hron, The fresh smell of ransomed coffee, https://decoded.avast.io/martinhron/the-fresh-smell-of-ransomed-coffee/. Last Accessed 25 Sept 2020
20. L. Fernandes, Global print security landscape (2019), https://quocirca.com/wp-content/uploads/2019/02/Quocirca-Print-Security-Feb-2019-Final-Web.pdf
21. E. Itkin, Y. Livneh, Y. Balmas, Faxploit: sending fax back to the dark ages, https://research.checkpoint.com/2018/sending-fax-back-to-the-dark-ages/. Last Accessed 12 Aug 2018
22. M.R. Warner, Internet of things cybersecurity improvement act of 2017, in *Proceedings 115th United States Congress* (2017), p. 1691
23. A. Rayes, S. Salam, IoT protocol stack: a layered view. *Internet of Things from Hype to Reality* (Springer, Cham, 2017). https://doi.org/10.1007/978-3-319-44860-2_5
24. T. Varshney, N. Sharma, I. Kaushik, B. Bhushan, Architectural model of security threats & their countermeasures in IoT, in *2019 International Conference on Computing, Communication, and Intelligent Systems (ICCCIS)* (IEEE, 2019), pp. 424–429

25. A. Singh, N. Chawla, J.H. Ko, M. Kar, S. Mukhopadhyay, Energy efficient and side-channel secure cryptographic hardware for IoT-edge nodes. IEEE Internet Things J. **6**(1), 421–434 (2019)
26. E. Fernandes, J. Jung, A. Prakash, Security analysis of emerging smart home applications, IEEE Secur. Privacy 636–654 (2016)
27. WeLiveSecurity, 10 things to know about the October 21 IoT DDoS attacks [Online] (2016). https://www.welivesecurity.com/2016/10/24/10-things-know-october-21-iot-ddos-attacks/
28. M. Nawir, A. Amir, N. Yaakob, O.B. Lynn, Internet of things (IoT): taxonomy of security attacks, in *2016 3rd International Conference on Electronic Design (ICED)* (IEEE, 2016), pp. 321–326
29. A.A. Cárdenas, S. Amin, Z.S. Lin, Y.L. Huang, C.Y. Huang, S. Sastry, Attacks against process control systems: risk assessment, detection, and response, in *Proceedings of the 6th ACM Symposium on Information, Computer and Communications Security* (2011), pp. 355–366
30. M. Albahri, R. Kirichek, A.A. Ateya, A. Muthanna, A. Borodin, Combating counterfeit for IoT system based on DOA, in *2018 10th International Congress on Ultra Modern Telecommunications and Control Systems and Workshops (ICUMT)*, Moscow, Russia, 2018, pp. 1–5, https://doi.org/10.1109/ICUMT.2018.8631257
31. H. Mohammed, S.R. Hasan, F. Awwad, Fusion-on-field security and privacy preservation for IoT edge devices: concurrent defense against multiple types of hardware Trojan attacks. IEEE Access **8**, 36847–36862 (2020)
32. W. Chen, X. Luo, A.N. Zincir-Heywood, Exploring a servicebased normal behavior profiling system for Botnet detection, in *Proceedings of International Symposium on Integrated Network Service Management (IM)*, May 2017, pp. 947–952
33. I. Hafeez, M. Antikainen, A.Y. Ding, S. Tarkoma, IoT-KEEPER: detecting malicious IoT network activity using online traffic analysis at the edge. IEEE Trans. Netw. Serv. Manage. **17**(1), 45–59 (2020)
34. Y. Liu, H.-H. Chen, L. Wang, Physical layer security for next generation wireless networks: theories, technologies, and challenges. IEEE Commun. Surv. Tutor. **19**(1), 347–376. 1st Quart (2017)
35. P. Williams, I. Dutta, H. Daoud, M. Bayoumi, Security aspects of internet of things–a survey, in *2020 IEEE 6th World Forum on Internet of Things (WF-IoT)* (IEEE), pp. 1–6
36. G. Vormayr, T. Zseby, J. Fabini, Botnet communication patterns. IEEE Commun. Surv. Tutor. **19**(4), 2768–2796. Fourthquarter (2017). https://doi.org/10.1109/COMST.2017.2749442
37. M.P.P. Yin et al., IoTPOT: analysing the rise of IoT compromises, in *Usenix Conference on Offensive Technologies* (USENIX Association, 2015), p. 9
38. D. Nandal, V. Nandal, Security threats in wireless sensor networks. **11**(01), 59–63 (2011)
39. K. Zhang, X. Liang, R. Lu, X. Shen, Sybil attacks and their defenses in the internet of things. IEEE Internet Things J. **1**(5), 372–383 (2014)
40. C. Kolias, G. Kambourakis, A. Stavrou, J. Voas, DDoS in the IoT: Mirai and Other Botnets. Computer **50**(7), 80–84 (2017)
41. V. Karagiannis, P. Chatzimisios, F. Vazquez-Gallego, J. Alonso-Zarate, A survey on application layer protocols for the internet of things. Trans. IoT Cloud Comput. **3**(1), 11–17 (2015)
42. G. Selander, F. Palombini, K. Hartke, Requirements for COAP end-to-end security. IETF, Fremont, CA, USA, July 2017 [Online]. http://www.ietf.org/internet-drafts/draft-hartke-coree2e-security-reqs-03.txt
43. G. Selander, J. Mattsson, F. Palombini, L. Seitz, Object security for constrained restful environments (OSCORE). IETF, RFC 8613, July 2019 [Online]. https://tools.ietf.org/html/rfc8613
44. S. Pérez, J.L. Hernández-Ramos, S. Raza, A. Skarmeta, Application layer key establishment for end-to-end security in IoT. IEEE Internet Things J. **7**(3), 2117–2128 (2019)
45. K. Zhang et al., Sybil attacks and their defenses in the internet of things. IEEE Internet Things J. **1**(5), 372–383 (2014)

46. A. Rajan, J. Jithish, S. Sankaran, Sybil attack in IOT: modelling and defences, in *2017 International Conference on Advances in Computing, Communications and Informatics (ICACCI)* (IEEE, 2017)
47. H. Yu, M. Kaminsky, P. Gibbons, A. Flaxman, SybilGuard: defending against Sybil attacks via social networks. IEEE ACM Trans. Netw. **16**(3), 576–589 (2008)
48. H. Yu, P. Gibbons, M. Kaminsky, F. Xiao, SybilLimit: a near optimal social network defense against Sybil attacks. IEEE/ACM Trans. Netw. **18**(3), 885–898 (2010)
49. A. Praseed, P. Santhi Thilagam. DDoS attacks at the application layer: challenges and research perspectives for safeguarding web applications/ IEEE Commun. Surv. Tutor. **21**(1), 661–685 (2018)
50. A.A. Zaid, M.H. Alalfi, A. Miri, Automated identification of over-privileged smartthings apps, in *2019 IEEE International Conference on Software Maintenance and Evolution (ICSME)* (IEEE, 2019)
51. SAINT Project, https://csl.fiu.edu/saint-project/saint-project-description/
52. F. Schmeidl, B. Nazzal, M.H. Alalfi, Security analysis for smartthings IoT applications, in *2019 IEEE/ACM 6th International Conference on Mobile Software Engineering and Systems (MOBILESoft)* (IEEE, 2019)
53. J. Ferdows et al., A comprehensive study of IoT application layer security management, in *2020 IEEE International Conference for Innovation in Technology (INOCON)* (IEEE, 2020)
54. K. Zhao, L. Geo, A survey on the internet of things security, in *International Conference on Computational Intelligence and Security (CIS)* (2013), pp. 663–667
55. N. Apthorpe, D. Reisman, N. Feamster, A smart home is no castle: privacy vulnerabilities of encrypted IoT traffic. arXiv:1705.06805 [cs] (2017)
56. W. Zhang, Y. Meng, Y. Liu, X. Zhang, Y. Zhang, H. Zhu, Homonit: monitoring smart home apps from encrypted traffic, in *Proceedings of the 2018 ACM SIGSAC Conference on Computer and Communications Security* (ACM, 2018), pp. 1074–1088
57. B. Park, Threats and security analysis for enhanced secure neighbor discovery protocol (SEND) of IPv6 NDP security. Int. J. Control Autom **4**(4) (2011)
58. H. Kim, Protection against packet fragmentation attacks at 6lowpan adaptation layer, in *2008 International Conference on Convergence and Hybrid Information Technology* (IEEE, 2008), pp. 796–801
59. R. Hummen, J. Hiller, H. Wirtz, M. Henze, H. Shafagh, K. Wehrle, 6LoWPAN fragmentation attacks and mitigation mechanisms, in *Proceedings of the Sixth ACM Conference on Security and Privacy in Wireless and Mobile Networks* (ACM, 2013), pp. 55–66
60. N. Park, N. Kang, Mutual authentication scheme in secure internet of things technology for comfortable lifestyle. Sensors **16**(1), 20 (2016)
61. C. Pu, S. Hajjar, Mitigating forwarding misbehaviors in RPL-based low power and Lossy networks, in *2018 15th IEEE Annual Consumer Communications Networking Conference (CCNC)* (2018), pp. 1–6
62. G.C. Kessler, An overview of steganography for the computer forensics examiner. Forensic Sci. Commun. **6**(3), 1–27 (2004)
63. D. Coppersmith, The Data Encryption Standard (DES) and its strength against attacks. IBM J. Res. Dev. **38**(3), 243–250 (1994)
64. P. Mahajan, A. Sachdeva, A study of encryption algorithms AES, DES and RSA for security. Glob. J. Comput. Sci. Technol. (2013)
65. A.A. Yazdeen, S.R. Zeebaree, M.M. Sadeeq, S.F. Kak, O.M. Ahmed, R.R. Zebari, FPGA implementations for data encryption and decryption via concurrent and parallel computation: a review. Qubahan Acad. J. **1**(2), 8–16 (2021)
66. DES Weak keys—SSLeay 0.9.0b—Jan 1999, http://www.umich.edu/~x509/ssleay/des-weak.html#:~:text=The%20semi%2Dweak%20keys%20are,4%20different%20subkeys%20of%2016
67. J.H. Moore, G.J. Simmons, Cycle structure of the DES with weak and semi-weak keys, in *Conference on the Theory and Application of Cryptographic Techniques* (Springer, Berlin, Heidelberg, 1986), pp. 9–32

68. J. Katz, Efficient cryptographic protocols preventing" man-in-the-middle" attacks. Columbia University (2002)
69. D. Coppersmith, D.B. Johnson, S.M. Matyas, A proposed mode for triple-DES encryption. IBM J. Res. Dev. **40**(2), 253–262 (1996)
70. S. Lucks, Attacking triple encryption, in *International Workshop on Fast Software Encryption* (Springer, Berlin, Heidelberg, 1998), pp. 239–253
71. R. Bhanot, R. Hans, A review and comparative analysis of various encryption algorithms. Int. J. Secur. Appl. **9**(4), 289–306 (2015)
72. P.C. van Oorschot, M.J. Wiener, A known-plaintext attack on two-key triple encryption. Eurocrypt '90, Springer LNCS **473**, 318–325
73. J. Daemen, V. Rijmen, AES proposal: Rijndael (1999). A. Kak, Lecture Notes on "Computer and Network Security". Purdue University (2015)
74. E. Tromer, D.A. Osvik, A. Shamir, Efficient cache attacks on AES, and countermeasures. J. Cryptol. **23**(1), 37–71 (2010)
75. D.J. Bernstein, Cache-timing attacks on AES (3) (2005)
76. C. Rebeiro, M. Mondal, D. Mukhopadhyay, Pinpointing cache timing attacks on AES, in *2010 23rd International Conference on VLSI Design* (IEEE, 2010), pp. 306–311
77. E. Biham, A. Shamir, Differential cryptanalysis of DES-like cryptosystems. J. Cryptol. **4**(1), 3–72 (1991)
78. David Ireland, RSA algorithm, https://www.di-mgt.com.au/rsa_alg.html
79. S.Y. Yan, *Cryptanalytic attacks on RSA* (Springer Science & Business Media, 2007)
80. V.B. Kute, P.R. Paradhi, G.R. Bamnote, A software comparison of RSA and ECC. Int. J. Comput. Sci. Appl. **2**(1), 43–59 (2009)
81. M. Seetha, A.K. Koundinya, Comparative study and performance analysis of encryption in RSA, ECC and Goldwasser-Micali cryptosystems. Int. J. Appl. Innov. Eng. Manag. (IJAIEM) **3**(1), 111–118 (2014)
82. The Blowfish encryption algorithm. Dr. Dobb's J. **19**(4), 38–40 (1994)
83. M.N.A. Wahid, A. Ali, B. Esparham, M. Marwan, A comparison of cryptographic algorithms: DES, 3DES, AES, RSA and blowfish for guessing attacks prevention. J. Comput. Sci. Appl. Inf. Technol. **3**(2), 1–7 (2018)
84. P.J.M. Finch, A study of the blowfish encryption algorithm Diss. City University of New York (1995)
85. T. Nie, T. Zhang, A study of DES and Blowfish encryption algorithm, in *Tencon 2009–2009 IEEE Region 10 Conference* (IEEE, 2009)
86. B. Schneier et al., in *The Twofish Encryption Algorithm: A 128-Bit Block Cipher* (Wiley, 1999)
87. P.S. Nandhini, B.M. Mehtre, Intrusion detection system based RPL attack detection techniques and countermeasures in IoT: a comparison, in *2019 International Conference on Communication and Electronics Systems (ICCES)* (IEEE, 2019)
88. E. Kfoury et al., A self organizing map intrusion detection system for RPL protocol attacks. Int. J. Interdiscip. Telecommun. Netw. (IJITN) 11(1):30–43 (2019)
89. G. Thamilarasu, S. Chawla, Towards deep-learning-driven intrusion detection for the internet of things. Sensors **19**(9), 1977 (2019)
90. I. Butun, P. Österberg, H. Song, Security of the internet of things: vulnerabilities, attacks, and countermeasures. IEEE Commun. Surv. Tutor. **22**(1), 616–644 (2019)
91. W.T. Zhu, J. Zhou, R.H. Deng, F. Bao, Detecting node replication attacks in wireless sensor networks: a survey. J. Netw. Comput. Appl. **35**(3), 1022–1034 (2012)
92. Z. Karakehayov, Using reward to detect team black-hole attacks in wireless sensor networks, in *Proceedings of Workshop Real World Wireless Sensor Network* (2005), pp. 20–21
93. A. Prathapani, L. Santhanam, D.P. Agrawal, Intelligent honeypot agent for blackhole attack detection in wireless mesh networks, in *IEEE 6th International Conference on Mobile Adhoc and Sensor Systems* (IEEE, 2009)
94. A. Prathapani, L. Santhanam, D.P. Agrawal, Detection of blackhole attack in a wireless mesh network using intelligent honeypot agents. J. Supercomput. **64**(3), 777–804 (2013)

95. R. Alattas, Detecting black-hole attacks in WSNs using multiple base stations and check agents, in *2016 Future Technologies Conference (FTC)* (IEEE, 2016)
96. S.N. Krishnan, P. Srinivasan, A QOS parameter based solution for black hole denial of service attack in wireless sensor networks. Indian J. Sci. Technol. **9**(38) (2016)
97. M. Tiwari, K.V. Arya, R. Choudhari, K.S. Choudhary, Designing intrusion detection to detect black hole and selective forwarding attack in WSN based on local information, in *Proceedings of IEEE 4th International Conference on Computer Sciences and Convergence Information Technology* ICCIT) (2009), pp. 824–828
98. M. Wazid, A. Katal, R.S. Sachan, R. Goudar, D. Singh, Detection and prevention mechanism for blackhole attack in wireless sensor network, in *Proceedings of IEEE International Conference on Communication and Signal Processing (ICCSP)* (2013), pp. 576–581
99. G.R.M.D.A. Patil, Data breaches as top security concern in cloud computing. Int. J. Pure Appl. Math. **119**(14), 19–28 (2018)
100. J. Hur, D. Koo, Y. Shin, K. Kang, Secure data deduplication with dynamic ownership management in cloud storage. IEEE Trans. Knowl. Data Eng. **28**(11), 3113–3125 (2016)
101. P. Yang, N. Xiong, J. Ren, Data security and privacy protection for cloud storage: a survey. IEEE Access **8**, 131723–131740 (2020)
102. A. Hughes, A. Park, J. Kietzmann, C. Archer-Brown, Beyond bitcoin: what blockchain and distributed ledger technologies mean for firms. Bus Horizons **62**(3), 273–281 (2019)
103. A. Al Sadawi, M.S. Hassan, M. Ndiaye, A survey on the integration of blockchain with IoT to enhance performance and eliminate challenges. IEEE Access **9**, 54478–54497 (2021)
104. J. Singh, T. Pasquier, J. Bacon, H. Ko, D. Eyers, Twenty security considerations for cloud-supported Internet of Things. IEEE Internet Things J. **3**(3), 269–284 (2015)
105. X. Li, Q. Wang, X. Lan, X. Chen, N. Zhang, D. Chen, Enhancing cloud-based IoT security through trustworthy cloud service: an integration of security and reputation approach. IEEE Access **7**, 9368–9383 (2019)

Reenie Tanya is currently working as Assistant Professor at SRM Institute of Science and Technology, Ramapuram, Chennai, India. She completed her Masters in Engineering from St. Joseph's College of Engineering and Bachelor's in Computer Science and Engineering from DMI College of Engineering. She has authored several research papers in reputed indexed journals as a beginning of her research journey. Her areas of interest also include Video summarization and Wireless transmission security. Ms. Reenie Tanya has 4 years of teaching experience and is currently pursuing her Ph.D. in the area of Internet of Video Things (Surveillance) in SRM University. She also served a stint at Cognizant Technology Solutions as a Programmer Associate and a Project Associate-I in CTDT, Anna University, Chennai working on Web Services Development.

Balika J Chelliah has an overall experience of 16+ years in the field of Computer Science both in Academics and Administration. He has obtained his Doctorate in Computer Science and Engineering from SRM University in the year 2018 and is currently working as Associate Professor in the Department of CSE, SRM Institute of Science and Technology, Ramapuram, Chennai, India. He has authored and Co-Authored 40+ Research publications in well reputed indexed journals and holds 8 Patents as his contribution to the research arena. His Research interests have been mainly in the field of Artificial Intelligence and Expert Systems. He is also currently guiding 6 Scholars in their Doctoral journey. He has also authored books on "Machine Learning" and "Introduction to Python" to share his expertise.

Wearable and Assistive Wireless Connectivity for Healthcare

Assistive Technology Strategy: Wearable Multi-Lingual Blind Technology for Persons with Impairment and Eye-Sight Disability Based on IoT and Cloud

Humayun Rashid, Aasim Ullah, Md. Mosaraf Hossain Khan, Md. Shahid Ullah, S. M. G. Mostafa, Mohammad Jalal Uddin, Abu Tayeb Noman, and Amran Hossain

Abstract People with challenged vision (both permanent and temporary) face several difficulties in their everyday life. A person having visual impairment may not differentiate between colors, which is an essential part of work in several industries such as Ready Made Garments (RMG) sector where sorting cloths based on color is essential. This work represents a more improved version of our previous work which was a demonstration of a talking color detecting device for blind people. Obstacle facing and fall occurrence are very important issues for visually challenged person that are addressed in this chapter. The proposed device uses the latest hardware components including upgraded Central Processing Unit (CPU) and sensors for IoT and cloud-based architecture that can detect color and obstacle efficiently. Moreover it gives notification regarding color and obstacle in multiple languages to visually challenged person. The device also sends fall notification through internet to the caretaker of the visually impaired user in case of fall detection, which is an added key feature of this work.

Keywords Alert generation · Cloud server · Color detection · Disability · Fall detection · IoT Assistive Device · Multilingual · Obstacle avoidance · Ready Made Garments (RMG) · Vision impairment

---

H. Rashid
Department of Electrical and Electronics Engineering, University of Turku, Turku, Finland
e-mail: hurash@utu.fi

A. Ullah (✉) · Md. M. H. Khan · Md. S. Ullah · S. M. G. Mostafa · M. J. Uddin · A. T. Noman · A. Hossain
Department of Electrical and Electronics Engineering, International Islamic University Chittagong, Chattogram 4318, Bangladesh
e-mail: aasim@kth.se

Md. S. Ullah
e-mail: shahideee04@iiuc.ac.bd

A. K. Pathan (ed.), *Towards a Wireless Connected World: Achievements and New Technologies*, https://doi.org/10.1007/978-3-031-04321-5_6

1 Introduction

Eye-sight impairment is a constraint in case of functioning of the eye. As mentioned in [1], the visual nerve system and persons having eye-sight impairment refers no vision in both eyes who has a visual acuity less than 6/60 or 20/200 in Snellen chart [2, 3]. According to a statistical research on blindness by WHO considerably more than 285 million people are struggling from visual impairment which includes blindness and low vision and unfortunately 87% of that total live in regions like Bangladesh [4].

The study also focuses on the opportunities of the visual impaired people that blind people usually regarded as burden to own family and to the outside world the disregard is out of world and unfortunately these people have extremely fewer possibilities to earn their very own livelihood. Generally this scenario has been observed in the underdeveloped countries while rich and developed countries provide at least more opportunities to them [5]. In case of children with visual impairment the sufferings are more noticeable as they are being excluded from the opportunity of getting education as well as other basic rights [6].

Color detection ability is an important privilege that seems not so important in a relative look which also can be a strength to some people in their daily life. Countries like Bangladesh, China, Vietnam, lot of people are gaining their livelihood through apparel along with clothing manufacturing garments in which cases visually impairment of a person can be a great hindrance while working, consequently making their livelihood. In these cases the opportunity to work in such a place reduces if they don't have the ability to detect and differentiate color. Also, color detection ability is benefited most for children for their education purposes.

Technological innovations have been becoming a hope for helping them in recent times in case of inability to color detection or even helpful for completely visual impaired person, although the concept of the assistive devices for the blind is actually not so traditional [7]. While reviewing the previously developed product, several blind assistive devices were found for color detection such as Color Talk [8], Coloresia [9], Colorino [8], Color Teller [8], Speech-master Talking Color with detection system [10], ColorTest 150 [8]. Some blind guidance systems have also been reviewed which are Sonar Glasses [11], iSonar [12], Smart walking stick [13], The GuideCane [14] and The NavBelt [15]. Some key limitation of these devices is not having speaking ability into multiple languages, IoT Connectivity and Wearability. Most importantly, no devices have been developed with both two features into one single wearable device with IoT connectivity. Also, there was no fall detection feature in any of the mentioned devices which is a very essential feature for any blind or disables assistive devices.

In authors' previous paper [7], a new architecture has been suggested in preceding research work [7] that can pronounce the name of numerous colors in both English and Bangla. This proposed architecture will help the blind people to recognize different colors in their real life and workstation. A new architecture has been proposed with current advancement of Internet of Things (IoT) and its wide variety of application in advance embedded system design. The suggested architecture incorporated with IoT and sensors to provide various features to identify 'fall' for the blind person. In addition, an obstacle avoidance unit has been added with proposed architecture to avoid collision and accident. The prime features of this proposed design is that all kind of warnings will be given as human vocal sound.

The book chapter arrangement has been organized in the following manners. The section discusses the architecture and methodology, the third and fourth section illustrates the software and hardware improvement, the fifth section represents the working principle and algorithm of the proposed system and the last two sections represent a comparison among earlier developed devices and the suggested system along with the conclusion following with the future scope of the work.

2 Review of Literature

Conducting activities of daily living for blind people, such as reading product labels and identifying currencies, can be difficult due to a variety of obstacles [16]. It demonstrates that a low-cost wearable device is accessible for anyone who wishes to take their phone with them. Dynamic multi-ability routing and an automated multi-lingual context are also in the works as are new innovations in dynamic routing [17]. Recent year research has also discovered that the relevant community is moving forward by providing new problems and scenarios ranging from multi-lingualism to driving. One may envision a blind person approaching an intelligent system who is knowledgeable in appropriate fields such as human–computer interactions, ubiquitous/wearable data processing, and so on, as depicted in [18]. Along with the growing diversity of wearable technologies, electronic food diaries and mobile applications enable users to keep a more detailed record of what they drink and eat. Yet there is a lot of scope for the future because of new technology in the health sector such as smartphone applications, smartwatches and sensors. Is the app accessible to disabled people (e.g., screen readers for blind people, closed captioning for deaf people) [19]. The current generation of smart assistants, which are embedded in home listening devices, smartphones and wearables, has developed over the last few years. The next generation of wearable devices straddles the boundaries between conversation and technology. Braille is a written language used by blind people to write and read, and it is similar to the language in which they communicate [20]. Recent technological advancements, such as deep learning-based speech identification, have enabled more accurate results to be obtained. The acoustic environment created by wearable gadgets was measured using a wrist-mounted device that was specially constructed for this purpose. As a result, both classifiers were unable to determine the group status of any of the participants [21]. Many surveys illustrate the relationship between humans (citizens) and computers (for example, wearables such as smartglasses). The survey allows users to connect to smart urban entities by utilizing augmented reality smartglasses (a sort of wearable computer). With these intelligent glasses, one could take his or her eyes off the road for a split second.

It is proposed that a variety of bus recognition systems be developed in the field of computer vision research, as well as in other domains; However, the majority of them employ sensors and active devices, such as RFID (Radio Frequency Identification), GPS tracking systems, Beacons, and so on. In the field of vision-based techniques, Wongta [8] developed a system that recognizes bus numbers by using MSER (Maximally Stable Extremal Regions). Instead than only finding the required texts in an image, their technique detects all of the texts in the image as well, for example the bus number. Guida et al. [9] developed a framework regarding bus route number that uses a number classifiers together including adaboost in order to determine any other items present at the very front of bus and some repairs are done to the recovered characteristics to get restored numerical value. The object is therefore transformed to Saturation, Hue, Value (HSV) colorspace, after which each numeric or digit value is partitioned.

At the end identification of digits was done via Optical character Recognition (OCR) and voice of the user is produced as the output. The proposed face detection framework by Viola and Jones [10] has three main contributions, namely integral images, boost and cascading classifiers, which is a simple and efficient way to detect faces from binary images. Bus detection system proposed by Pan et al. [11] helps the visually challenged people. It uses Histogram of the oriented gradients to extract features from bus images and for detecting the bus facade in frames of windows via a bus classifier a cascading Support Vector Machine model is implemented. To support the visually impaired person Tsai and Yeh [12] introduced a process for bus detection. The functionalities of that process contain detecting the moving bus as well as the detection of bus panel and also detecting text from the text region of the bus panel. The process showed high accuracy when features were extracted from different frames of the video as it uses the method of MAFD (Modify Adaptive Frame Differencing).

Bouhmed employed an ultrasonic sensor and camera in a walking cane to sense hazards in the path and communicate this information to the blind in [13]. The module creates output via voice. Zahir et al. [14] implemented a prototype of a wearable head-mountable device by adapting Virtual Reality glasses with ultrasonic sensors and HC-SR04 since it takes the least amount of time to detect and can discover hazards over a longer distance. Arduino is used to creating the prototype. Ani et al. [15] established a voice-assisted text-reading system that works with eyeglasses. A camera is incorporated into the eyeglass to obtain an image, and text is retrieved from the image using Tesseract-OCR. For TTS, Open Software E-speak is used. Khan and Khusro [16] presents a method for end-to-end real-time scene text localization and recognition, with "false positives" demonstrating the resilience of the suggested method against noise and poor character contrast. Murdoch et al. [17] introduces a Convolution Neural Networks (CNN) model for detecting English and Thai text from natural scene photos with higher accuracy than earlier approaches. Dengel et al. [18] describes an approach for extracting multi-script text from natural scenes that uses the collaboration of proximity and similarity rules to generate text-group predictions. Chawla et al. [19] investigates the topic of finding correspondences between two photographs taken from various perspectives. To support the blind, a variety of procedures and unique concepts have been offered. The majority of these systems are dynamic, which makes them easier to use in real-time. Our proposed solution is designed for real-time application and relies on video recognition rather than image capture. Because the bus arrival and waiting time are continually changing, every frame must be checked rather than just single image snaps, which would be less effective in detecting the bus and bus board. The suggested system can be connected with any existing arrangement with components that are nearly identical, or it can be constructed as a standalone system with any additional route number characteristics added.

3 Methodology and Architecture

The system has two main units. Each unit has several sub-units. The main two units are listed as:

1. Sensors and IoT section
2. Cloud server section.

A 8-bit micro-controller is used to develop the system in the previous design [7]. The design adopts simplicity in order to develop the efficiency. In latest studies [21, 22], several experiments are utilized with 8-bit based micro-controller. It is understandable why Atmega 328P is not worked as efficient device for the systems. As BLE and WiFi plays an important role which misses out in the chip. The missing features leads to no connectivity function with IoT base architectures. A supplemental Classic Bluetooth module is recommended in the study as well in order to use in previous research [7]. The purpose is to deliver the Bluetooth connectivity. Although it cause greater power consumption that cause a big negative aspect for any wearable device. 8-bit micro-controller has its additional supplemental constraint. It demands additional Wi-Fi module in order to supply Wi-Fi connectivity. The feature eventually enhances the power consumption along with the cost of the production. In this chapter, a micro-controller featuring Wi-Fi and BLE has been propositioned so that the elimination of an auxiliary Bluetooth and Wi-Fi module can be achieved as well as a gentle cut in the power consumption through the adaptation of this chip.

One of the significant challenges for visually impaired users is the perception of the colors that have been presented briefly in our earlier paper [7]. A color sensor had been suggested to not only recognize the colors but also to adapt the name of them in dual language from the pre-stored audio data in the storage of the system.

A color sensor with a similar concept has been utilized in this chapter to identify the colors and transcribe them in multiple languages with the help of various linguistic data packs stored in the cloud. The alternative language pack can be updated using a smartphone that will necessitate the participation of a caretaker but will allow users from other countries and languages to use the gadget in their native language. One of many significant features is the ability to listen to the audio using both wired and wireless headsets as well as existing sound systems.

Our earlier design could only detect colors with the use of a color sensor and had no other capabilities that could assist a blind person. When working on an earlier version of this chapter, the author suggested and created a blind assistive robot that employed an ultrasonic sensor to identify obstacles and allow blind people to walk beside the robot. A speech alarm, as well as vibration alerts, have been used to implement this obstacle avoidance system, which makes use of an ultrasonic sensor and a small motor driver in conjunction with the motor to create vibration.

Falling while walking or moving is a regular occurrence among blind people, and it can be observed in many different situations. Some falls can be dangerous, and recovering from them may necessitate support from others. As a result, those responsible for the care of blind people should be aware of their deteriorating condition. A fall detection feature has been introduced to this chapter, and the caretaker will be notified whenever a fall occurs. The basic block diagram of our suggested system is depicted in Fig. 1.

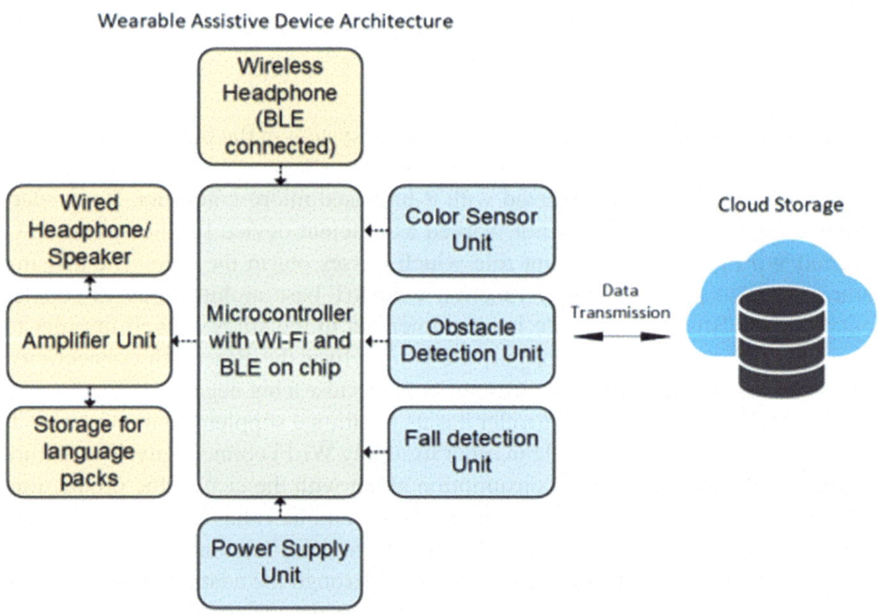

Fig. 1 The block diagram for the prospective system's architectures

4 Hardware Development

4.1 Microcontroller Unit

A 32-bit based micro-controller known as Node MCU ESP-32S has been proposed to be used as core micro-controller for our proposed system. NodeMCU ESP-32S is considered as a high-performance micro-controller for developing IoT based system as it is equipped with industrial grade specifications and able to perform efficiently for better integration, wireless transmission, lower power consumption and better network connectivity [22]. It is powered with ESP32 chip that is considered as a scale-able and adaptive chip. A principal feature of NodeMCU is that it has two CPU cores that can be controlled individually along with the capability to adjust the clock frequency from 80 to 240 MHz Another main feature of this micro-controller is that it can be operated without having any additional power supply unit and featured with ultra-low power consumption [3]. The module is also featured with traditional Bluetooth, Bluetooth low energy and Wi-Fi (802.11n @ 2.4 GHz up to 150 Mbit/s) along with other popular data transmission protocol of I2C, SPI, and UART that makes it very suitable to construct IoT based system.

4.2 Sensor Unit

Different sensorz require to be interfaced with the micro-controller unit to achieve different goal of the proposed concept. The three main goals are to detect colors, obstacle and fall. Assistive Navigation Application for Blind People using a White Cane Embedded System and sensor unit is also reported in [22].

4.3 Color Sensor

The TCS3200 colour is a Programmable colour light-to-frequency converter that has ability to convert light intensity to frequency and it allows optimized output range. The operating process of TCS3200 depends on Photodiodes which can detect colour when Light from the LED of the sensor is shone above the subject placed in front of the sensor. Three different values of red, green and blue colors are obtained by generating a square wave of 50% duty. Different logic permutations for three colors value are utilized along with various pulse provided from the micro-controller [4] to detect color. Similar color act with A finger wearable audio-tactile device using customized color tagging is reported in [23].

4.4 Obstacle Avoidance

HC SR04 is a ultrasonic distance sensor that need to be interfaced with ESP32 to detect obstruction in front of the user by with assistance of non-contact measurement functionality. The main feature of this sensor is that accuracy is up to 2–400 cm by utilizing Trig and Echo, but the ranging accuracy can be reached up to 3 mm by transmitting a ultrasonic signal and detect incoming output [5]. The sensor module is featured with additional control circuit to prevent inconsistent "bouncy" data that may cause false distance measurement.

4.5 Fall Detection

MPU-6050 is one of the first Motion Tracking devices that has been constructed with feature of low power consumption along with the ability to provide high-performance. It consists of 3-axis accelerometer, a 3-axis gyroscope and a motion processor that has capability to establish communication through I2C communication protocol. The main feature of this module is that it has capability to carry out complicated 6-axis motion detection that is required to detect the fall of the user [6]. A Fine-Grained Indoor Fall Detection System With Ubiquitous Wi-Fi Devices is found in [24], Contact-less fall detection for the elderly in [25]. Also, A Distributed Fall Detection Architecture Using Ensemble and machine Learning is mentioned in [26–28] (Fig. 2).

4.6 Alert Generation in Multiple Languages

The proposed system will have multilingual supports and the function to acquire the multilingual support requires to have a cloud server. The language files will be stored and updated to the clod server from where it required to be downloaded and and stored into a SD card storage that will be interfaced with the system.

The process to download and update the languages audio files requires the support of a an app that can be installed and run to an android powered smartphone. The app should have the capability to connect with hardware with the support of lowe power Bluetooth. Another approach can be developed to directly download the audio files to the assistive device with the help of built-in WI-FI of ModeMCU module. Also, a different methodology can be developed where audio language files will be downloaded to the smartphone with the help of the application first and sync up the assistive device with the latest audio files through Bluetooth. SD Card Module will be attached with the assistive device via SPI communication protocol [7].

Vibration alert will be generated when the assistive device will detect any kind obstacle and a vibrating Mini Motor Disc need to be utilized with ESP32 to achieve

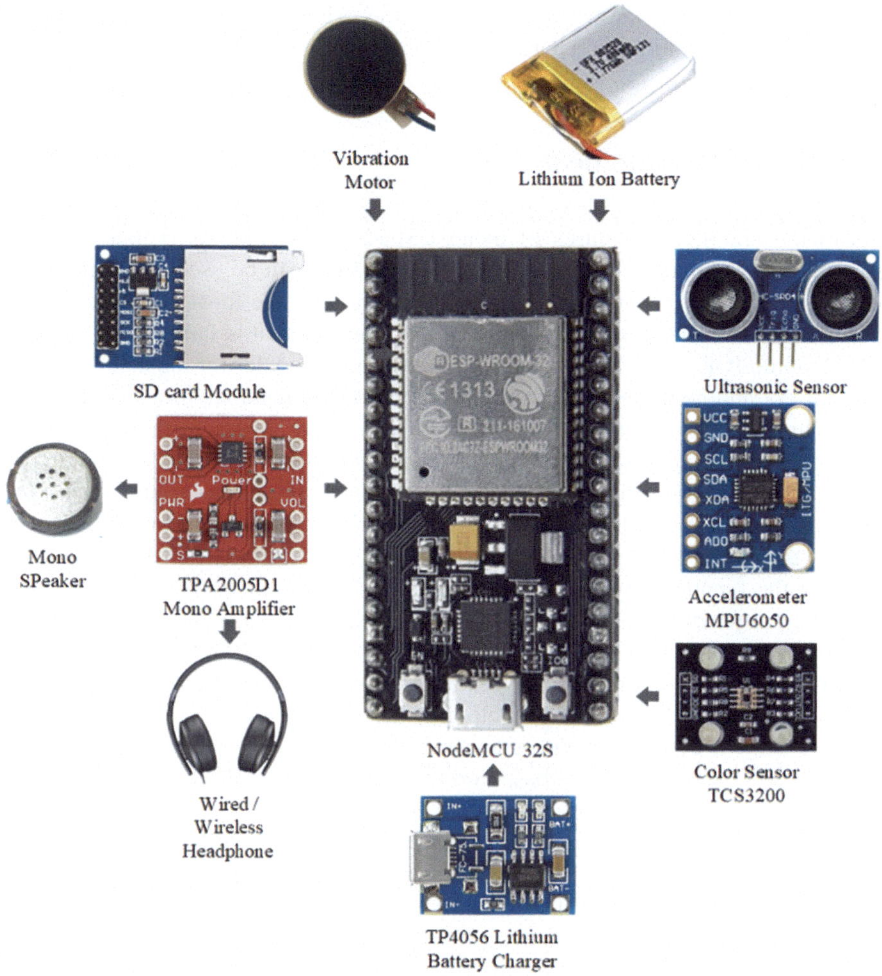

Fig. 2 The diagram of hardware settings for the proposed system

this goal. Low powered motor needs to be interfaces that can be operated with 2 V current (40 mA) to ensure higher operating time by the assistive device.

4.7 *Power Supply Unit*

lM7805 & LE33 were used to build the power supply unit for our previous development. But the need of using a separate power supply has been omitted for our current proposed assistive device as we are using NodeMCU featured with built-in power supply to convert 5 V into 3.3 V for different sensors and other components. A lithium changer module known as TP4056 requires to be interfaced to have projection against Dual Functions and recharge the battery.

5 Software Development

One of the significant features of using ESP32 based NodeMCU that different programming languages like Lua, Python, Java, Ruby can be used to program the device. To achieve our goals for our proposed system, the most convenient is to utilize C++ based Integrated development environment that is familiar as Arduino IDE. Arduino IDE has rich collection of libraries and modules for most of the available sensors and components. It removes the need to create additional library for each component from scratch. Another option is to use Micro Python an object-oriented powerful programming language based on Python for NodeMCU but it, but C/C++ based firmware can ensure faster compilation compare to Micro-Python for IoT based software development. Programming flowchart of the proposed system is demonstrated in the figure. The hardware requires to be initialized at starting to ensure the functionality of all components. Following the initialization process, the system requires to verify the selected policy by the user, which is determined through a select switch for choosing a specific policy among from several policies. The selected policy will run and determine the system functionality. The program requires to check and verify if the the policy has been changed by the users and in case no change is detected, the program will continue running the same policy. The program can change the policy based on the user selection. Policies are described below:

Policy 1: The policy 1 has been developed to run with full capability thus it utilizes higher power consumption. All the features are enabled that causes to run all the sensors and modules as weak as allow to enable all features. The user can detect color, obstacle, and fall. Performance efficiency of the proposed assistive device will be at the highest level although power efficiently will reduce significantly compare to the other policies. It can also be configured as the default policy when the assistive device system will be operated without any user's specific selection.

Policy 2: The policy 2 will be developed to have higher power efficiency by reducing power consumption through enabling selected features of obstacle avoidance and fall detection. The policy has been design with consideration for those users who don't requires to utilize the color detection always. This policy will allow to have lower power consumption compare to policy 1 as well the performance accuracy will be lower than he policy 1. Policy 2 will have higher operating time compare to policy 1.

Policy 3: Policy 3 will allow t user to use only color detection and thus it will have lowest power consumption compare to policy 1 and policy 2. The main reason to design such policy is to provide user flexibility to use only color detection with higher operating time. As the power consumption will be lowest, it will also have significant affects on performance accuracy (Fig. 3).

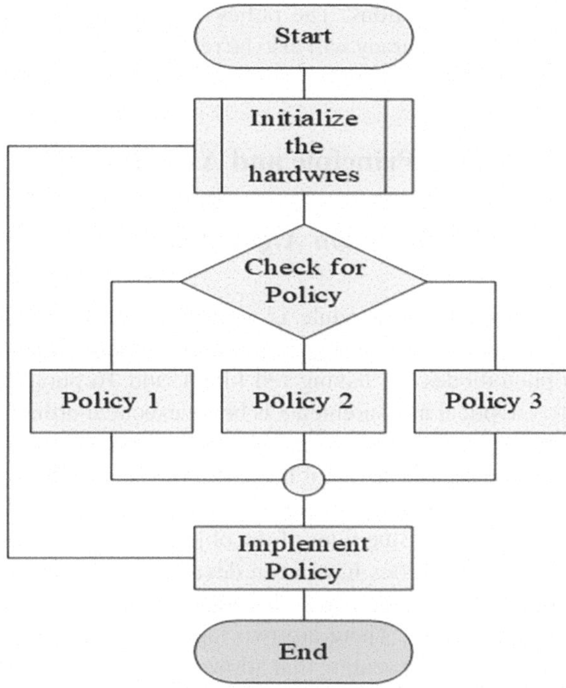

Fig. 3 The flowchart for the programming of MCU

The flowchart of the programming is illustrated in the following figure. In the initial stages, the hardware will be initialized. After the initialization procedure, the system will verify which policy has been chosen. There is a configurable selective switch for a different policy. According to the specified policy, the program will run the policy. During the process, the program will always check whether the policy has been updated or not. If it is not modified by the user, it will continue to run, or else if the policy change has been identified, it will alter the policy according to the user's choices.

Policy 1: In policy 1, all the components are active, and all the functions may be concurrently performed. The user may detect color, impediment and fall. Power consumption will be high but the performance efficiency of the device will be at the optimum standard. It will be also the default policy when the system would commence without any user's option.

Policy 2: In policy 2, only obstacle avoidance will be triggered together with fall detection. As perceived that users may not always wish to utilize color detection rather it is easy to use the device for obstacle avoidance and fall detection. This strategy would generate medium power usage combined with medium performance accuracy.

Policy 3: In this policy, the user will be allowed/permitted to Turn/switch on the color detection alone and the rest of the two features will be turned off. The user may activate this policy when he needs color detection just and does not require for other two features/functions. The policy will create reduced power consumption as the performance accuracy will also be relatively less compared to the other two policies.

6 Working Principle and Algorithm

6.1 Color Detection Algorithm

The colour sensor module is equipped with 8 × 8 array of photodiodes, where 16 photodiodes are having blue filters, 16 photodiodes are having green filters and 16 photodiodes are having red filters, and 16 photodiodes are having clear or no filters. Optical measurements is being executed utilizing small-angle incident radiation with assistance of an optical source. The basic principle of working functionality of a colour sensor is that it has capability to converts different colour into different frequencies. TCS3200 module is utilized to get the frequencies generated due to Green, Red and Blue filter of the object by allowing specific filter actives. Different logic and conditions have been developed to detect different colours. TCS3200 is featured with 5.6 mm lens that can operate with better functionality for the area of 1" and 1 1/16. There are two logic inputs known as S0 and S1 for controlling Output-frequency scaling that allows the output range to be optimized for a different kind of measurement method. The functioning procedure of TCS3200 module has been shown in Fig. 4. Duty cycle, time period and frequency are determined from the below-mentioned equation [4].

$$\text{The duty cycle of square wave T, TH} = \text{TL} \tag{1}$$

$$\text{Time Period, T} = 2\ \text{TH} \tag{2}$$

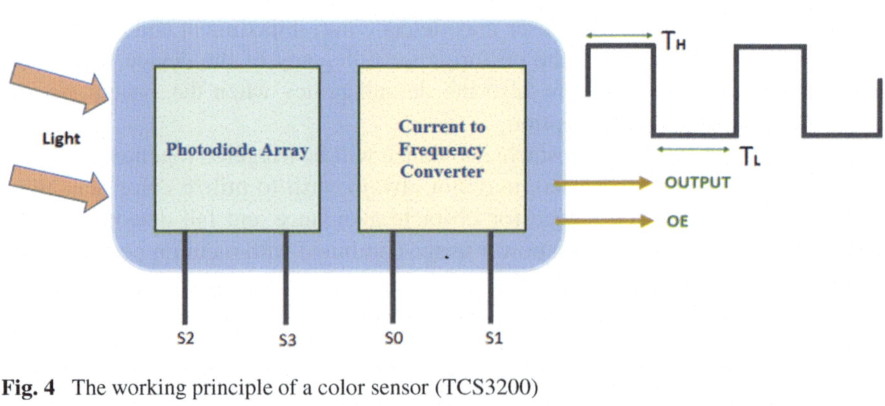

Fig. 4 The working principle of a color sensor (TCS3200)

$$\text{Frequency} = \tfrac{1}{2}\,\text{TH} \tag{3}$$

NodeMCU will detect the colour based on the logic created and the assistive device will find the selected language file and generate the sound of the name of the color in that selected languages. However, if the selected languages files are not available in the SD card module, the system will try to found it from cloud server through the application and if the sound file is found, it will be download it from the cloud server. The output audio sound can be heard into the attached wired or wireless headphone.

6.2 Obstacle Avoidance Algorithm

An ultrasonic sensor attached with the assistive system to transmit a high-frequency sound pulse to detect the distance between the user and the obstruction. The main working functionality depends on detecting the transmission time and receiving time of the ultrasonic sound. A short 10 uS pulse is required to enable the triggering of transmitting an 8 cycle burst of ultrasound at 40 kHz to detect the ranging. NodeMCU processes the timing of the echo and reflection of the sound for duration into a distance conversation. The processed value is compared with a pre-defined parameter to obtain the result with the help of following equation [19]:

$$\text{Distance} = (\text{high level time} \times \text{velocity of sound}(340\text{M/S})/2 \tag{4}$$

6.3 Fall Detection Algorithm

The algorithms for fall detection has been developed using a variation of acceleration and angular motion during fall occurs. Sum Vector Magnitude (SVM), the angle between x and z-axis and differential SVM (DSVM) are determined to detect if the fall has occurred (Fig. 5). The equation can be expressed as follow where *i* represent the sample number [11]:

$$SVM_i = \sqrt{{x_i}^2 + {y_i}^2 + {z_i}^2} \tag{5}$$

$$\delta = \arctan\left(\frac{\sqrt{{x_i}^2 + {z_i}^2}}{x_i}\right) * \frac{180}{\pi} \tag{6}$$

$$DSVM_i = \sqrt{(x_i - x_i - 1)^2 + (y_i - y_i - 1)^2 + (z_i - z_i - 1)^2} \tag{7}$$

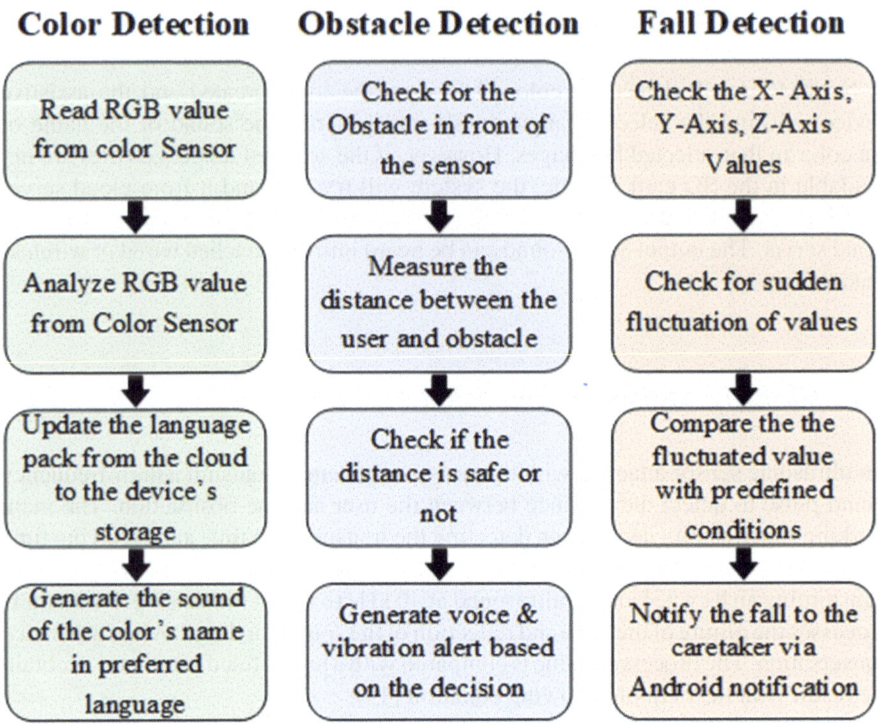

Fig. 5 Algorithm for color, obstacle and fall detection

6.4 Language File from Cloud Server and Notification to the Caretaker About Fall Detection

A cloud platform needs to be integrated with a android app to develop multilingual support and notification service. The development platform of the app is MIT app inventor that allows to develop an android smart app using drag and drop logic and functions. Multilingual support requires to have different languages pack stored in a cloud server. The concept of the notification is that a fall detection will generate a notification to the user's caretaker along with information of fall occurrence time and magnitude of the fall that will allow the caretaker to be able to find out the location of the user and arrange emergency services to reduce health hazard of the user.

6.5 Language File from Cloud Server

Several languages pack will be available to a cloud server. The languages file can be updated via system by the user. The user can select his own preferred language pack from the cloud and the language pack will be updated through the Wi-Fi of the device. Another approach is that the language file will be updated to the user's phone and where the system will be connected with the user's phone, it can update the features from the smartphone also. The app has been developed using MIT app inventor which offers an easy solution to build an android smart app using logic and functions.

7 Discussion

TCS3200, a color sensor is used to convert the color into frequency. In proposed model this TCS3200 module will be used to collect the frequency acquired with Green, Red and Blue filter of the object through enabling corresponding filter actives. In this case, several conditions are applied in order to detect each color. Generally, four white LEDs are attached at the front side of the TCS 3200 color sensor which effects in the spectrum by making a deviation whenever the sensor is initialized in the device, illustrated in Fig. 6 [11].

The Figs. 6 and 7 displays relative response vs. wavelength curves showing four different shades red, green, blue and black which plotted in the graph, represents four photo-diodes of red, green, blue and clear.

Three-axis data of x, y, z is collected from the Accelerometer which are pre-processed and analyzed with previously defined values to detect fall. Variation of Sum Vector Magnitude (SVM) and angle between x between z-axis with the processed

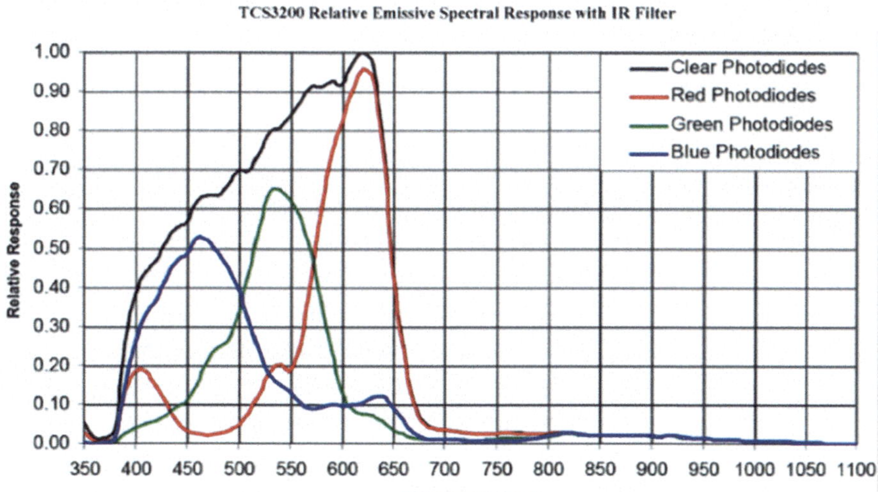

Fig. 6 The curve comparison of Relative response and Wavelength with IR Filter [9]

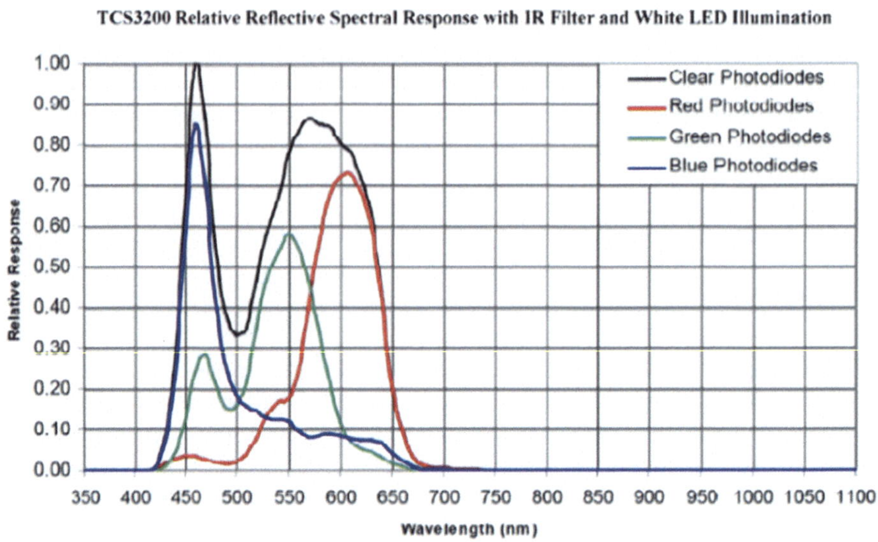

Fig. 7 Relative response versus Wavelength with IR filter and LED illuminations [9]

data can be shown as well to determine the threshold as well as a violation of threshold values which indicates a fall.

Different components have different power consumption that have been demonstrated in the table. When adopting a different policy, different power consumption has been noticed which has been documented in Table 1. The three different policy has been developed in a way that they can be employed for power efficiency. Table 2 has been showing the details of the three policy.

Table 1 Hardware specifications of proposed system

Components	Model	Operating voltage
Microcontroller	NodeMCU 32S	3.3–5.5 V
Color sensor	TCS3200	2.7–5.5 V
Ultrasonic sensor	HC-SR04	5 V
Accelerometers	MPU6050	3.3 V
SD card module	SparkFun microSD	3.3 V
Mono-amplifier	TPA2005D1	3.3 V
Vibration motor	Disk vibrating motor	−3.3 V

Table 2 Power consumption & performance analysis for different policy

Policy name	Color detection	Obstacle detection	Fall detection	Power consumption	Performance
Policy 1	On	On	On	High	High
Policy 2	Off	On	On	Medium	Medium
Policy 3	On	Off	Off	Low	Low

8 Comparison Study with Earlier Research

Table 3 illustrates the product functionalities that have been incorporated in the systems are given with the features and comparison of ten old products. The components are mainly categories according to some preferred terms such as- Obstacle Detection ability, Color Detection ability, speaking capability, Fall Detection ability, IoT Connectivity, and Wearability. From the contrast, it is quite clear that Color Talk [8], Colorino [8], Color Teller [8], Speech Master Talking Color [8], Color Test 150 [8] and Coloresia [9] are presenting same functionalities. All of these products can identify color and have the speaking capability. But these are not suitable for obstacle and fall detection, multiple language recognition capability or having IoT connectivity. Most significantly these components are not wearable either.

Obstacle detection is possible for Sonar Glasses [11] along with wearable functionalities. However, detection of color, detection of fall, IoT connectivity, speaking and multiple languages are not possible by this device [12]. The identical constraints have been found in Smart walking stick [10] and GuideCane [9]. In case of a blind assistive system iSonar is worth nothing if the above abilities are not found in this device. The NavBelt [11] has navigation capability and wearability by obstacle detecting unit. But it does not pose other features compared to the proposed architectures.

The architecture and the product which is suggested in this study is designed such a sensible way that all kinds of usable features including fall detection, color detection, obstacle detection, speaking ability, multiple language recognition ability are added in proposed product. As well as the most demanding 'wearability' has included for such case of study. The distinct feature in this proposed is that it can be linked and controlled through the Internet of Things (IoT) which is not available in the devices that are discussed above.

The market price of Coloring and Speech-Masters Talking Color is expensive enough. Obviously the suggested product will be economical comparatively other available devices in the market. The costs of proposed device are about $80 which is comparatively very low than the other available products. Additionally, the unavailability of all the mentioned features are present in the existing proposed devices.

Table 3 Comparison of previously developed system with our proposed system

No	Product	Color detection	Obstacle detection	Fall detection	Speaking Ability	Multiple language Ability	IoT connectivity	Wear-ability
1	Color Talk [8]	✓	X	X	✓	X	X	X
2	Coloresia [9]	✓	X	X	X	X	X	X
3	Colorino [8]	✓	X	X	✓	X	X	X
4	Color Teller [8]	✓	X	X	✓	X	X	X
5	Speech-master Talking Color [10]	✓	X	X	✓	X	X	X
6	ColorTest 150 [8]	✓	X	X	✓	X	X	X
7	Sonar Glasses [11]	X	✓	X	X	X	X	✓
8	iSonar [12]	X	✓	X	X	X	X	X
9	Smart walking stick [10]	X	✓	X	X	X	X	X
10	The GuideCane [9]	X	✓	X	X	X	X	X
11	The NavBelt [11]	X	✓	X	X	X	X	✓
12	Proposed Device	✓	✓	✓	✓	✓	✓	✓

9 Future Work and Conclusion

The Proposed device that we have discussed in this chapter overcomes major shortcomings of our previous designed system. A user ready version will be very beneficial for the consumer even though there is still plenty of scope for future improvements. By incorporating power efficient micro-controller with Bluetooth 5 and GPS Module, the current position as well as the directional navigation of the user can be accessed by the caretaker. By including more sensors with the proposed three sensors vital health status of the user can be monitored. The cheaper locally available sensors that are used here can be replaced with more efficient sensors for better performance. The color sensor and ultrasonic sensor can be replaced with a camera which can more efficiently detect variations of color and obstacles. The Proposed device interfaced with a machine learning algorithm to predict the user's priority and making the system more user-friendly will provide much better experience for the consumer.

References

1. A. Cashin-Garbutt, What is visual impairment? News-Medical.net (2012). [Online]
2. B. Punani, N. Rawal, *Visual Impairment Handbook* (Blind People's Association, India, 2000), pp. 1–10
3. Disability in Bangladesh: Prevalence and pattern, *Population Monograph of Bangladesh*, vol. 5 (2015)
4. World Health Organization, *Global Data On Visual Impairments 2010* (2012), pp. 1–17
5. Bangladesh fights to end blindness, *The Guardian* (2010). [Online]. Available: https://www.theguardian.com/world/2010/sep/28/bangladeshvolunteers-childhood-blindness-treatment
6. Help for the visually impaired in Bangladesh, Ft.com, (2016). [Online]
7. H. Rashid, A.S.M. Rabbi Al-Mamun, M.S.R. Robin, M. Ahasan, S.M. Taslim Reza, Bilingual wearable assistive technology for visually impaired persons, in *2016 International Conference on Medical Engineering, Health Informatics and Technology (MediTec)* (2016)
8. P. Wongta, T. Kobchaisawat, T.H. Chalidabhongse,An automatic bus route number recognition, in *2016 13th International Joint Conference on Computer Science and Software Engineering (JCSSE)* (IEEE, 2016)
9. C. Guida, D. Comanducci, C. Colombo, Automatic bus line number localization and recognition on mobile phones—a computer vision aid for the visually impaired, in *International Conference on Image Analysis and Processing* (Springer, Berlin, Heidelberg, 2011), pp. 323–332
10. S. Pattanayak, C. Ningthoujam, N. Pradhan,A survey on pedestrian detection system using computer vision and deep learning, in *Advanced Computational Paradigms and Hybrid Intelligent Computing* (Springer, Singapore, 2022), pp. 419–429
11. M. Vardar, P. Sharma,An optimized object detection system for visually impaired people, in *Second International Conference on Sustainable Technologies for Computational Intelligence* (Springer, Singapore, 2022), pp. 25–38
12. R. Priyatharshini, R. Senthil Kumar, M. Sanjay Sivakumar, A. Mathumathi, N.S. Johnson,A wearable assistive device for safe travel using transfer learning and IoT for visually impaired people, in *Advanced Soft Computing Techniques in Data Science, IoT and Cloud Computing* (Springer, Cham, 2021), pp. 3–26
13. E. Bouhamed, I. Kallel, D.S. Masmoudi, New electronic cane for visually impaired people for obstacle detection, in *Proceedings of the IEEE International Conference on Vehicular Electronics and Safety* (2012)

14. E. Zahir, K. Hossain, K. Balachander, C. Venkatesan, R. Kumar,Safety driven intelligent autonomous vehicle for smart cities using IoT. Int. J. Pervasive Comput. Commun. (2021)
15. R. Ani, E. Maria, J. Jameema Joyce, Smart specs: voice assisted text reading system for visually impaired persons using TTS method, in *IEEE International Conference on Innovations in Green Energy and Healthcare Technologies* (2017)
16. A. Khan, S. Khusro, An insight into smartphone-based assistive solutions for visually impaired and blind people: issues, challenges and opportunities. Univ. Access Inf. Soc. **20**(2), 265–298 (2021)
17. T. Murdoch, T. Pey, E. Brooks,A step towards truly independent access for everyone, everywhere. Assis. Technol. 1–5 (2021)
18. A. Dengel, L. Devillers, L.M. Schaal,Augmented human and human-machine co-evolution: efficiency and ethics, in *Reflections on Artificial Intelligence for Humanity* (Springer, Cham, 2021), pp. 203–227
19. S. Chawla, J.K. Sabharwal, B. McCarthy, R. Erhardt,Technology acceptance, social marketing and the design of a mobile health app to support active ageing amongst senior citizens in the Asia-Pacific region, in *Broadening Cultural Horizons in Social Marketing* (Springer, Singapore, 2021), pp. 239–261
20. D. Sayers, R. Sousa-Silva, S. Höhn, L. Ahmedi, K. Allkivi-Metsoja, D. Anastasiou, Š. Beňuš et al.,The Dawn of the Human-Machine Era: A forecast of new and emerging language technologies (2021)
21. B. Little, O. Alshabrawy, D. Stow, I. Nicol Ferrier, R. McNaney, D.G. Jackson, K. Ladha et al.,Deep learning-based automated speech detection as a marker of social functioning in late-life depression. Psychol. Med. **51**(9), 1441–1450 (2021)
22. L.-H. Lee, T. Braud, S. Hosio, P. Hui, Towards augmented reality driven human-city interaction: current research on mobile headsets and future challenges. ACM Comput. Surv. (CSUR) **54**(8), 1–38 (2021)
23. A. Mocanu, V. Sita, C. Avram, D. Radu, A. Aştilean,Assistive navigation application for blind people using a white cane embedded system, in *2020 IEEE International Conference on Automation, Quality and Testing, Robotics (AQTR)* (IEEE, 2020), pp. 1–5
24. C. Rui, Y. Liu, J. Shen, Z. Li, Z. Xie,A multi-sensory blind guidance system based on YOLO and ORB-SLAM, in *2021 IEEE International Conference on Progress in Informatics and Computing (PIC)* (IEEE, 2021), pp. 409–414
25. Y. Wang, S. Yang, F. Li, W. Yue, Y. Wang, Fall viewer: a fine-grained indoor fall detection system with ubiquitous wi-fi devices. IEEE Internet Things J. **8**(15), 12455–12466 (2021)
26. M. Nahian, M.H. Raju, Z. Tasnim, M. Mahmud, M.A.R. Ahad, M. Shamim Kaiser, Contactless fall detection for the elderly, in *Contactless Human Activity Analysis* (Springer, Cham, 2021), pp. 203–235
27. C.-C. Chang, Y.-C. Chen, B.-H. Sieh, Y.-M. Ooi,A distributed fall detection architecture using ensemble learning, in *2021 IEEE 4th International Conference on Knowledge Innovation and Invention (ICKII)* (IEEE, 2021), pp. 81–84
28. C.R. Kumar, M. Kaleel Rahman, E. Derrick Gilchrist, R. Lakshmi Pooja, C. Sruthi, Smart band for elderly fall detection using machine learning. NVEO-Nat. Volatiles Essential Oils J. NVEO 8269–8285 (2021)

Maximum Power Design and Simulation for a Low Return Loss Wearable Microstrip Patch Antenna

Syed Zahidur Rashid, Abdul Gafur, Aasim Ullah, Md. Akbar Hossain, Sultan Shah Mamun, and Md. Qudrat-E.-Alahi Majumder

Abstract A single band rectangular microstrip patch antenna is proposed for wearable applications in this book chapter. The design inspectorates radiating patch on one segment of the substrate and a ground plane on the other segment of the substrate. The designed antenna size is about 39×47 mm^2 which is very small compared to the resonant frequency's wavelength. The primary feature of this proposed antenna is its very low return loss. It also features decent gain and bandwidth of 58.84 dB, 2.85 dB and 95 MHz respectively along with a low Specific Absorption Rate (SAR) value which makes it safe for off body implementations. The recommended rectangular patch antenna is designed and simulated using the CST Microwave studio. In the studio an operating frequency of 2.45 GHz is maintained. FR-4 (4.3) material is used as substrate which is normally used for low frequencies. It is also a light-weight, low cost and available material for antenna fabrication. The gain and return loss of the antenna is further improved by selective optimization. Overall, the antenna holds great significance in wearable application.

Keywords Directivity · ISM band · Off body application · Patch antenna · Radiation pattern · Return loss · SAR · Single band · VSWR · Wearable

---

S. Z. Rashid · A. Gafur · S. S. Mamun · Md. Q.-E.-A. Majumder
Department of Electronic and Telecommunication Engineering, International Islamic University Chittagong, Chittagong, Bangladesh
e-mail: szrashidcce@yahoo.com

A. Ullah (✉)
Department of Electrical and Electronic Engineering, International Islamic University Chittagong, Chittagong, Bangladesh
e-mail: aasim@kth.se

Md. A. Hossain
Auckland University of Technology, Auckland, New Zealand
e-mail: akbar.hossain@aut.ac.nz

A. K. Pathan (ed.), *Towards a Wireless Connected World: Achievements and New Technologies*, https://doi.org/10.1007/978-3-031-04321-5_7

1 Introduction

The use of Wireless Body Area Network (WBAN) in various application like health-care monitoring, sports, military, and industry is increasing rapidly. The wearable antenna is used in a wearable device to monitor the health of people of different ages, learning human signs, physical exercising, tracking, training and emergency rescue solutions etc. Thus, the antenna being required to be flexible, compact, lightweight, robust and as efficient as possible. Here the patch antenna becomes a good option. The micro-strip patch antenna can be regarded as an essential device in body area network. It claims to be low cost, low profile, easy to fabricate as well as provide unidirectional radiation pattern [1, 2].

The antenna polarization is normally required to be in regular direction to the body surface in case of on/off communication regarding to the body. To make the antenna as efficient as possible it is required to consider the effects caused due to the human body [3]. The large ground located under the patch is used to minimize the mutual coupling between the human body and the antenna. Consequently, it increases forward radiation at the bore sight [2].

There are numerous possibility of the antenna to be designed based on the fact that how it will be integrated as wearable. It can be the antenna that acts as the transmitter or receiver for the wearable device. Or it can be an antenna that relay sensor-base information to a nearby device. Even an antenna can be fabricated as part of a garment and can be worn on the human body [4]. Based on those possibility the antennas can be designed in so many ways. As the proposed antenna is intended as part of a wearable it is been designed using FR-4 substrate which allows it to be compact, low cost and easy to fabricate.

In case of frequency, a number of frequency bands happen to be designated for WBAN systems. For example, the Medical Implant Communication System (MICS) which is around 400 MHz band. Also, the Industrial Scientific Medical (ISM) band which is around 2.4 GHz and 5.8 GHz. The Ultra-wideband (UW) [5–7] is around 3–10 GHz. The 2.45 GHz frequency of ISM band is popular due to its high readable range, fast reading speed, large information capacity and low-cost [8].

An additional significant feature of the wearable antenna system is Specific Absorption Rate (SAR). It is an essential assessment aspect when an antenna functions on or near the human body. It specifies the total energy emitted from the electromagnetic field that is absorbed in human body. Antennas used in WBAN's necessitate a low SAR in order to not cause any harm to the human body from radio wave exposure [2, 9]. But reducing the value of SAR without affecting the antenna parameters is also a challenge. The standard value of SAR given by IEEE Std C95.1-1999 is 1.6 W/Kg per 1 g tissue model [10].

In paper [11], a SRR loaded antenna is designed where Teflon is used as substrate, return-loss is 32 dB while gain and directivity is extremely low and VSWR is 1.37. In paper [12], the antenna was designed using a paper substrate where the antenna gain and directivity was 2.6 dB, 4.2 dBi respectively and with a low return loss and bandwidth but VSWR were not given specificly. In paper [1] the Rogers Ultralam 3850 substrate and co-planar waveguide (CWP) being utilized where the return-loss and bandwidth are 36 dB,110 MHz correspondingly while the gain is only 1.13 dB and directivity is not stated. Same as ours, the author in [13] used FR-4 substrate for antenna fabrication. Though the antenna directivity is 5.4 dB, the return loss is 14 dB and VSWR value is 2.10 which is over the standard measurement. Even though the gain is not mentioned in the paper, the return loss and VSWR indicated poor gain along with bulk antenna size. In paper [14] FR-4 substrate uses for antenna design the gain of the antenna is 5.11 dB but return-loss and directivity are less and VSWR is high.

This work presents a high gain then [1, 11, 12] with low return-lost then [1, 11, 12, 14] and low SAR wearable microstrip patch antenna for an ISM band, which is designed and simulated by using antenna simulator software. The low SAR value has been proposed placing the antenna at an optimum distance. and the SAR value is lower than the Theoretical standard.

2 Design Specification

To design the microstrip patch antenna for a specific resonant frequency the size of the antenna is needed to be measured using the theoretical analysis. Using the transmission line model [15] the theoretical width and length of the patch antenna is calculated which is then optimized to obtain the proposed antenna's width and length. The equations derived in transmission line model are used here to calculate the antenna width and length [16].

$$Width\,(W) = \frac{c}{2f_r}\sqrt{\frac{2}{\epsilon_r + 1}} \tag{1}$$

where

c = velocity of light (3×10^8 m/s).

f_r = resonant frequency.

ϵ_r = dielectric constant of the substrate ($\epsilon_r = 4.3$).

The length of the patch can be calculated by using the following formula as shown in Eq. (2)

$$Length\,(L) = L_{eff} - \Delta L \tag{2}$$

where L_{eff} = effective length of the patch and ΔL = patch length extension in mm.

Here we need to calculate the effective length to obtain the patch length as follows

$$L_{eff} = \frac{c}{2f_r\sqrt{\epsilon_{eff}}}$$

where ϵ_{eff} the effective dielectric constants as are stated below

$$\in_{eff} = \frac{\in_r + 1}{2} + \frac{\in_r - 1}{2}\left(1 + \frac{10h}{W}\right) \tag{3}$$

where h is height of the patch in mm, W is the width of the patch in mm.

Now extension length ΔL is as follows

$$\Delta L = 0.412h \frac{(\in_r + 0.3)(\frac{W}{h} + 0.264)}{(\in_r - 0.258)(\frac{W}{h} + 0.8)} \tag{4}$$

where ΔL = patch length extension in mm, h = patch height in mm and W = patch width in mm.

The antenna is simulated using these equations given above and the simulation is done using CST Microwave studio.

Figure 1 shows the geometric view of the proposed microstrip patch antenna after proper optimization. The theoretical measurements are changed through optimization to obtain a better efficient result. The patch and the ground of the antenna are constructed using a 0.035-mm-thick copper [2, 17]. As shown in Table 2, the substrate is designed by using dimensions of 38.64 mm × 48.16 mm. The substrate used is FR-4 which has dielectric constant of 4.3 and the substrate height used in this antenna is 1.7 mm. Thus, the overall size of the antenna is about 38.64 × 48.16 × 1.77 cubic mm which is relatively smaller the most compared one bellow. The ground plane used in this antenna works as the shield for the back-ward radiation. Thus, through optimization the desired frequency is achieved and the efficiency is improved which is done by varying the width, length of the patch and the feed line. Table 1 shows the different dimensions of the proposed antenna such as height, length, width etc. which are achieved after proper optimization (Figs. 2 and 3).

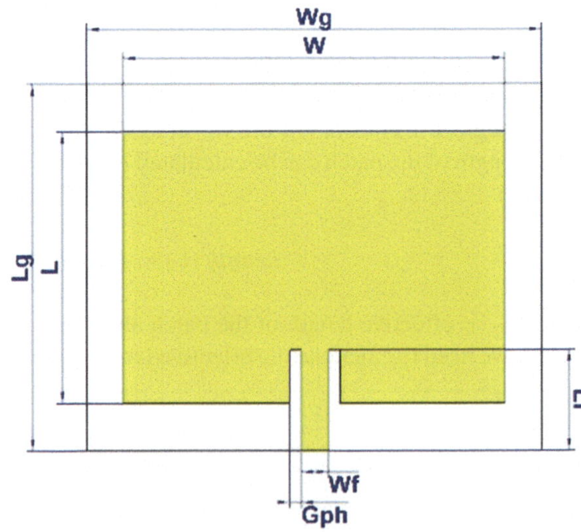

Fig. 1 Proposed antenna geometry

Table 1 The designed antenna parameters

Symbols of parameters	Parameters description	Values in (mm)
L	Patch length	28.62
W	Patch width	38.64
H_p	Patch height	0.035
L_g	Ground length	38.65
W_g	Ground width	47.18
W_f	Feedline width	2.85
L_f	Feedline length	10.65
G_{ph}	Gap between feedline and patch	0.855

Table 2 Characteristics of the substrate

Parameters symbols	Description of parameters	Value (mm)
$L_s = L_g$	Substrate length	38.64
$W_s = W_g$	Substrate width	47.18
h	Substrate height	1.6
$\in_r$	Relative permittivity	4.3

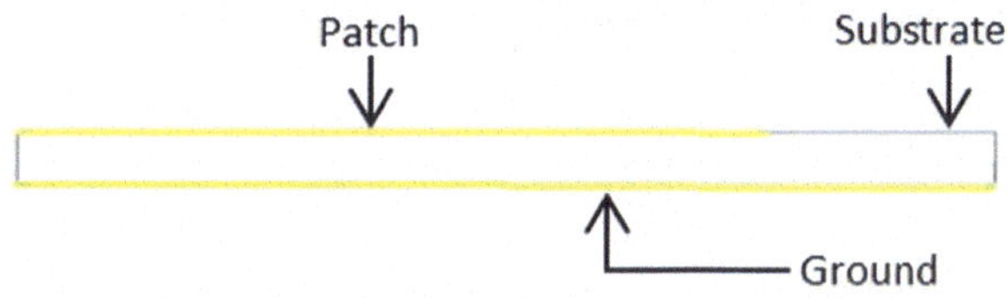

Fig. 2 Side view

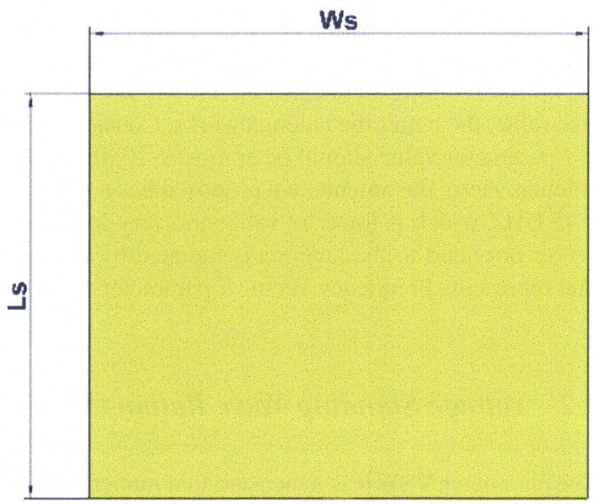

Fig. 3 Ground view

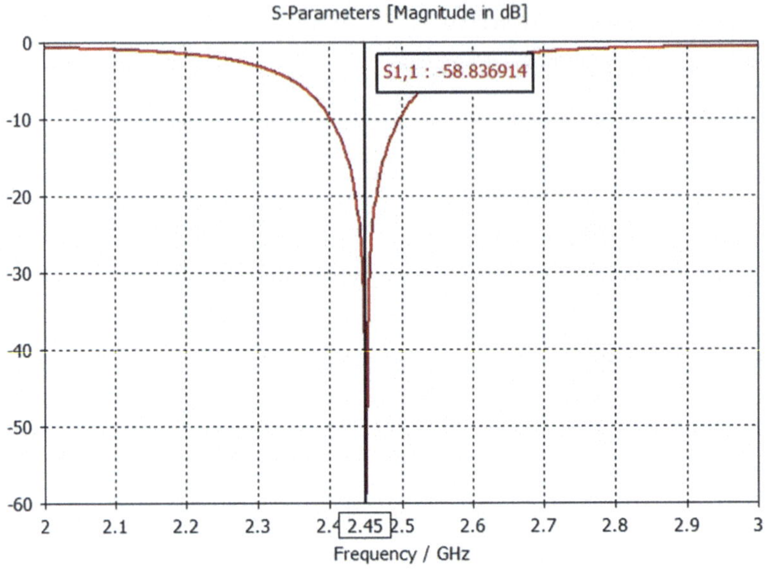

Fig. 4 Frequency versus S parameter (in dB)

3 Simulated Result

3.1 Return Loss

Return loss generally known as s11 parameter refers to the reflection coefficient denoted by (S11). It describes the relationship between ports input output and represents how much power the antenna reflects. The plot of the return loss also provides data on how much matched the feed-line is to the antenna. If s11 parameter is 0 dB, the antenna reflects all the energy, nothing is radiated. Thus the lower the s11 parameter value, the better the antenna works. Considering the real world applications, the s11 parameter value should be at most −10 dB to give efficient performance by the antenna. Here, the antenna we proposed has a value of −58.83 dB at a frequency of 2.45 GHZ which is lower in value and very low loss of power. Thus the maximum power provided to the antenna is radiated by the antenna. Figure 4 shows the plot that represents Frequency versus S parameters in (dB).

3.2 Voltage Standing Wave Ration (VSWR)

The parameter VSWR is a measure that numerically describes how well the antenna is impedance matched to the radio or transmission line it is connected to. It also represents how much power is reflected from the antenna. The VSWR is always a real and positive number for antennas. The smaller the VSWR is, the better the

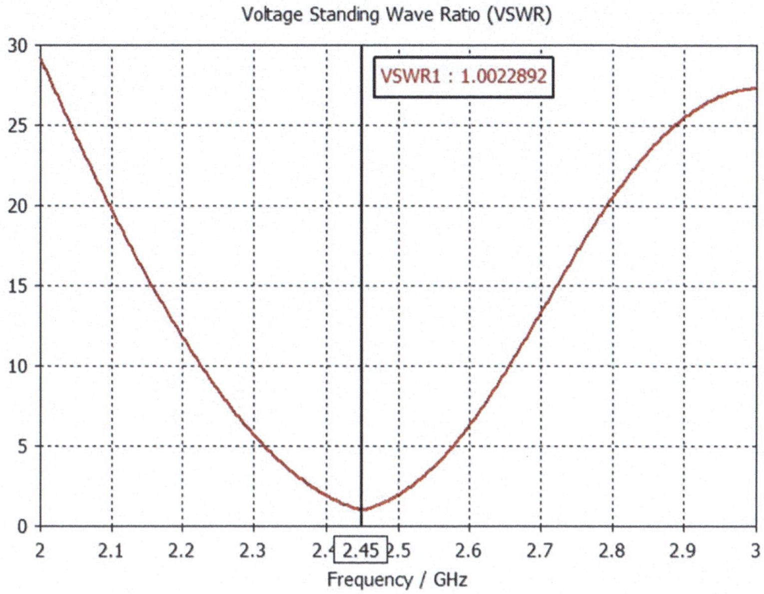

Fig. 5 Frequency versus VSWR

antenna is matched to the transmission line and the more power is delivered to the antenna. The minimum VSWR is 1.0. In that case, no power is reflected from the antenna, which is ideal. The plot that represents Frequency versus VSWR is shown in Fig. 5. The figure shows that proposed patch antenna has a peak value VSWR of 1.002 at 2.45 GHZ. The required standard value of VSWR is between 1 and 2. Thus the proposed antenna is considered very efficient and almost all power is radiated.

3.3 Radiation Pattern

Radiation pattern refers to the directional dependence of the strength of the radio waves from the antenna. In other word, it represents the amount of energy that is radiated by the antenna. 2D radiation pattern of the antenna indicates that the designed antenna provides a good radiation pattern and very narrow beam width. The required half circular radiation pattern is obtained as shown in the figure. The respective radiation pattern can be seen in 2D as shown in Fig. 6.

Whereas the 3D pattern is usually measured at a sufficient distance from the antenna known as the far-field. Just like the 2D radiation pattern a good antenna should also maintain its 3D radiation pattern throughout the frequency range of operation. From 3D radiation pattern makes it easier to observe the amount of power delivered to a specific direction. The 3D radiation pattern of the proposed antenna is shown in Fig. 7.

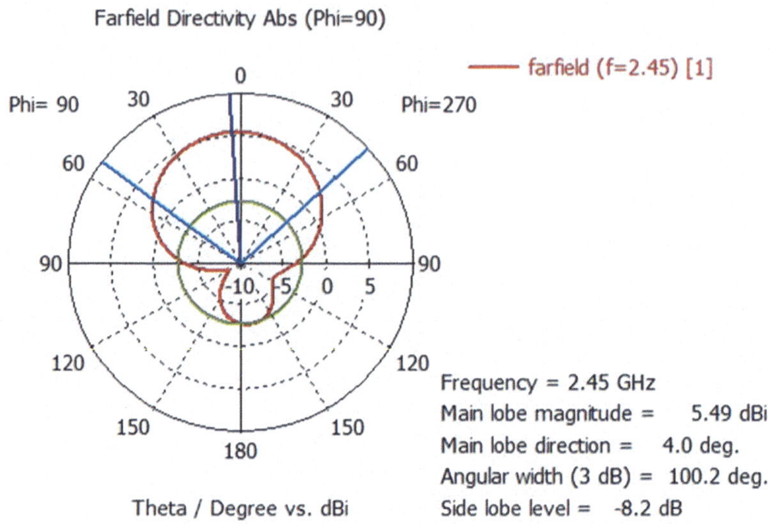

Fig. 6 2D radiation pattern

Fig. 7 3D radiation pattern

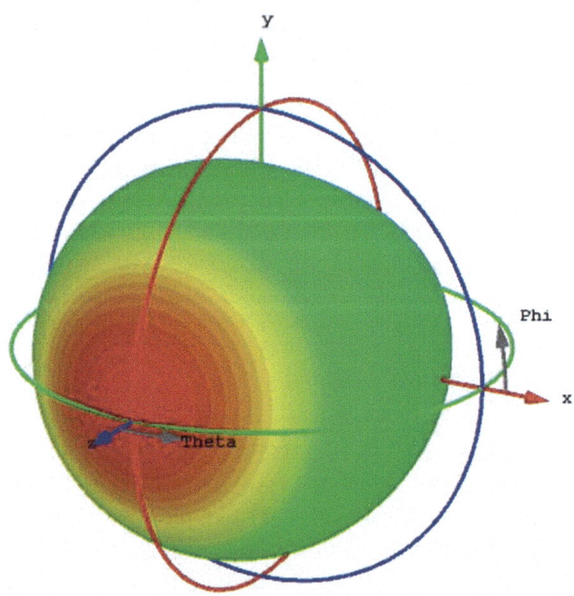

3.4 Gain Pattern

The gain of the antenna is closely related to the directivity, it is a measure that takes into account the efficiency of the antenna as well as its directional capabilities. The 2D radiation pattern of the proposed patch antenna is shown in Fig. 8a and b for both phi = 0° and phi = 90° respectively.

3.5 Specific Absorption Rate (SAR) Analysis

Since the antenna is intended for wearable purposes, it is necessary to analyse the Specific Absorption Rate (SAR) primarily for human health safety issues. The SAR is measured for a 100 mm long human hand model where the radius of bone, muscle, fat and skin are respectively 15, 10, 3 and 2 mm. A value of 1.4 W/kg is obtained while the antenna is 2 mm apart from the model hand. The IEEE C95.3-2002 decides the standard of the SAR calculation is shown in Fig. 9. The average SAR value for 1 g should not exceed 1.6 W/kg as set out in the rules of ICNIRP.

Table 3 shows the comparison between the proposed antenna with some recent published works in terms of return loss, gain, directivity, bandwidth, VSWR etc. As shown, the proposed antenna exceeded them in almost every aspect which makes it more suitable for the proposed applications.

4 Future Scope

The proposed antenna shows improved performance compared to the other antenna of similar field. It is a common knowledge that wearable needs to be efficient, lightweight and safe for human body too. Considering the off body health monitoring devices and wearable that are used in modern age, all the components used in them are also required to be small in size and efficient. On design perspective the proposed antenna can be further shrinked while maintaining this performance or even improving it. Further more as it operates in ISM band, it can be integrated in different types of wearable devices of various fields such as fitness, health monitoring, aiding and prosthetic devices etc. For a wearable device of specific application in those various fields, the antenna can provide a way to operate in wireless environment. In future, we intend to do further research on reducing the antenna in size and implementing it in a health monitoring device such as blood sugar monitoring device.

(a)

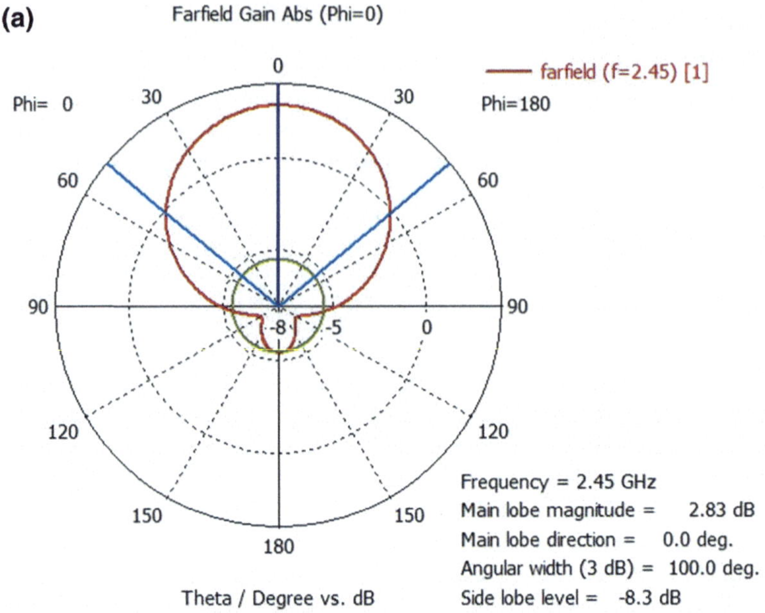

(b)

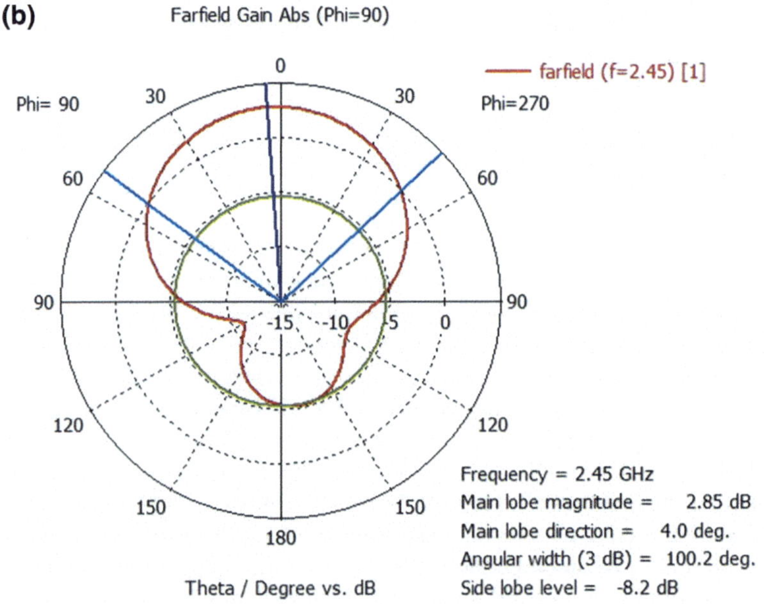

Fig. 8 **a** The 2D pattern of the gain at Phi = 0. **b** The 2D pattern of the gain at Phi = 90

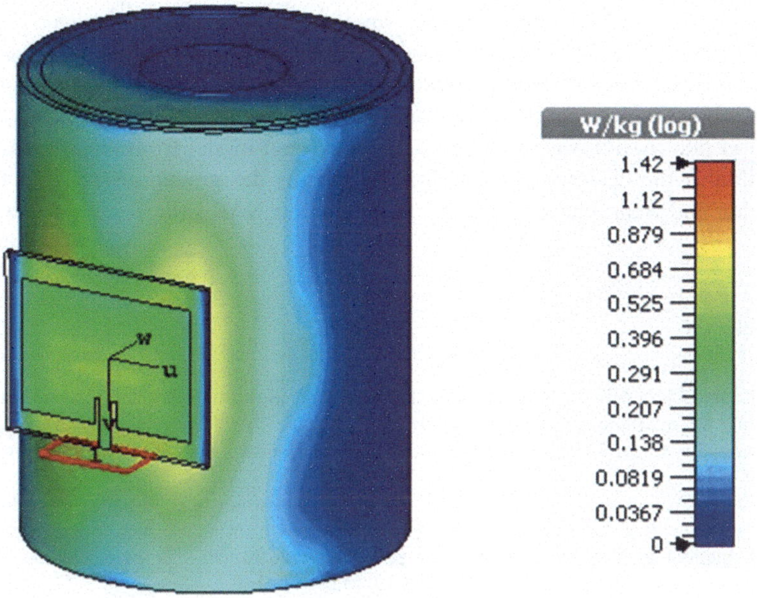

Fig. 9 Analysis of specific absorption rate (SAR)

Table 3 The comparison of our proposed antenna with existing wearable antennas

Reference	S11 (dB)	Gain (dB)	Directivity	10 dB-BW (MHz)	Dielectric materials	VSWR
Proposed antenna	58.84	2.85	5.49 dBi	95	FR-4	1.002
[11]	32	−10	2.33 dBi	80	Teflon	1.37
[12]	12*	2.6	4.2 dBi	30	Paper	–
[1]	36	1.13	–	110	Rogers Ultralam 3850	–
[13]	14	–	5.4 dB	–	FR-4	2.10
[14]	14.90	5.11	3.68 dB	–	Denim	1.51

* Approximate

5 Conclusion

We designed and simulated our suggested wearable microstrip patch antenna using CST Microwave Studio for Healthcare monitoring application to operate at 2.45 GHz band. The results of our simulation show that the antenna has a gain of 2.85 dBi which is greater than the most antenna in term of size and similar application as shown in the Table 3 comparison. The suggested antenna has a return loss of 58.84 dB which implies that the antenna reflects a very small quantity of energy and most of the energy is radiated to the antenna. Our antenna introduced has a 10 dB-bandwidth of 95 MHz. The VSWR of the simulated antenna is almost 1 through the whole frequency band which maintain the standard to exist between 1 and 2. The specific absorption rate (SAR) was evaluated using the voxel model from the antenna. The SAR value of the design is 1.42 W/kg. Hence this design can be used for health monitoring and wearable applications due to the low SAR value.

References

1. M. Kadry, M.E. Atrash, M.A. Abdalla, Design of an ultra-thin compact flexible dual-band antenna for wearable applications, in *2018 IEEE International Symposium on Antennas and Propagation & USNC/URSI National Radio Science Meeting* (Boston, MA, 2018), pp. 1949–1950
2. M. Faisal, A. Gafur, S.Z. Rashid, M.O. Shawon, K.I. Hasan, M.B. Billah, Return loss and gain improvement for 5g wireless communication based on single band microstrip square patch antenna, in *2019 1st International Conference on Advances in Science, Engineering and Robotics Technology (ICASERT)* (IEEE, 2019, May), pp. 1–5
3. D.M. Pozar, *Microwave Engineering*, 3rd edn. (Wiley, Hoboken, NJ, USA, 2005)
4. P. Schilingovski, V. Vulfin, S. Sayfan-Altman, R. Shavit, Wearable antennas design for wireless communication, in *2017 IEEE International Conference on Microwaves, Antennas, Communications and Electronic Systems COMCAS 2017*, vol. 2017–Novem (2018), pp. 1–3
5. K. Kwon, J. Choi, Antennas for wireless body area network, in *7th European Conference on Antennas and Propagation (EuCAP)* (2013), pp. 375–379
6. Z. Wang, E.G. Lim, T. Tillo, F.Z. Yu, Review of the wireless capsule transmitting and receiving antennas, in *Wireless Communications and Networks—Recent Advances*. Intech Open. Chapter 2, March 2012, pp. 27–46. ISBN 978-953-51-0189-5
7. E.G. Lim, Z. Wang, M. Leach, R. Zhou, K.L. Man, N. Zhang, Compact size of textile wearable antenna, in *Lecture Notes in Engineering and Computer Science: Proceedings of The International MultiConference of Engineers and Computer Scientists 2014*, IMECS 2014, 12–14 March, 2014, Hong Kong, pp. 870–873
8. M.A. Rahman, A. Shaikat, I.S. Iqbal, A. Hassan, Microstrip patch antenna design and performance analysis for RFID applications at ISM band (2.45 GHz), in *2013 2nd International Conference on Advances in Electrical Engineering (ICAEE)*, Dhaka, 2013, pp. 305–308
9. U. Kim, J. Choi, Design of a microstrip patch antenna with enhanced FIB for WBAN applications. Trans. IEICE I135–1141 (2011)
10. IEEE standard for safety levels with respect to human exposure to radio frequency electromagnetic fields, 3 kHz to 300 GHz, in *IEEE Std C95.1*, 1999 Edition, pp. 1–83, 16 April 1999. https://doi.org/10.1109/IEEESTD.1999.89423
11. K. Sajith, J. Gandhimohan, T. Shanmuganantham, Multilayered and dual-band CB-CPW fed wearable antenna for healthcare monitoring applications, in *2018 International Conference on Current Trends towards Converging Technologies (ICCTCT)*, Coimbatore, 2018, pp. 1–5
12. T.H. Nguyen, T.L.H. Nguyen, T.P. Vuong, A printed wearable dual band antenna for remote healthcare monitoring device, in *2019 IEEE-RIVF International Conference on Computing and Communication Technologies (RIVF)*, Danang, Vietnam, 2019, pp. 1–5.
13. R. Jamal, I.K. Singh, R.K. Singh, Development of 2.45 GHz patch antenna and measurement of SAR over Human head model. 2019 Int. J. Appl. Eng. Res. **14**(2) (2019) (Special Issue). ISSN 0973-4562

14. R. Darwin, S. Vinodhini, Design of multiband wearable rectangular slot antenna for WIMAX and WLAN applications. Int. J. Innov. Technol. Explor. Eng. (IJITEE) **8**(2S2) (2018). ISSN: 2278-3075
15. G. Kaur, Wearable antennas for on—body communication systems, **6**, 568–575 (2014)
16. M. Muthusamy, Int. J. Emerg. Technol. Adv. Eng. www.ijetae.com (ISSN 2250-2459, ISO 9001:2008 Certified Journal, 3(12), December (2013) Reduction of sags and swells in a distributed generation system based on environmental characteristics. Int. J. Emerg. Technol. Adv. Eng. **3**, 382–385
17. M.N. Suma, P.C. Bybi, P. Mohanan, A wideband printed monopole antenna for 2.4-GHz WLAN applications. Microw. Opt. Technol. Lett. **48**(5), 871–873 (2006)
18. D. Mitra, S. Das, S. Paul, SAR reduction for an implantable antenna using ferrite superstrate, in *2019 International Workshop on Antenna Technology (iWAT)* (Miami, FL, USA, 2019), pp. 1–4

Syed Zahidur Rashid received the M.Sc. degree in Computer Science and Engineering from Daffodil International University, Bangladesh. Previously he received the B.Sc. degree in Computer and Communication Engineering from International Islamic University Chittagong, Bangladesh. He is currently working as the Chairman of the Department of Electronics and Telecommunication Engineering, Faculty of Science and Engineering, International Islamic University Chittagong, Bangladesh. He has contributed numerous research articles in various international refereed journals and conferences. His main research interests include free-space optical communications, wireless communications, blockchain technology, neural networks, enabling technologies for IoT, machine learning, etc. He can be contacted at email szrashidcce@yahoo.com.

Abdul Gafur received the M.Sc. degree in Electrical Engineering (Telecommunications) from Blekinge Institute of Technology, Sweden. He is currently working as an Associate Professor in the department of Electronic and Telecommunication Engineering (ETE) under the Faculty of Science and Engineering (FSE) of International Islamic University Chittagong (IIUC), Bangladesh. He has contributed to numerous technical articles in different international journals and conferences as an author and co-author. His main research interests include coherent optical transmission systems, free space optic and optical wireless communication systems. He can be contacted at email engr.abdul.gafur@gmail.com.

Aasim Ullah (aasim@kth.se) received his Ph.D. from the School of Engineering, Computer and Mathematical Sciences at Auckland University of Technology, Auckland, New Zealand. He is currently working as an Assistant Professor at International Islamic University Chittagong (IIUC), Bangladesh. Previously, he held M.Sc. in Power Engineering from Norwegian University of Science & Technology (NTNU)Trondheim, Norway.

Md. Akbar Hossain (akbar.hossain@manukau.ac.nz) received his Ph.D. from the School of Engineering, Computer and Mathematical Sciences at Auckland University of Technology, Auckland, New Zealand. He is currently working as a Senior Lecturer at Manukau Institute of Technology Ltd. Previously, he held M.Sc. in Telecommunication Engineering from the University of Trento, Italy. He received his Bachelor's degree in Electrical and Electrical Engineering from the Islamic University of Technology, Bangladesh. Between his Masters and Ph.D. studies, he was awarded a prestigious Marie Curie Early Stage Research fellowship to work as a researcher in the different European Union (EU) funded projects in Germany and Greece. He is active in wireless networks, the Internet of Things, and Cognitive Radio Ad-Hoc Networks research. The current research project includes D2D communication for post-disaster communication, a Technology-assisted Medical emergency support system, and Cognitive internet of Things Smartlife solutions. He also served as a TPC member for many international conferences and a reviewer for several IEEE transactions. He currently serves as a lead guest editor for EURASIP Journal of Wireless Communication and Networking for a special issue on Cognitive UAVs in Critical Missions: IoT-based Applications, Protocols, and Deployments and Associate Editor for International Journal of Computers and Applications.

Sultan Shah Mamun has completed his B.Sc in Electronics and Telecommunication Engineering from International Islamic University Chittagong, Bangladesh. He completed his graduation in 2019 alongside been served as a teaching and research assistant till the end of 2019. Currently, he is employed as Core Network Engineer in a Private IT company that provides internet all over Bangladesh. Along working in the practical field he is interested in IT technologies, wireless technologies, network architecture, Cloud computing, machine learning, etc.

Md. Qudrat-E-Alahi Majumder completed his B.Sc in Electronics and Telecommunication Engineering from International Islamic University Chittagong, Bangladesh in 2019. He stated his carieer as Assistant Support Engineer in 2021 and currently he is working as a Assistant Network Engineer in a Private IT company. His fields of interest are Wireless technologies, Network architecture, Cloud computing, Data mining and ETL.

Intelligent Wireless Communication Issues

Context-Aware Cognitive Communication for Sustainable Digital Twins

Zhihan Lv and Liang Qiao

Abstract In order to study the application of sustainable digital twins (DTs) technology and context-aware computing in the field of network communication, this study puts forward the concept of DTs network aiming at network security in industrial environment by studying the application of DTs technology in intelligent industrial environment; the network framework of intelligent industrial environment based on context-aware reasoning is established; aiming at the current security problems of wireless sensor network (WSN) in industrial network environment, a trusted data aggregation algorithm based on context-aware and data density correlation is raised and verified by a practical case. The results show that in the unreliable environment, compared with the traditional algorithms, the data aggregation quality of the adaptive scheduling algorithm under time and energy constraints in the fixed data aggregation tree (DAT) increases with the increase of the threshold, and the performance is improved by at least 1%, up to 33 and 23% on average. The proposed trusted data aggregation method has more accurate perceived trust and greater system throughput when the perception and the link are unreliable. The missing report rate is 6.7%, and false positive rate is 1.4%, which is much inferior than the traditional methods. The data aggregation method based on context-aware has better performance in the direction of network communication security.

1 Introduction

Digital twins (DTs) sustainable development has become a hot technology in the field of industrial Internet and is widely used in industrial design, industrial manufacturing, and other fields [1]. With the continuous deepening of case analysis and simulation level and the support of emerging technologies such as Internet of things

Z. Lv (✉)
Department of Game Design, Faculty of Arts, Uppsala University, Visby 62167, Sweden
e-mail: lvzhihan@gmail.com

L. Qiao
College of Computer Science and Technology, Qingdao University, Qingdao 266071, China

A. K. Pathan (ed.), *Towards a Wireless Connected World: Achievements and New Technologies*, https://doi.org/10.1007/978-3-031-04321-5_8

(IoT), cloud computing, and big data, people have a deeper and deeper digitization of the physical world, virtual-real interaction and virtual-real integration, and DTs have a wider range of application scenarios and values [2]. "DTs" is a way of digital representation of the physical world, which originates from specific application requirements and is limited by people' understanding of the physical world. Industrial communication network is an expensive and fragile "system", while industrial field is a business field that focuses on cost and efficiency. Therefore, there are great challenges in the evaluation and verification of security system for industrial network [3]. In addition, it is very difficult to carry out attack and defense verification for the expensive and fragile system of industrial network. On the one hand, using the industrial network entity for attack and defense verification may cause irreparable damage to the industrial network security; on the other hand, it is very expensive to copy a complete industrial network with high fidelity [4]. DTs can perform low-cost verification for this expensive and fragile system [5]. The DTs of industrial Internet is established and different DTs are connected to jointly build the DTs space of industrial security, which can provide guarantee for security system of industrial network [6].

To meet the industrial applications, a secure network architecture is indispensable. Qian et al. [7] pointed out that only a few studies have solved the relationship between passion and personality, most of which are carried out outside the working environment. Wireless sensor network (WSNs) often used in industrial communication networks are usually in an unattended and unreliable communication environment and are vulnerable to attacks [8]. Traditional security technology is privacy protection or encryption technology, which can't deal with attacks from internal nodes of the network. Therefore, trusted computing technology based on context-aware provides a new idea to solve attacks from internal networks [9]. With the rapid development of the IoT, the scale of sensor nodes also shows explosive growth. Aggregating all nodes in the network will take a long time and can't meet more and more application needs. More and more applications restrict delay of data aggregation [10]. Context-aware computing enables the system to perceive the environment context information to provide users with relevant services and computing resources [11]. Deploying WSNs in the intelligent industrial environment to collect data, the sensors in WSNs communicate with each other, and combined with context-aware analysis and processing for industrial network security analysis can increase the accuracy of sensor network identification [12].

In this study, the DTs technology is applied to the intelligent industrial environment by means of literature research and algorithm model. Aiming at the network security problem of industrial environment, a DTs network architecture is proposed. For the current security threats in WSNs, a data aggregation analysis method based on context-aware computing is presented, which can obtain better data aggregation quality and network throughput, and can provide a new idea for WSN management.

2 Research Status of DTs in Intelligent Industrial Environment

Research in the field of intelligent environment technology is very extensive, such as context-aware computing, wireless sensor related technologies, activity recognition related technologies [13]. The DTs technology is to make full use of physical model, sensor update, operation history, and other data, integrate the multi-disciplinary, multi physical quantity, multi-scale, and multi probability simulation process, and complete the mapping in the virtual space, so as to reflect the whole life cycle process of the corresponding physical equipment [14]. Obschonka et al. [15] highlighted the role of feedback seeking in linking empowered leadership to mission performance, accountability, and voice, based on social communication theory. DTs consists of three parts: physical product of physical space, virtual product of virtual space, connection data and information between virtual and real. It emphasizes the two-way connection between digital world and physical world to realize the synchronization and feedback between physical twins and the DTs [16]. This makes the original digital simulation no longer an isolated and static model, but can change with the physical world, interact with physical world, and even affect the twins of physical world. The change not only increases the authenticity of digital simulation, but also makes the DTs play a better role [17]. The combination of DTs and context-aware computing in industrial network communication awareness can provide a better guarantee for network security.

Gehrmann and Gunnarsson [18] pointed out how to use DTs model and corresponding security architecture to allow data sharing and critical control safety processes. The design-driven security requirements based on DTs data sharing and control are determined, and other security components of high-level design and evaluation structure are given. New security framework lays a foundation for future research in this new field. Jiang et al. [19] pointed out that the industrial IoT has brought value-added services and has become an important business and technical model in industry 4.0 era. In addition, the combination of DTs and the IoT can promote the integration of the real and the virtual, which has become the key to give full play to the value of the IoT. With the continuous expansion of the scale of the IoT network, how to optimize the network, allocate limited resources, and provide high-quality services is still a major issue of concern. At present, the work in this field mainly depends on the model with low practical value for the IoT network with limited resources, and it is difficult to simulate the dynamic system in real time. Lu et al. [20] studied the integration of DTs and edge networks, and came up with digital double edge networks to fill the gap between physical edge networks and digital systems.

There are many researches on the application of DTs in various industries, but there are few researches in the direction of industrial network security. In this study, context-aware technology is applied to the data aggregation of WSN communication security in industrial environment, and a data aggregation analysis method based on context-aware computing is proposed, which can greatly improve network throughput and embankment sensing error rate, and fill the research gap in the direction of hidden network security dangers caused by unattended communication network in industrial network security.

3 Data Aggregation Analysis of DTs Sensor Networks Based on Context-aware

3.1 Industrial Network Communication Security Based on DTs

One of the important application fields of DTs is industrial manufacturing, and an important link in industrial manufacturing is network communication. The demand for industrial network security capability is still inseparable from systematic ability of detection, defense, and response. The specific contents include functional requirements such as asset identification, traffic monitoring, log audit, boundary protection, link encryption, situation awareness, and emergency response [21]. The display, diagnosis, verification, prediction, and decision-making functions of DTs can just serve the construction of industrial network security system. Through display function, it can build a global network topology and asset display, and diagnosis and verification function can support the evaluation and verification of security protection capability to realize security threat early warning and automatic and intelligent emergency response of industrial communication network [22]. DTs has natural advantages in visual display. Through real-time connection with physical entities, it can achieve more real traffic replay and business simulation, and form the industrial Internet security situational awareness ability of global awareness and panoramic display [23]. DTs gives new meaning to traditional network data (Fig. 1).

The establishment of high-fidelity industrial Internet range through DTs can simulate controllers and production equipment from a digital perspective, to truly realize the integration of physical space and network space, and provide high-simulation, low-cost, and scalable range environment. Moreover, the high-simulation industrial network DTs based on DTs is a natural dense network environment. Combined with the corresponding attack deception technology, it can easily induce attackers to enter the wrong attack path, protect the physical network entities, and provide the basis for traceability and counterattack. Giving the influence of network space and physical space, a global system including system composition, business carrying, and personnel behavior of industrial network is established. By opening up the connection between physical and digital world, the DTs of the physical entity is reconstructed in the digital space, and the industrial DTs network including equipment of production, control, network, and computing is established. Industrial network security issues also pose new demands on computer technology (Fig. 2).

Combined with the DTs and attack verification methods of industrial network established by DTs technology, repeated tests and verification of network security

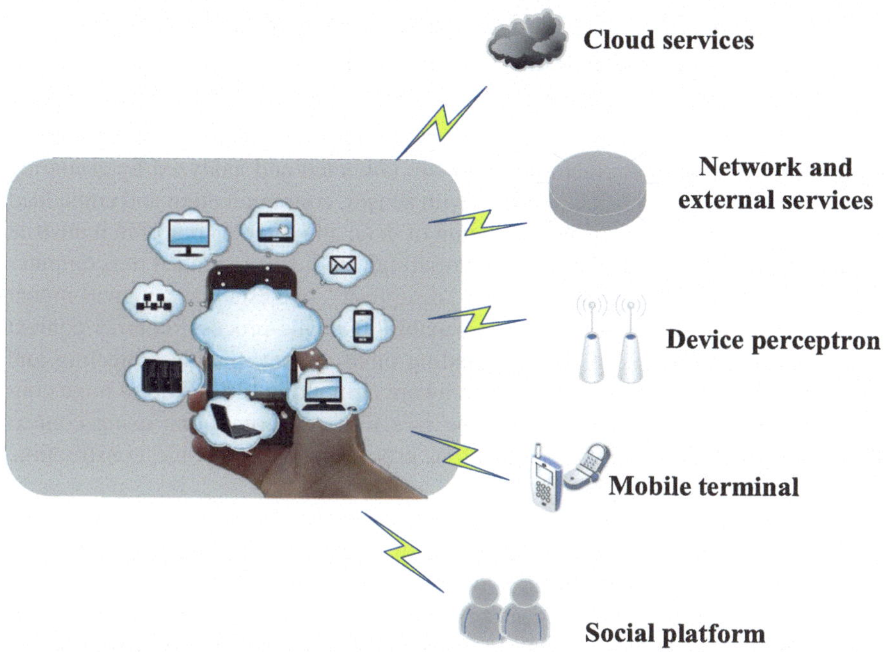

Fig. 1 Data transformation in DTs

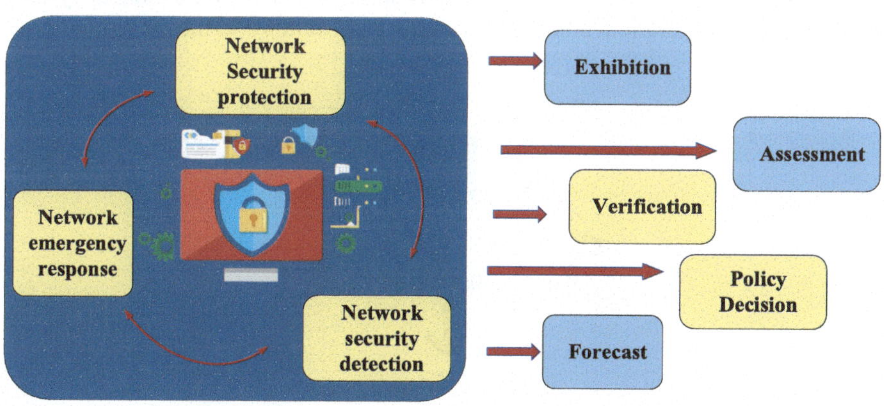

Fig. 2 Analysis of the capability needed for industrial network security

can make industrial enterprises clearly understand their networks' defense ability against different attacks and the scope of influence after being subjected to network attacks. Therefore, a global, on-demand security defense system is established by adjusting the network architecture and deployment according to demand, and such deployment is also economical.

3.2 Hierarchical Structure Design Based on Rule Reasoning in Intelligent Industrial Environment

In the intelligent industrial environment formed by application of IoT technology in the field of industrial manufacturing, data are collected and analyzed by arranging various sensing devices [24]. Combined with sensor, communication and embedded technology, sensing, monitoring, prediction of data, and remote control in industrial manufacturing are realized. Pervasive computing [25] needs to embed the computer into the environment, make the computer "disappear", and establish an environment full of computing and communication capabilities. This process requires context reasoning and perception of the surrounding environment, mainly collecting and processing data through sensors. Context-aware computing [26] involves huge content. Context reasoning aware computing is a mode of computing using context information, including context acquisition, processing, and module construction. The perception process is given in Fig. 3.

The environmental state and activity scene in intelligent industrial environment are changing at any time with the progress of industrial manufacturing activities, and are extremely complex. In view of the problem that the user needs can't be met and the actual needs change at any time in the industrial environment, it is necessary to establish an intelligent industrial environment framework with strong configurability and adaptability to different industrial environment transformations. In order to better shield the underlying details and improve the efficiency of system modeling and configuration, combined with the IoT framework, the data processing

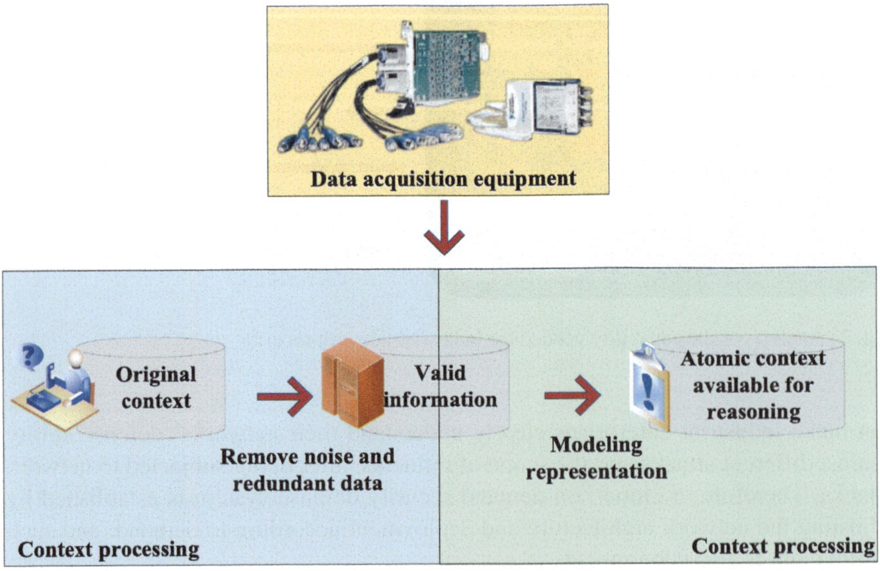

Fig. 3 Acquisition and processing of context in context reasoning and perception

layer is established on the physical layer, and the original context data collected by the underlying equipment are processed to convert them into the corresponding parameters that can be introduced into fact base, and then reasoning is carried out in the reasoning engine. After inference engine obtains the result, the data processing layer is transformed into the parameters that can be used to control the equipment into the industrial control equipment to achieve the purpose of shielding the details of the hardware environment.

Because of the complexity of intelligent environment, context-aware computing technology is introduced into the framework. Context-aware provides users with implicit interactive services by sensing, decision-making, and reasoning of context information. Therefore, context-aware computing is the basis for human-computer interaction and one of the core technologies of pervasive computing. Since context data perception is heterogeneous, it is divided into context acquisition, context processing, and context reasoning perception. Therefore, context reasoning perception module is formed by combining rule reasoning with intelligent layer, which can integrate context acquisition and context processing into physical layer and data processing layer, so as to establish a context reasoning perception framework based on rule reasoning (Fig. 4).

Figure 4 suggests that the framework includes physical, data processing, intelligent, and application layers. The physical layer contains basic components, sensors, controllers, and wireless communication modules required by the smart node. As the core module of the framework, the intelligent layer is an independent layer that integrates service function of the industrial environment. The rules can be independent, so that in the installation and deployment of the industrial environment, the staff can dynamically configure the rules according to the needs of the enterprise, without considering the differences of the underlying equipment, so as to realize the rapid deployment of the industrial environment and the flexible adjustment when the function changes. Since this network framework is designed in context-aware computing, the three functions of context acquisition, modeling representation processing, and context reasoning perception are allocated to the physical, data processing, and intelligent layers. Therefore, the physical layer defines the hardware of the node. After the data processing layer reads the original data from the sensor in the physical layer, it is introduced into fact base in the intelligent layer. The intelligent layer shields the complexity of the underlying and realizes the basic content of the reasoning framework. The output conclusion is transmitted to a control device in the application layer.

3.3 *Trusted Data Aggregation Based on Context-aware and Data Density Relevance*

Due to the limitation of low power, the communication between sensors in WSNs under industrial environment can only be short-range communication, resulting that WSNs are high-density, multi-hop, self-organizing networks. Adjacent sensors often detect the same event, which will make the perceived data have a lot of redundant

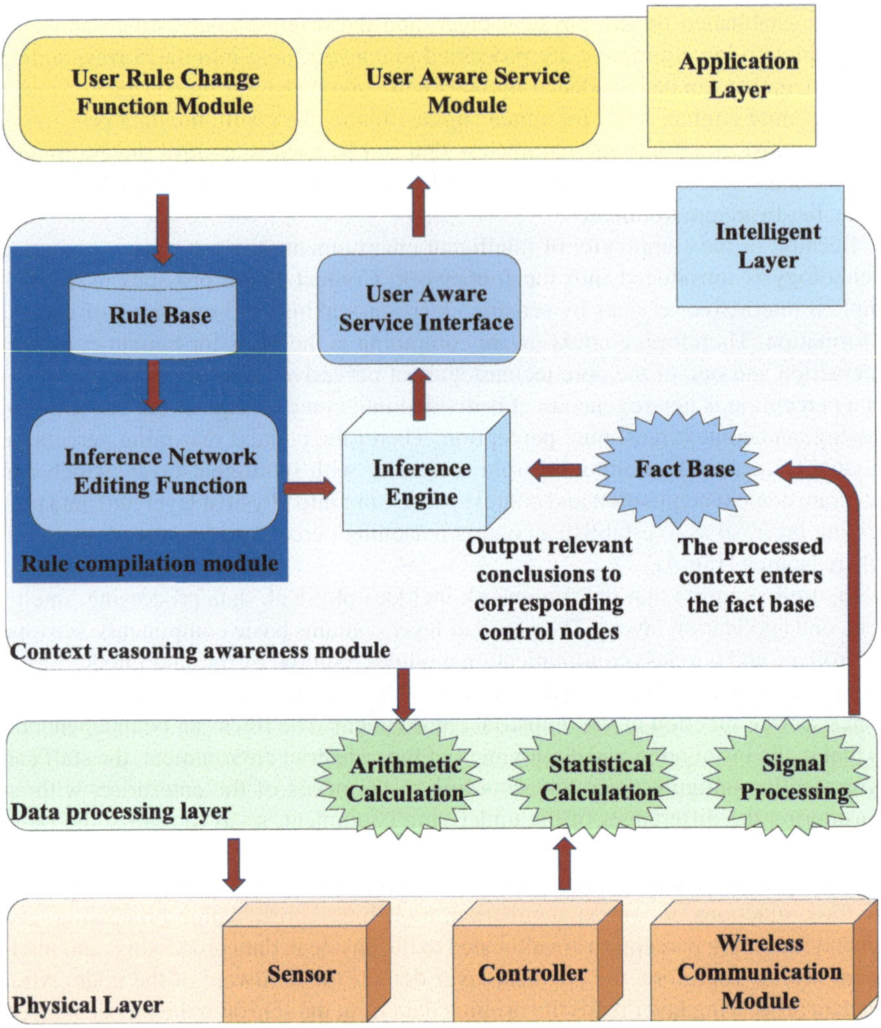

Fig. 4 Context reasoning-based intelligent industrial environment network framework

information. This will lead to the waste of energy and communication bandwidth, reduce life cycle of communication network, increase the probability of collision between signals when transmitting data, and reduce channel utilization rate and communication efficiency as well as the timeliness and accuracy of information collection. During data transmission, data aggregation technology can't directly transmit the data received by each intermediate node to the subsequent node, but aggregate the multiple data received first, remove redundant information, and integrate it into a more accurate and effective data that meets the needs of users and has smaller data volume, which ensures information perception and efficient data transmission. Data aggregation includes cluster aggregation, tree aggregation, and mixed aggregation.

With the continuous expansion of sensor scale, unreliable communication links in sensor networks usually cause data transmission failure. Re-uploading failed data not only wastes time but also consumes energy, affecting the operation cycle of the entire network. This study considers the problem of maximizing the quality of data aggregation under time and energy constraints in unreliable environments.

The network system is modeled as a graph $G = (V \cup \{S\}, E)$, V represents the set of sensor nodes, S represents the sink node, and E represents the set of links among all nodes. If there are N nodes in the system, $|V| = N$, then the source node senses target and generates source data, and all nodes can aggregate the data. In this way, data aggregation is carried out on a G network generation tree. It is assumed that one unit of time slot is required to transmit a single packet and that each communication link is single-capacity. In an unreliable environment, creating a data aggregation tree (DAT) with the greatest aggregation quality under time and energy constraints can be regarded as a problem for combinatorial network optimization.

$$Z: \max_{\psi \in Tr(G)} \left(\max_{\overrightarrow{t^{\psi}}, \overrightarrow{w^{\psi}}}, \sum_{i \in V} w_i^{\psi}(j) \right) \quad (1)$$

G represents network topology diagram, ψ presents generation tree of network G, $Tr(G)$ represents generation tree collection of network G, $w_i^{\psi}(j)$ denotes information by node i receiving from node j in DAT ψ, t^{ψ} represents number of time slot for sending data in ψ, w^{ψ} denotes time slot when sending data packet in ψ.

$$s \cdot t \cdot t_i^{\psi} \in \{0, 1, \ldots, T_i^{\psi}\}, \forall i \in V \quad (2)$$

T_i^{ψ} indicates the maximum number of transmission data time slots assigned to a node i in ψ.

$$e_i^{\psi} \cdot E_T + \sum_{j \in c^{\psi}(i)} e_i^{\psi} \cdot E_R \leq E_t - ES, \forall i \in V \quad (3)$$

The equation indicates that the sum of energy of node sensing, receiving data, and sending data during a data aggregation process can't exceed the agreed energy constraint. e_i^{ψ} represents the energy unit allocated to the node i to send data in ψ, E_T shows the energy required for a single data transmission, E_R is the energy required for a single data reception, and ES is the energy required for a single perception and generation of data.

$$\sum_{j:j \in C} t_i^{\psi} \leq W_i^{\psi} - \min_{j:j \in C} W_j^{\psi} \quad (4)$$

It indicates that the sum of the time required to send data by the brother nodes participating in data aggregation is not greater than the distinction of sending time slot of parent node and the minimum that of brother nodes participating in data aggregation. t_i^{ψ} denotes the number of time slots for sending data assigned to i in ψ.

$$W_i^{\psi} \in \{0, 1, \ldots, D-1\}, \forall i \in V \quad (5)$$

W_i^{ψ} represents the time slot when i sends data packet in ψ.

$$W_S^{\psi} = D \quad (6)$$

W_S^{ψ} represents the time slot when a sink node sends data packet in a DAT ψ.

$$\psi \in Tr(G) \quad (7)$$

In the single-hop interference model, within the same time slot, only one node in all brother nodes is allowed to send data to the parent node. When this brother node is sending data to parent node, other brother nodes can choose to receive the data sent by their child nodes. The number of generation trees in the network means the exponential level of the number of nodes. Then, the maximum clique problem (MCP) is defined as: S represents the base set, $U = \{S_1, S_2, \ldots, S_m\}$ represents the set of base subsets. $k(0 \leq k \leq m)$ is an integer, MCP means selecting at most k elements from U to maximize the sum of elements of the selected l subsets.

$$\begin{aligned} &\max |U_{i \in 1,2,\ldots,l} S_i'| \\ &s \cdot t \cdot S_i' \in U, l \leq k, k \leq m \end{aligned} \quad (8)$$

Revenue version of MCP is each S_i corresponding to a revenue $Pro(S_i')$, revenue version of the MCP target is to maximize the selected l subsets of the revenue.

$$\begin{aligned} &\max | \sum_{i \in 1,2,\ldots,l} Pro(S_i')| \\ &s \cdot t \cdot S_i' \in U, l \leq k, k \leq m \end{aligned} \quad (9)$$

Maximum clique problem group (MCPG) is a variant of MCP problem. In MCPG problem, elements in U are divided into k groups, $G_1, G_2, \ldots, G_k$, and at most one element is selected from one group. The MCPG expression of the revenue version is as follows.

$$\begin{aligned} &\max | \sum_{i \in 1,2,\ldots,l} Pro(S_i')| \\ &s \cdot t \cdot S_i' \in U, l \leq k, |S_1', S_2', \ldots, S_l'| \cap G_j \leq 1, \forall j \in \{1, \ldots, k\} \end{aligned} \quad (10)$$

Markov chain is a stochastic process with Markov property, and its feature is memoryless. The state of the system at $t+1$ is only related to that of the current t. The system can predict the next state of random variables according to the current state.

$$P\{x_{t+1} = b_{t+1} | x_t = b_t, x_0 = b_0, \ldots, x_0 = b_0\} \tag{11}$$

A network is composed of user set R and network configuration set F. Each network configuration is composed of each user using their own local configuration. In a network configuration f, each user can get a profit $x_r(f)$. The revenue of the system is maximized by selecting the optimal configuration.

$$MWC : \max_{f \in F} \sum_{r \in R} x_r(f) \tag{12}$$

In creating the maximum quality DAT, each ψ corresponds to f. Generation tree set $Tr(G)$ corresponds to F.

$$x_\psi = \sum_{r \in R} x_r(\psi) = \max_{\overrightarrow{t^\psi}, \overrightarrow{w^\psi}} , \sum_{i \in V} w_i^\psi(j) \tag{13}$$

$$MWC : \max_{\psi \in Tr(G)} x_\psi \tag{14}$$

$$\begin{aligned} MWC - EQ &: \max_{p \geq 0} \sum_{\psi \in Tr(G)} P_\psi \cdot x_\psi \\ s \cdot t \cdot &\sum_{\psi \in Tr(G)} P_\psi = 1 \end{aligned} \tag{15}$$

P_ψ means the probability that ψ is adopted. According to the log-sum-exp approximate function, the following equation is deduced.

$$g_\beta(x) = \frac{1}{\beta} \sum_{\psi \in Tr(G)} \exp(\beta \cdot x_\psi) \tag{16}$$

β is a normal quantity.

$$\begin{aligned} &\max_{p \geq 0} \sum_{\psi \in Tr(G)} P_\psi \cdot x_\psi - \frac{1}{\beta} \sum_{\psi \in Tr(G)} P_\psi \cdot \log(p_\psi) \\ &s \cdot t \cdot \sum_{\psi \in Tr(G)} P_\psi = 1 \end{aligned} \tag{17}$$

Equation (16) is used as MWC and $MWC - EQ$ approximation function, and an information entropy is obtained.

$$-\frac{1}{\beta} \sum_{\psi \in Tr(G)} P_\psi \cdot \log(p_\psi) \tag{18}$$

The approximation accuracy is controlled by β. Since Eq. (17) is a closed convex function, the optimal solution p_{ψ}^{*} is obtained by solving Karush-Kuhn-Tucker.

$$p_{\psi}^{*} = \frac{\exp(\beta \cdot x_{\psi})}{\sum_{\psi \in Tr(G)} \exp(\beta \cdot x_{\psi})} \tag{19}$$

The upper limit of the difference between the approximate value obtained by bringing the optimal solution and the approximate difference of the optimal value is as follows.

$$-\frac{1}{\beta} \sum_{\psi \in Tr(G)} P_{\psi} \cdot \log(p_{\psi}) \tag{20}$$

When β approaches to infinity, the approximate difference is 0. However, the real β is not likely to be very large, and too large β value will lead to too fast convergence of Markov chain into local optimum.

This study establishes two simulation cases on MATLAB platform, and compares the proposed algorithm with the optimal solution and similar algorithms in other literatures. If $E_R = 1$, $E_T = 2$, $ES = 1$, each node allows up to seven units of energy consumption in a single data aggregation. When the network scale is small, the optimal solution can be obtained by the exhaustive method. Here, a large-scale network with 100 nodes evenly distributed in a square region with a width of 300 m is established. 80% nodes are randomly selected as the source nodes, and other nodes are used as non-source nodes. The communication range of all nodes is 75 m.

Trusted computing technology is a reliable method to solve network internal attacks. For the security problems in WSNs, trust and reputation system plays an important role by evaluating the credibility of participants. At present, the existing data fusion based on trusted computing mostly compares the data with all its neighbor node data to calculate the node's perceived trust, usually without considering the data correlation among nodes. Moreover, the sensor location is often in an unsupervised environment. The uncertainty of terrain and communication conditions often leads to the non-correlation among neighbor nodes. In this study, a perceptual trust calculation method is proposed. The clustering algorithm can be used to cluster and exclude irrelevant nodes to obtain a value with more trust. It is assumed that the data of the sensor is close to the data of a specific number of adjacent nodes, then the sensor is the core sensor node.

It is supposed there is a sensor node as v, its n neighbors are the $v_1, v_2, \ldots, v_n$ representing $D_1, D_2, \ldots, D_n$. If there is m data in $D_1, D_2, \ldots, D_n$ and distance between the data with D is less than ε, and there is $\min\ p \leq m \leq n$, where ε and min p represent the data threshold and the quantity threshold, the data density correlation of node v is as follows.

$$sim(v) = \begin{cases} 0, m < \min p \\ a_1(1 - \frac{1}{\exp(N - \min p)}) + a_2(1 - \frac{d_{\Delta}}{\varepsilon}) + a_3(1 - \frac{d}{\varepsilon}), m \geq \min p \end{cases} \tag{21}$$

d presents the mean distance between m data and D, d_{Δ} means the distance between data center of m data and D, a_1, a_2, a_3 are the weight values, and $a_1 + a_2 + a_3 = 1$.

WSN is a network that regards data as the center and needs to consider trusts of perception, link, and node. This study puts forward a trust model based on context-aware. This model also considers the energy factor, and selects different coping measures for different context trust problems.

It is supposed the network system is a graph $G = (V \cup \{S\}, E)$, where V is a set of nodes, S is a sink node, and E is a set of edges. Network contains a sink node and N nodes, when i and j within the communication range of each other, the two nodes are adjacent. The trusted data aggregation framework of this study firstly clusters network nodes based on data correlation. Then, node trust, perceived trust, and link trust are calculated. Finally, according to different contexts, different processing methods are adopted to improve system throughput and robustness. If the node is untrusted, it can't be adopted as the aggregation and the forwarding nodes, so all the untrusted child nodes select the parent node again. It is assumed that the link is not credible, only the child nodes on the link need to re-select the parent node. If the node perception is not credible, the node-aware data is discarded, without any node to select the parent node.

Perceived trust is adopted to evaluate data consistency and fault tolerance. The data perceived by sensors are related in time and space, namely, in different time periods, the data perceived by the same category of sensors are similar. If the data are in accordance with normal distribution, the probability density function is expressed below.

$$f(x) = \frac{1}{\sigma\sqrt{2\pi}} e^{-\frac{(x-\mu)^2}{2\sigma^2}} \tag{22}$$

μ is mean and σ is variance. The closer the value and the mean μ are, the higher the trust is. For any i node, perceived trust value calculation is expressed in Equation (23).

$$T_{data}(i) = 2(0.5 - \int_{\mu}^{v} f(x)dx) = 2\int_{\mu}^{v} f(x)dx \tag{23}$$

v means the perceived data value for the i node. To prevent malicious data attacks, the median absolute deviation (MAD) is utilized here to remove abnormal values. The median and MAD obtained after eliminating the abnormal values are utilized as mean and variance to calculate the data trust value.

The value of link trust is determined by packet error rate and packet loss rate. Packet error rate is usually calculated according to bit error rate. Bit error rate P_{ber} is calculated by Nakagami-m channel fading model.

$$P_{ber} = \frac{1}{2} erfc(\sqrt{SNR}) \tag{24}$$

Because there is signal interference, forward error correction is used. Then, the expression of packet error rate is as follows.

$$P_{ber} = 1 - (1 - P_{ber})^n \tag{25}$$

n indicates the number of effective loads in the packet.

Packet loss rate P_{plr} is calculated by packet reception rate P_{prr}. The expression of packet reception rate is as follows.

$$P_{prr} = \frac{n_{rec}}{n_{sen}} \tag{26}$$

n_{rec} presents the number of packets received successfully, and n_{sen} represents the sum of packets sent. It is important to distinguish packet loss due to unreliable links. Because the wireless communication channel has memory ability and there is temporary correlation in the lost grouping, the autoregressive integrated moving average model (ARIMA) is needed to predict the packet loss because of the unreliable communication channel. Based on packet error rate and reception rate, the link quality calculation expression is shown in Eq. (27).

$$Lq = (1 - P_{per})P_{prr} \tag{27}$$

Node trust includes direct trust T_{direct} and recommendation trust. T_{direct} is the principal's observation of the trustee. It is based on communication behavior and does not consider the impact of node residual energy on node credibility. Therefore, when calculating node direct trust, node residual energy is also regarded as the evaluation index of node trust. Recommendation trust is indirect trust from third-party recommendation. Communication behavior includes successful and unsuccessful communication. Malicious nodes and unreliable communication links will also cause communication failure. Therefore, when calculating the direct trust of nodes, it is necessary to eliminate packet loss and modification caused by unreliable communication links.

$$\begin{aligned} T_{direct} &= \left(\frac{s + (P_{plr} + P_{per}) \cdot (s + f)}{s + f}\right) \cdot Rel \\ &= \left(\frac{s + (1 - P_{prr} + P_{per}) \cdot (s + f)}{s + f}\right) \cdot Rel \end{aligned} \tag{28}$$

s shows the number of successful communications, Rel represents residual energy, and f is the number of failed communications. The recommended trust calculation for i is shown below.

$$T_{rec} = \sum_{j=1}^{n} T_{j,i} \tag{29}$$

There is trust value calculation expression of the node.

$$T_{node} = w_{direct} \cdot T_{direct} + (1 - w_{direct}) \cdot T_{rec}$$
$$0 \leq w_{direct} \leq 1 \tag{30}$$

w_{direct} means the weight of direct trust. Then, attenuation function is adopted to update the trust value.

$$w_{decay} = exp(-\delta \cdot (t - t_0)) \tag{31}$$

δ indicates adjustment factor, t is the computation time of current trust, and t_0 indicates that of the last trust value. The process of data aggregation algorithm based on context-aware and data density correlation is illustrated in Algorithm 1.

Algorithm 1 Data aggregation algorithm

Start
Clustering network nodes.
Compute perceived trust: $T_{data}(i) = 2(0.5 - \int_{\mu}^{v} f(x)dx) = 2\int_{\mu}^{v} f(x)dx$
Calculate link trust value: $Lq = (1 - P_{per})P_{prr}$
Calculate node direct trust value:

$$T_{direct} = \left(\frac{s + (P_{plr} + P_{per}) \cdot (s + f)}{s + f}\right) \cdot Rel$$
$$= \left(\frac{s + (1 - P_{prr} + P_{per}) \cdot (s + f)}{s + f}\right) \cdot Rel$$

Calculate recommended trust value: $T_{rec} = \sum_{j=1}^{n} T_{j,i}$
Calculate the total trust value of the node:

$$T_{node} = w_{direct} \cdot T_{direct} + (1 - w_{direct}) \cdot T_{rec}$$
$$0 \leq w_{direct} \leq 1$$

According to $w_{decay} = exp(-\delta \cdot (t - t_0))$ update the perceived trust value, link trust value and node trust value.
if the perception of the node i is untrusted **then**
    only the data generated by the node i is discarded, but the node i itself can still aggregate the data from other nodes as an intermediate node i and forward it to the node's parent node.
end if
if the link of the node i is not trusted. **then**
    The node i replaces a feasible node through a trusted link.
    if i failed to replace parent node **then**
        All child nodes update the parent node, and i is discarded.
    end if
end if
End

The performance of this context-aware algorithm is verified by actual case analysis. The platform adopts MATLAB, and the data set is the public data set collected by Inter Berkeley Research Room. There are 54 sensors to measure temperature, humidity, light, voltage, etc. every 31 s. By the method in this study, the sensors are clustered every 30 times, and the data threshold and quantity thresholds are set to 0.4

and 2, respectively. In order to simulate abnormal data, the error data is randomly injected into the data set, and then six nodes are randomly selected as malicious data generation nodes, and then put with a probability of 0.8. This method is compared with the traditional perceptual trust method (the method without data density aggregation, distinguishing node trust, link trust, and perceptual trust).

4 Results and Discussion

4.1 Data Aggregation Results Under Time and Energy Constraints in Unreliable Environment

The performance of the method in this study is compared with the optimal solution and similar methods in literatures [27–30] (Fig. 5a). In order to compare with the optimal solution, it is given that all nodes are source nodes and the link is reliable, $\alpha = 0.2$, $\beta = 1$. In addition, a case study of the algorithm in this study in the network environment with large-scale uniformly distributed is presented in Fig. 5b, $\alpha = 0.2$, $\beta = 0.5$, and time constraint $D = 20$.

Figure 5a shows that comparison results between the algorithm and other similar methods, the data aggregation quality obtained by this algorithm at different time points is the highest, and it is closest to the optimal solution. In Fig. 5b, aggregation quality of the algorithm is better than that of similar algorithms, and the data aggregation quality increases with the network size.

Figure 6 are data aggregation quality comparison under different time constraints D, data aggregation quality comparison of different algorithms under different link

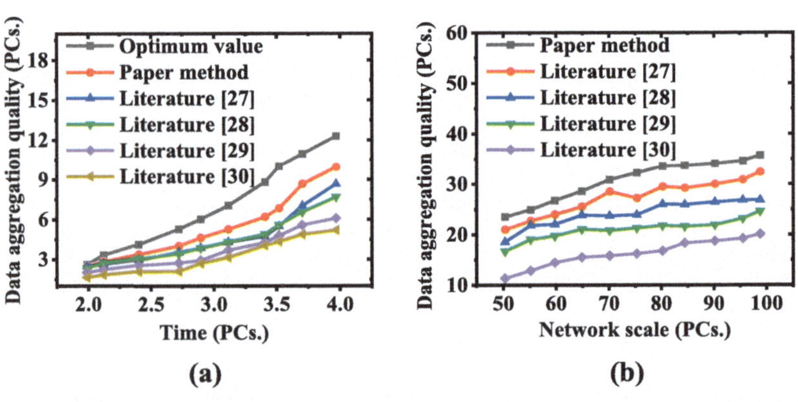

Fig. 5 Comparison of the performance of the algorithm proposed in this study and other algorithms (**a** the performance comparison of the algorithm and the optimal solution; **b** the comparison of the aggregation quality of the algorithm with different network sizes under evenly distributed scenarios)

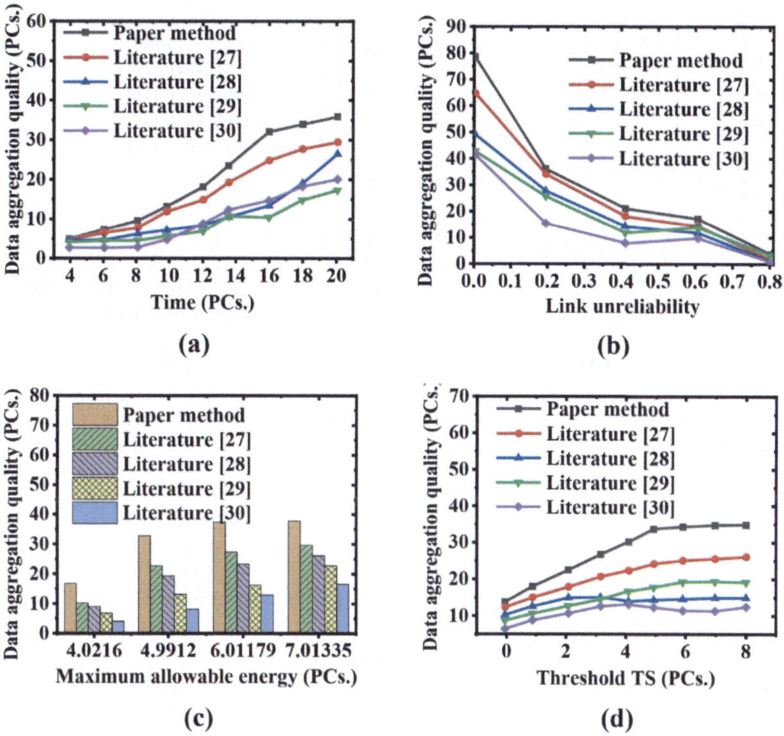

Fig. 6 Comparison of performance of different algorithms under uniform distribution scenario (**a** comparison of data aggregation quality under different time constraints D; **b** comparison of data aggregation quality under different link reliability; **c** comparison of data aggregation quality under different energy constraints; **d** comparison of data aggregation quality under different thresholds)

reliabilities, data aggregation quality under different energy constraints and different thresholds.

Figure 6a indicates that the data aggregation quality of the algorithm in this study is better than that of similar algorithms in other literatures under different time constraints, and the performance is improved by at least 3%. The value of link unreliability in abscissa represents the random value (Fig. 6b). When the link unreliability increases, the packet loss rate will increase, then the amount of data transmitted to the receiver will decrease, and the amount of data aggregation will decrease, and the algorithm performance is improved by an average of 50% compared with other methods. It is found in Fig. 6c that the data aggregation quality also increases with the increase of the value of the maximum allowable energy. Since the increase of the value will make the sensor have more energy to transmit the lost data packets again, the data aggregation quality will increase correspondingly. The algorithm performance is improved by an average of 58% compared with other methods. It can be observed in Fig. 6d that the quality of data aggregation increases with the increase

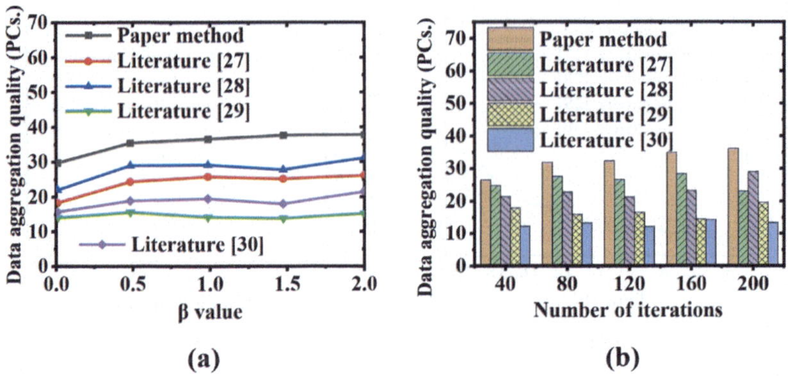

Fig. 7 Comparison of data aggregation quality under different values of parameters β and different iteration times (**a** the effect of different parameters β on data aggregation quality; **b** comparison of data aggregation quality under different iteration times)

of threshold TS, and the performance of the algorithm improves by at least 1% and at most 33%, with an average improvement of 23% compared with other methods.

The effect of different algorithm parameters β on the quality of data aggregation under different values and different iteration times is given in Fig. 7.

From Fig. 7a, the quality of data aggregation increases with the β increasing, and when $\beta = 1$, the quality of data aggregation tends to be stable, because when approaching the optimal solution, the improvement of aggregation quality will become smaller and smaller. The algorithm in this study tends to be stable earlier than other methods and the trend is smooth and almost free of fluctuations, indicating that the algorithm performance in this study is more stable. Figure 7b suggests that each iteration means that a timer ends to start a new aggregation number. The data aggregation quality increases with the number of iterations. When the number of iterations reaches 200, data aggregation quality gradually converges and finally tends to a global optimal solution.

4.2 Data Aggregation Results Based on Context-aware Computing

The calculation results of perceived trust of six malicious nodes are given in Fig. 8.

This data is implausible malicious data when the trust value is below 0.5 (Fig. 8). The missing report rate of the proposed method is 0.067, while that of the traditional method is 0.53. This study greatly reduces the omission rate based on the context-aware method.

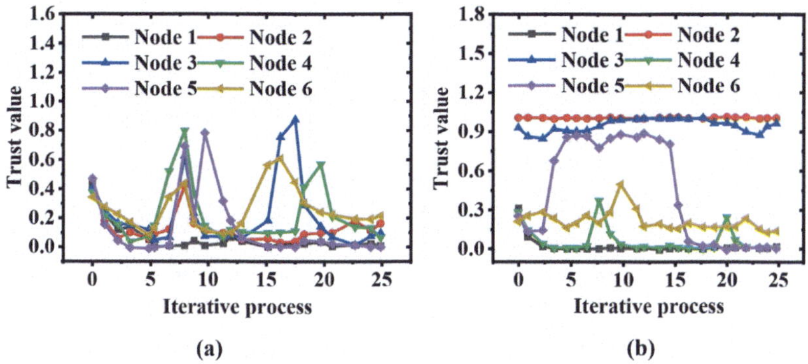

Fig. 8 Perceived trust of different malicious nodes (**a** the perceived trust of the method in this study; **b** the perceived trust of traditional methods)

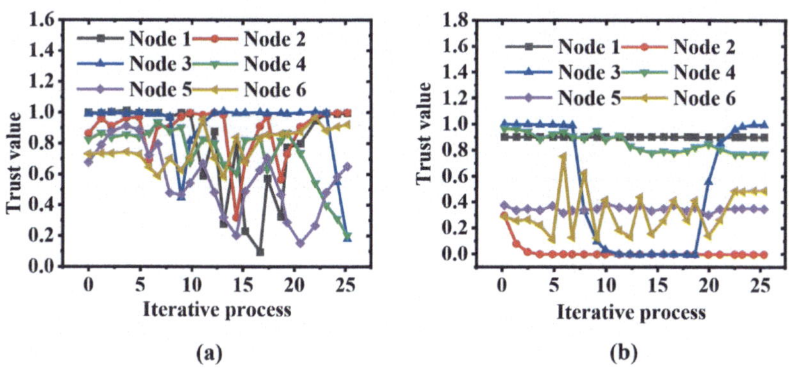

Fig. 9 Perceived trust of non-malicious nodes (**a** the perceived trust of method in this study; **b** the perceived trust of traditional methods)

Six nodes without malicious data are randomly selected to test the false alarm rate results (Fig. 9). It is not difficult to find that the error rate of warning of the proposed method is 0.14 much lower than that of the traditional method (0.42).

Comparison of the number of nodes unable to connect to the network under different number of unreliable links and under nodes with different number of sensing anomalies are illustrated in Fig. 10. The network throughput between this method and traditional methods as well as similar methods is compared.

From Fig. 10a, b, the number of unreliable links increases, and the number of nodes not connected to the network will also increase. Both the present method and the traditional methods have different degrees of node discarding. However, the discard amplitude of this algorithm is small and the curve remains stable. From Fig. 10c, d, as the number of sensing abnormal nodes increases, the number of nodes that can't be connected into the network increases. This method should choose to discard the

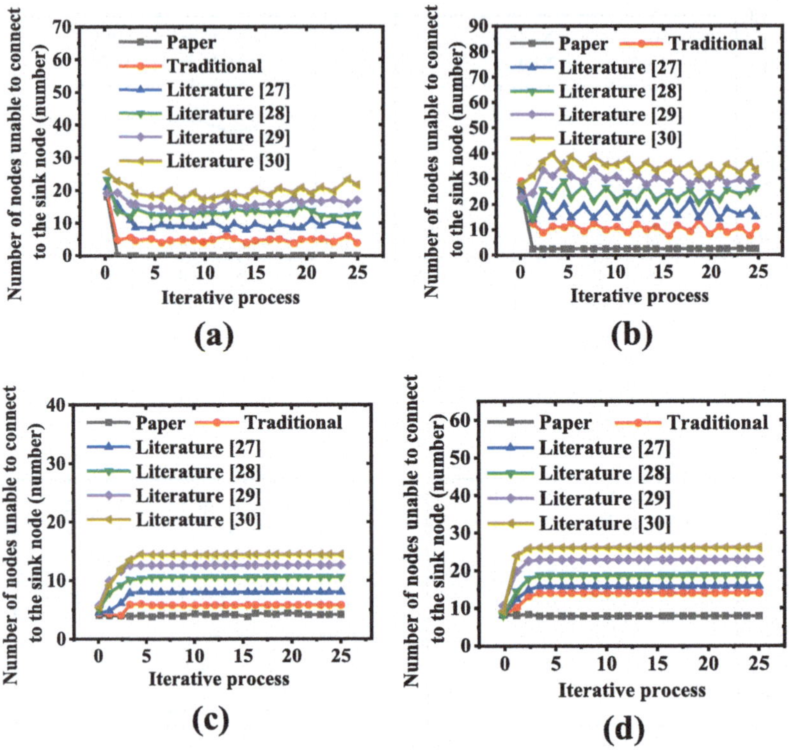

Fig. 10 Comparison of throughput under unreliable links and under sensing anomalies (**a** 8 unreliable links; **b** 16 unreliable links; **c** 4 sensing anomaly nodes; **d** 8 sensing anomaly nodes)

sensed data of unreliable nodes, but it can still aggregate and transmit the data from other nodes, so it improves the network throughput.

In addition to the industrial network data transmission scheme proposed in this chapter, DTs will play a more important role in the future. Based on the basic state of physical entities, DTs will make a highly realistic analysis of the established model and collect data in a dynamic and real-time way for monitoring, predicting, and optimization of physical entities. In addition, DTs as edge side technology can be effective connection device layer and network layer, become the industry the Internet platform of knowledge extraction tool, will constantly in the industrial system fragmentation knowledge transfer to the industrial Internet platform, DTs body of different maturity, the industry knowledge of different granularity reassemble, through the industrial APP calls. Therefore, the industrial Internet platform is the incubation bed of DTs, and DTs is an important scene of the industrial Internet platform.

Besides, accurate mapping of virtual models and fast feedback control of physical entities are the key to realizing DTs. The accuracy of the virtual model, the fast feedback control ability of physical entities, and the interconnection of massive physical devices put forward higher requirements on the data transmission capacity, transmission rate, and transmission response time of DTs. 5G communication technology has high speed, large capacity, low delay, and high reliability. It can meet the data transmission requirements of DTs, meet the mass data low-delay transmission of virtual model and physical entity, and meet the interconnection of many devices to promote the application of DTs better.

5 Conclusion

With the development of DTs technology and the background of other computer technologies and the IoT, the application field of intelligent environment is also expanding, but there are accompanying problems, such as weak network configurability, poor real-time performance, and privacy protection. The sustainable development of industrial manufacturing is inseparable from DTs technology. The network security problems existing in the intelligent industrial environment need to be solved by more effective technology. By establishing the DTs model of industrial network environment, this study establishes the hierarchical network framework of intelligent industrialization environment. By studying data aggregation technology, it attempts to solve the problems of node redundancy, data aggregation delay, and security caused by the explosive growth of node size in the IoT. A trusted data aggregation model based on context and data density correlation is proposed. Clustering nodes can improve the accuracy of network perceived trust. It is verified that the proposed method achieves better system throughput than other traditional methods in the process of network cognitive communication. The disadvantage is the article studies the maximization of aggregation quality of single sink node network. In the future, it is necessary to further study the maximization of data aggregation quality of multiple sink nodes.

References

1. Y. Dai, K. Zhang, S. Maharjan, Y. Zhang, Deep reinforcement learning for stochastic computation offloading in digital twin networks. IEEE Trans. Ind. Inform. **17**(7), 4968–4977 (2020)
2. W. Sun, S. Lei, L. Wang, Z. Liu, Y. Zhang, Adaptive federated learning and digital twin for industrial internet of things. IEEE Trans. Ind. Inform. **17**(8), 5605–5614 (2020)
3. K. Židek, J. Pitel', M. Adámek, P. Lazorík, A. Hošovský, Digital twin of experimental smart manufacturing assembly system for industry 4.0 concept. Sustainability **12**(9), 3658 (2020)
4. C. Zhang, G. Zhou, H. Li, Y. Cao, Manufacturing blockchain of things for the configuration of a data-and knowledge-driven digital twin manufacturing cell. IEEE Internet Things J. **7**(12), 11884–11894 (2020)
5. P. Jia, X. Wang, X. Shen, Digital-twin-enabled intelligent distributed clock synchronization in industrial IoT systems. IEEE Internet Things J. **8**(6), 4548–4559 (2020)
6. D. Wang, Z. Zhang, M. Zhang, M. Fu, J. Li, S. Cai, X. Chen, The role of digital twin in optical communication: fault management, hardware configuration, and transmission simulation. IEEE Commun. Mag. **59**(1), 133–139 (2021)

7. J. Qian, B. Song, Z. Jin, B. Wang, H. Chen, Linking empowering leadership to task performance, taking charge, and voice: the mediating role of feedback-seeking. Front. Psychol. **9**, 2025 (2018)
8. W. Wang, Z. Deng, J. Wang, Enhancing sensor network security with improved internal hardware design. Sensors **19**(8), 1752 (2019)
9. T. Alam, A middleware framework between mobility and IoT using IEEE 802.15. 4e sensor networks. J. Online Informatika, **4**(2), 90–94 (2020)
10. B. Cao, J. Zhao, Y. Gu, S. Fan, P. Yang, Security-aware industrial wireless sensor network deployment optimization. IEEE Trans. Ind. Inf. **16**(8), 5309–5316 (2019)
11. S. Otoum, B. Kantarci, H.T. Mouftah, On the feasibility of deep learning in sensor network intrusion detection. IEEE Netw. Lett. **1**(2), 68–71 (2019)
12. V. Bhasin, S. Kumar, P.C. Saxena, C.P. Katti, Security architectures in wireless sensor network. Int. J. Inform. Technol. **12**(1), 261–272 (2020)
13. T.G. Nguyen, T.V. Phan, B.T. Nguyen, C. So-In, Z.A. Baig, S. Sanguanpong, Search: a collaborative and intelligent nids architecture for sdn-based cloud iot networks. IEEE Access **7**, 107678–107694 (2019)
14. A. Fuller, Z. Fan, C. Day, C. Barlow, Digital twin: enabling technologies, challenges and open research. IEEE Access **8**, 108952–108971 (2020)
15. M. Obschonka, J. Moeller, M. Goethner, Entrepreneurial passion and personality: the case of academic entrepreneurship. Front. Psychol. **9**, 2697 (2019)
16. H.X. Nguyen, R. Trestian, D. To, M. Tatipamula, Digital twin for 5G and beyond. IEEE Commun. Mag. **59**(2), 10–15 (2021)
17. D. Chen, D. Wang, Y. Zhu, Z. Han, Digital twin for federated analytics using a Bayesian approach. IEEE Internet Things J. (2021)
18. C. Gehrmann, M. Gunnarsson, A digital twin based industrial automation and control system security architecture. IEEE Trans. Ind. Inf. **16**(1), 669–680 (2019)
19. Z. Jiang, Y. Guo, Z. Wang, Digital twin to improve the virtual-real integration of industrial IoT. J. Ind. Inform. Integr. **22**, 100196 (2021)
20. Y. Lu, X. Huang, K. Zhang, S. Maharjan, Y. Zhang, Communication-efficient federated learning and permissioned blockchain for digital twin edge networks. IEEE Internet Things J. **8**(4), 2276–2288 (2020)
21. J. Moyne, Y. Qamsane, E.C. Balta, I. Kovalenko, J. Faris, K. Barton, D.M. Tilbury, A requirements driven digital twin framework: specification and opportunities. IEEE Access **8**, 107781–107801 (2020)
22. M. Liu, S. Fang, H. Dong, C. Xu, Review of digital twin about concepts, technologies, and industrial applications. J. Manuf. Syst. **58**, 346–361 (2021)
23. B. He, K.J. Bai, Digital twin-based sustainable intelligent manufacturing: a review. Adv. Manuf. **9**(1), 1–21 (2021)
24. B. Cao, X. Wang, W. Zhang, H. Song, Z. Lv, A many-objective optimization model of industrial internet of things based on private blockchain. IEEE Netw. **34**(5), 78–83 (2020)
25. K. Kolomvatsos, C. Anagnostopoulos, A deep learning model for demand-driven, proactive tasks management in pervasive computing. IoT **1**(2), 240–258 (2020)
26. F. Farahbakhsh, A. Shahidinejad, M. Ghobaei-Arani, Context-aware computation offloading for mobile edge computing. J. Ambient Intell. Human. Comput. 1–13 (2021)
27. J. Song, Y. Liu, J. Shao, C. Tang, A dynamic membership data aggregation (DMDA) protocol for smart grid. IEEE Syst. J. **14**(1), 900–908 (2019)
28. G. Zhu, J. Xu, K. Huang, S. Cui, Over-the-air computing for wireless data aggregation in massive IoT. IEEE Wireless Commun. **28**(4), 57–65 (2021)
29. J. He, L. Cai, P. Cheng, J. Pan, L. Shi, Consensus-based data-privacy preserving data aggregation. IEEE Trans. Automatic Control **64**(12), 5222–5229 (2019)
30. S.A. Dehkordi, K. Farajzadeh, J. Rezazadeh, R. Farahbakhsh, K. Sandrasegaran, M.A. Dehkordi, A survey on data aggregation techniques in IoT sensor networks. Wireless Netw. **26**(2), 1243–1263 (2020)

Zhihan Lv received the Ph.D. degree from Ocean University of China in 2012. He has been an assistant professor with the Shenzhen Institutes of Advanced Technology, Chinese Academy of Sciences, from 2012 to 2016, an associate professor with Qingdao University, China, from 2017 to 2021. He was with CNRS, France, as a research engineer, with Umea University, Sweden, as a postdoctoral research fellow, with Fundacion FIVAN, Spain, as an experienced researcher, with University College London, UK, as a research associate, with University of Barcelona, Spain, as a postdoctor. He is currently an associate professor with Uppsala University, Sweden.

Liang Qiao received the bachelor's degree from Qingdao University, he is currently working toward the graduate degree in the College of Computer Science & Technology at Qingdao University. His research interests include machine learning, blockchain, and virtual reality. In 2019. In 2018, he won the second prize of the National Software and Information Technology Competition in China. He has rich experience in algorithm design.

Cognitive Mobile Computing for Cyber-Physical Systems (CPS)

Akramul Azim and Md. Al Maruf

Cyber-Physical Systems (CPS) are combined with sensor networks that have embedded computing ability to sense, monitor, and control the physical environment. Due to the growth of sensor data traffic and mobile applications (e.g., object detection, cameras, and security), the infrastructure of cyber-physical systems has become more complex. These applications require computational intensive machine algorithms to make intelligent decisions. Therefore, we define the term as cognitive mobile computing, when a machine can intelligently sense the required data and compute efficiently in making the right decision based on the human thought process in any complex situation. The main characteristics of cognitive mobile computing are pervasive computing, extensive networking, and the degree of automation during wireless communications without any human supervision. However, mobile computing devices connected over a network face challenges for computational power constraints, communication delay and physical interactions. Therefore, the future of cognitive mobile computing shows the necessity of implementing an efficient framework in selecting the right computing platforms, machine learning algorithms, and data analytic models considering mobility.

A. Azim (✉) · Md. Al Maruf
Ontario Tech University, Oshawa, Canada
e-mail: akramul.azim@ontariotechu.ca

Md. Al Maruf
e-mail: md.maruf@ontariotechu.ca

A. K. Pathan (ed.), *Towards a Wireless Connected World: Achievements and New Technologies*, https://doi.org/10.1007/978-3-031-04321-5_9

1 Introduction

Cognitive computing is a subfield of artificial intelligence (AI), and it refers to a platform that computes diverse tasks with reasoning and understanding to facilitate human intelligence [1]. Cognitive mobile computing provides support to compute tasks leveraging machine learning to extend the ability of processing data.

Cognitive Computing: Cognitive computing is a self-learning system that employs machine-learning methods to perform complex tasks intelligently like a human. A system is considered cognitive if it holds three important concepts: contextual insight from the model, hypothesis generation, and continuous learning from the sensor data [2].

Mobile Computing: Mobile computing refers to a technology that is used to perform a wide variety of tasks in remote computing platforms or mobile (non-static) environments. Mobile computing uses wireless communication (e.g., Wi-Fi, cellular networks) to transmit data to mobile computing devices (e.g., smart vehicles, cloud).

Cognitive Mobile Computing: The computing process uses AI, Machine learning, context analysis, and natural language processing to solve any problem. Therefore, when cognitive computing is conducted on any mobile platform in the cyber-physical systems, we define it as cognitive mobile computing. Mobile devices and platforms are connected through a wireless network in cognitive mobile computing. Moreover, it has the ability to manage dynamic resource allocation to achieve the trade-offs between mobile devices and wireless resources. It enables the facility to compute data wirelessly from one device to others considering varying application constraints.

Cognitive computing possesses some unique properties to differentiate itself from the other learning approaches such as environmental awareness, adaptiveness, and intelligence. It is similar to human cognition that involves real-time analysis of the environment, context, and intent like the human brain to solve problems. Moreover, it focuses on the following aspects [1]:

- *Interactive*: To obtain the capabilities of cognitive computing, the technology needs to be dynamically interactive with different processes, computing devices, platforms like cloud, fog and edge networks.
- *Adaptive*: This should have the ability to learn over time and adapt to changes in new environments. Nowadays, cyber-physical systems process a wide variety of data in real-time. These dynamic data need to be understood by the system to make decisions to change the learning mechanism by itself.
- *Iterative and stateful*: Another important attribute of cognitive computing is identifying the problem requirements and updating the state by asking questions or analyzing given input data. The system constantly updates its learning state for solving a problem from similar situations that occurred in the past.
- *Contextual*: It should be aware of the situation and context to perform a task. Examples of context are task requirements, domain, time, location, background, source of information, and goal of a task.

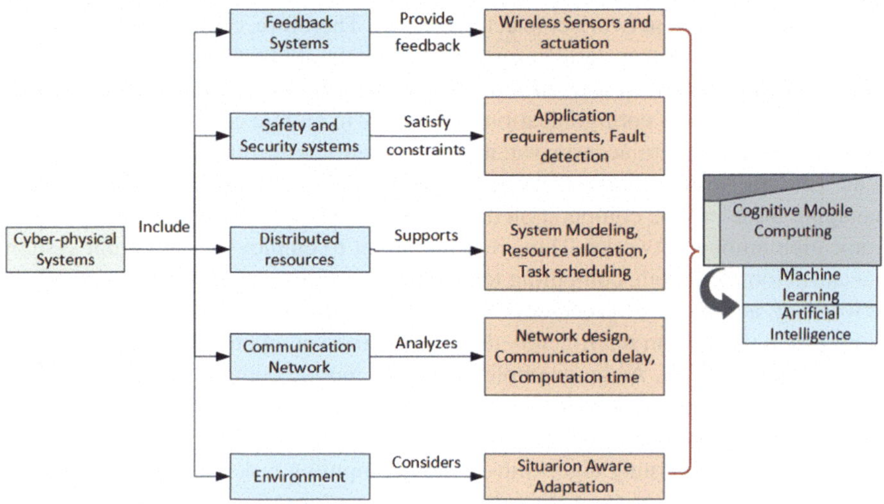

Fig. 1 Cognitive computing components in cyber-physical systems (CPS)

To understand the associated components in cognitive mobile computing in CPS, a list of components are shown in Fig. 1. The cyber-physical systems include different components such as feedback systems, safety systems, wireless networks, and physical environments controlled through different machine learning algorithms. Cognitive mobile computing provides the facility to make the decision of different tasks within each component intelligently.

2 Cognitive Mobile Computing for CPS

A cyber-physical system integrates systems with varying software (e.g., machine learning applications) and hardware (e.g., sensors) components connected over the Internet. A CPS generates a large amount of sensor data that are used in making decision for different Internet of Things (IoT) applications such as speech recognition and Google voice typing. The computational intensive applications use different machine learning algorithms to make intelligent decisions. These algorithms have become very popular in different industry's services, including healthcare, transportation, automotive, education, and many more. Cognitive mobile computing empowers the machine intelligence capability with the ability to self-learn and improve its learning from experiences automatically.

With computing becomes ubiquitous, different resource-constrained devices are required to compute different types embedded applications (e.g., automobiles, healthcare devices). Traditional system solutions are unable to offer adequate facilities to handle resource-constrained devices, which are limited by computational power, memory, storage, and network connectivity [3, 4]. Therefore, cognitive mobile computing in cyber-physical systems can greatly benefit the application users to overcome different challenges. Moreover, different computing devices are connected to a network through wireless communication technology in mobile computing. It provides services to end-users to access available computational resources and information in the network while mobile. Cyber-physical systems introduce new challenges in managing these mobile computation devices to ensure faster computation and communication among networks. Thus, it is important to employ an intelligent system that can automate mobile computing technology providing support for computation, networking, and physical processes. We refer to the term as cognitive mobile computing when the system intelligently decides where to compute and what to process to serve user requests. As an example, cognitive mobile computing can be a good solution for autonomous self-driving vehicles. Autonomous vehicles require to process raw sensory data (e.g., images, videos, text, and speech) and learn over time by applying machine learning algorithms. These computing tasks need to be managed smartly to ensure the safety of the cyber-physical systems.

Cognitive mobile computing can also take advantages of cloud-based solutions to provide support for resource management, data storage, security and access to different software applications. However, cloud-based approaches are not always feasible for applications that have constraints such as response time and cost. Moreover, cloud service is not always a preferred solution considering downtime, communication delay privacy, limited control and flexibility [5]. The cognitive mobile computing can benefit to run time sensitive applications to the nearest computing platforms(e.g., vehicular fog computing, road side units (RSUs)) to provide better performance. Furthermore, these mobile computing platforms are still required to investigate different issues such as hardware architectures, run time, communication mechanisms, wireless networking, and application development on the constrained devices, making applications seamlessly run across diverse platforms [6].

3 Use Cases

Depending of the scope of different applications, many use cases exist for cognitive mobile computing in the CPS domain. A few areas of specialization are discussed next, associated with fog computing, transfer learning, situational awareness, runtime monitoring, and context-based learning.

3.1 Cognitive Fog Computing for CPS

Cognitive mobile computing supports a wide range of applications where the computing platforms play a huge role in being adaptive while making decisions in real-time. Cloud is a popular computing platform for the IoT and CPS applications. How-

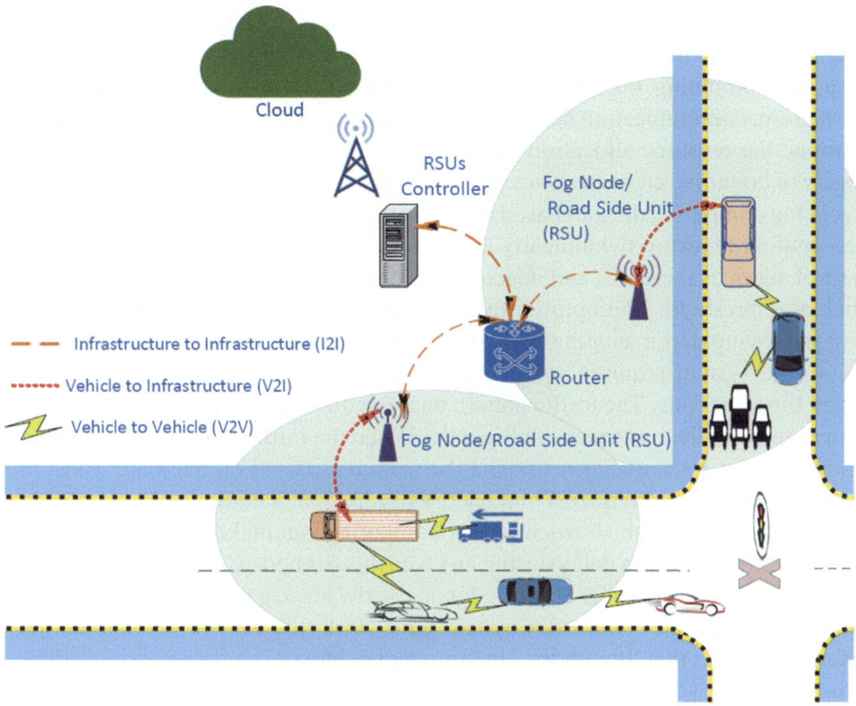

Fig. 2 Fog computing architecture in cyber-physical systems

ever, advanced machine learning applications require intelligent computing services where the decision can be made based on the application requirements. To meet the necessity, fog computing is a promising option for time sensitive computation (e.g., vehicular over-the-air (OTA) updates), as it can offer enhanced network durability and lower communication delays, as compared to the cloud [7]. To understand the fog computing architecture, Fig. 2 shows the connected components like vehicles, fog computing node or road side unit (RSU), cloud and communication medium (e.g., wireless). The fog computing nodes are placed besides the roads to make it more closer to the vehicles. This helps vehicles get connected quicker than the cloud with faster computation and update.

Although the current network communication latency and resilience are improved enough, the fog computing platform has not been explored to support cyber-physical systems due to the absence of intellectual ability. There is a large scope of research for making the fog computing platform fit for cognitive mobile computing. For example, vehicular OTA updates demand faster computation during a software update or object detection. In this scenario, fog computing can facilitate better services (e.g., high-quality video) and high-speed data transfer over cloud computing that possesses high latency. Therefore, we need an intelligent infrastructure of fog nodes management to tackle the extended demand of fog computing for resource allocation.

The recent research works [8, 9] show a potential to allocate resources in distributed fog computing architectures to minimize the OTA software update time.

In a fog computing infrastructure, the important factors are to understand the traffic pattern, communication networks, and each fog node's computational capacity to examine the resource allocation and faster computation. Therefore, it shows the necessity of cognitive characteristics that can make decisions to allocate resources in different fog computing devices based on the traffic density and service demands. The fog computing platform dynamically expands its computational capacity when the request of services increases and deactivates different fog nodes while the demands are below a threshold. The optimal resource assignment can potentially lower the cost of deploying and managing fog nodes in an extensive scale. To make a smart decision, the system requires analyzing the real-traffic loads and their pattern at different time intervals. The traffic pattern analysis and their prediction will help the system to manage the resources intelligently. When the traffic loads vary at different locations, the resource manager assigns the resources based on the load. Thus, the resource in terms of running cost and energy dissipation is minimized.

Moreover, the research shows how fog computing can make smart decisions on predicting communication delay during an Over-the-Air update. The communication delay prediction using transfer learning for any software size for a particular location can provide insight into total OTA update time. With this insight, the vehicle may take further actions that fit with the situation and context.

This automation in selecting the required number of computational nodes considering traffic load and predicting OTA update time when to compute makes fog computing more attractive as a cognitive computing platform. Moreover, adding cognition on how to compute tasks can lead to better resource utilization and user experiences.

3.2 Transfer Learning for Mobile CPS Applications

Transfer learning refers to a supervised learning technique that encodes the knowledge from a previously trained model and uses the knowledge to solve a different or similar problem in another domain [10]. In mobile CPS applications, transfer learning can play a vital part in reusing one domain knowledge to another. Many machine learning applications depend on hardware components to interact and collect data from sensors in real-time. To build a model for this kind of application is expensive and time-consuming in the real world. In such a situation, simulation of any system can benefit different stakeholders to gain knowledge for real-world implementation. Therefore, applying the simulation knowledge to the real world is an instance of transfer learning techniques. Alternatively, the real-world knowledge can be adapted in a simulation tool to verify and validate the system before running the system in the real environment. This is also another form of transfer learning to consider. As an example, the CARLA [11] simulator supports different environments to support training and validation in autonomous driving. In transfer learning, the pre-trained

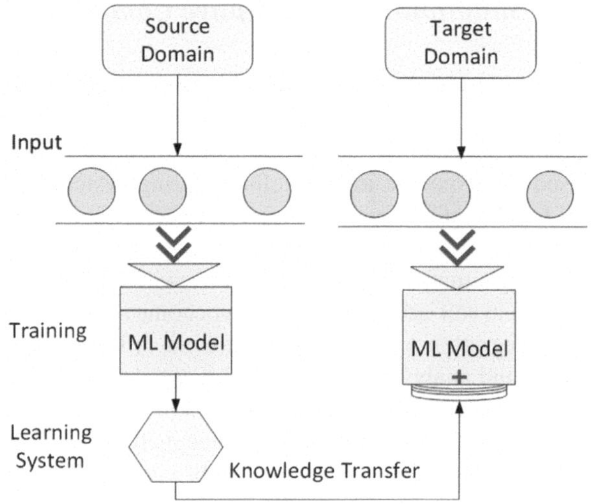

Fig. 3 Transfer learning concept in cognitive computing

machine learning model holds its domain knowledge and is used in another domain to solve a new problem. Figure 3 demonstrates how transfer learning transmits one domain knowledge to another. Therefore, transfer learning has the ability to solve different tasks in new environments.

There are different scenarios where transfer learning has impacts on mobile CPS applications. For example, many resource-intensive machine learning applications need larger training time for training the model. Once the model is trained, you can reuse this trained model for solving another problem where it can potentially avoid re-training the new model incorporating the pre-trained model. This transfer learning drastically reduces the training time for training another model for solving a problem. Feature extraction is a well-known approach to implement transfer learning. It assists in extracting the required features from the learned model and incorporating those features into a much smaller model for evaluation.

In a recent work [9], the authors predict the communication delay using the transfer learning approach. The approach initially set the goal to calculate the communication delay of different geographical areas given in the NYC taxi dataset. These location points do not have communication delay information. Therefore, a pre-trained model of another location is used to predict the delay for the NYC taxi dataset locations. The pre-trained model uses WiFi hotspot signal strength dataset [12] and a 5G dataset [13]. The experimental results show that transfer learning using a deep neural network has better accuracy in predicting communication delay than traditional approaches. This model training can be conducted in the fog computing platform, applying to transfer learning knowledge. Therefore, it can benefit the cyber-physical systems to achieve faster execution in both training and testing phases.

3.3 Situation-Aware Adaptive Computing for CPS

In cyber-physical systems, system components interact with different objects from the environment. Each component needs to satisfy different constraints such as safety and performance. To satisfy these constraints, the system is required to be operated based on a specific model defined during system design. Many real-time systems are highly interactive and need to respond correctly in a changing environment. Therefore, it shows the necessity of cognitive learning that can intelligently handle changing situations and monitor the environment for taking further decision.

System self-adaptation [14] has become an attractive research area with the extensive integration of real-time systems in different environmental situations. System design and analysis are performed in the development and testing phases in the traditional approach. However, the system requires maintaining the safety and tasks schedulability at runtime to guarantee its functional and timing behavior. For example, a transportation system needs to be knowledgeable to understand its surroundings using video stream data mining. The video mining approach captures different temporal events analyzing system runtime behavior and patterns. After that, it makes a knowledge base from those events. This knowledge base is used to train the model, and the model adapts itself, executing a set of pre-defined tasks based on the current situation.

The cognitive approach considers the self-adaptation factors before executing the tasks at runtime. In a dynamic environment, the system may require to execute additional tasks along the existing tasks to adapt to the changes. However, executing such additional tasks through a task scheduler overlooks the important factors like release delay, task preemption, context switches, and cache effects. These factors have an impact on the runtime performance. Therefore, the authors [14] propose a situation-aware adaption framework that verifies the task's constraints before execution so that it does not violate the constraints. The proposed approach mines the environmental inputs (e.g., sensor video data) and extracts different events training a machine learning model. The learning model learns from the events and adapts its system configurations to satisfy different system constraints. Figure 4 shows the cognitive computing approach for adapting situations based on current inputs. The adaptation system always updates itself from time to time based on the environmental context and situation. The experimental results compare system failure probability between the with and without situation-aware adaptation framework. In addition, the results show that the self-adaptation framework can improve the system runtime performance by reducing the scheduling overhead and response time.

3.4 Runtime Monitoring for Self-adaptive CPS

The cyber-physical systems (e.g. autonomous cars) can have different modes (e.g., safety and performance) of operations where a system can switch to different modes to satisfy the system requirements at runtime. Therefore, the cyber-physical systems

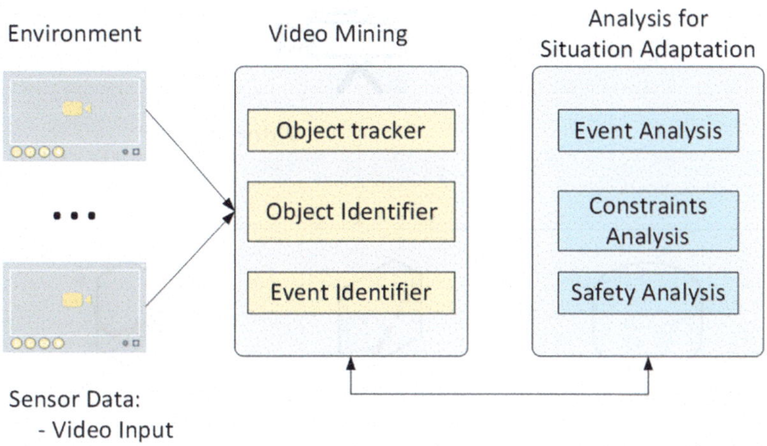

Fig. 4 Situation-aware adaptation using cognitive learning

require cognitive learning to identify the expected mode at runtime and verify the model. The existing works propose different approaches to assure the runtime behavior with varying requirements for self-adaptive systems. For example, in paper [15], the authors consider the feature modeling technique to adapt the reusable configurations at design time. Feature modeling extracts varying features and their requirements for developing a new system from the existing reusable systems. The proposed method presents two dynamic variability dimensions: environment variability and structural variability. The environmental variability identifies the constraints of cyber-physical systems under which they should adapt themselves to change their execution mode. On the other hand, the structural variability finds the system configurations required to be adapted. The experimental results find that the feature-based self-adapted systems require less configuration time and adaptation time compared to traditional approaches. Moreover, it shows that it has less probability of failure to satisfy the constraints in any changing environment.

In another approach [16], the authors present a system monitoring module that observes the environment and identifies events that occur at runtime. An event-driven dataset is created for failure analysis. Based on the probability of failure, the system makes decisions for each environmental situation. Moreover, the proposed approach extracts real-time and non real-time properties from the events. These findings further help to predict future events or system behavior. For example, the early collision probability prediction among two cars can determine the correct mode selection at runtime. The integrated verification and validation models check the failure probability. A threshold value named SV determines whether a system is safe/unsafe for deciding during the mode switches. If the calculated failure probability value is less than the threshold SV value, the system is verified as safe; otherwise, it requires switching into a safe mode. The workflow for assuring runtime behavior of

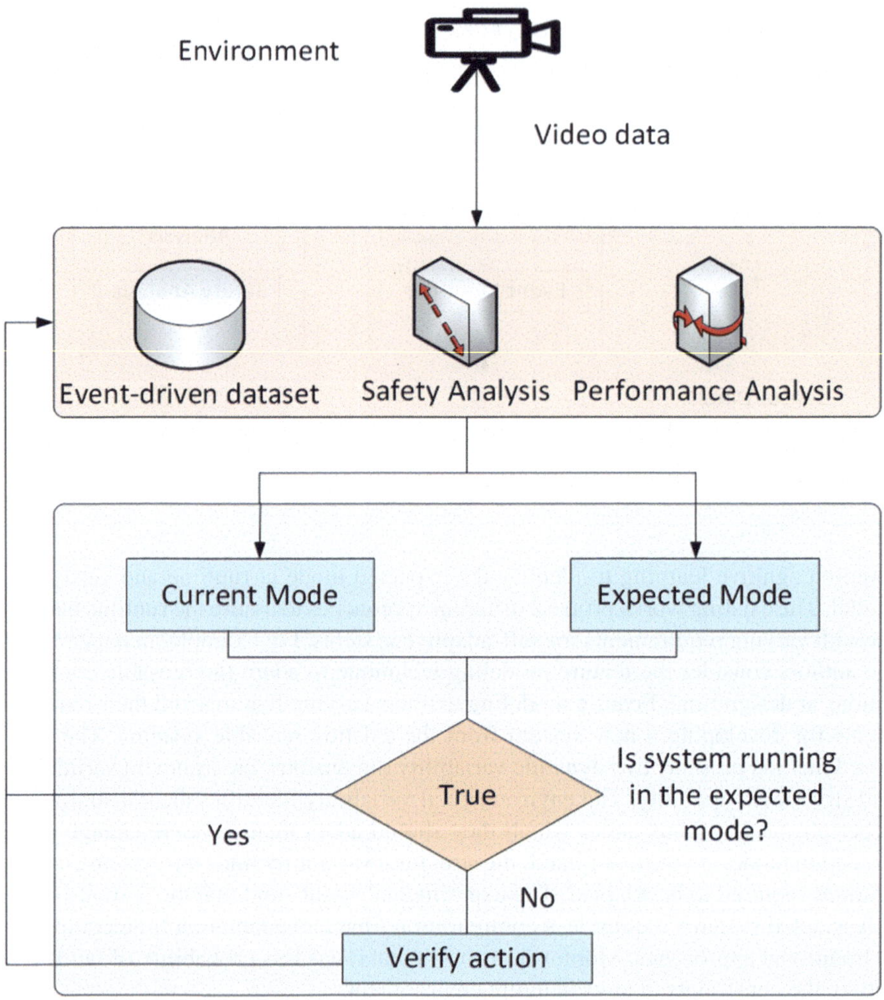

Fig. 5 Runtime behavior of self-adaptive cyber-physical systems

self-adaptive systems is shown in Fig. 5. The system model collects the input data from the environment and trains the model to analyze different requirements (e.g., safety, performance). The training model predicts the runtime behavior and helps the self-adaptive system in updating its configurations to meet requirements.

To some extent, the other existing works predict future vehicle trajectories in real-time to avoid collisions. For example, in autonomous driving, pedestrian or vehicle trajectory prediction can potentially prevent traffic injuries and improve safety [17]. However, future pedestrian trajectory prediction is challenging as it involves real-time interactions, pedestrian behaviors, and dynamic environments. Existing approaches process real-time video images combining convolution neural network (CNN) and long term-short memory (LSTM) to predict the trajectory or future movement [18]. The CNN performs image processing and segmentation for feature extraction and the LSTM predicts the future sequence of pedestrian movements.

3.5 *Context-Based Learning for CPS*

Context-based learning is a part of cognitive learning and it uses real-life experiences to teach different concepts. In this scenario, a concept is a thought process derived from a context where a context is the specific series of events or occurrences of a system. To enable context-awareness learning in autonomous vehicles, machine learning algorithms are used to predict all the events in prior.

Cyber-physical systems like autonomous cars need to extract context from the environment to learn from the partially observable complex environments. There are different context-awareness learning algorithms, but Deep Recurrent Q-Network (DRQN) approaches are great options to use [19]. It is challenging to train a model from its surrounding environments extracting all the information such as the number of surrounding vehicles and their speeds and trajectories at different times. Therefore, autonomous cars solely depend on the sensor data (e.g., obstacles ahead, speeds, and positions) to train their model.

To make the context-based learning simple, this work [19] proposes to combine the Deep Recurrent Q-Network (DRQN) and Long-term Short Memory (LSTM) to develop a reward-punishment mechanism. In this approach, agents with decision-making and context-awareness ability will be rewarded for correct prediction and punished for the wrong prediction. As the algorithms are different forms of reinforcement learning, the system will learn from the environment from the feedback control loop. The experimental results show that the DRQN algorithms perform better than DQN, considering a higher reward rate. Therefore, this approach shows how the system cognition capability can be further improved to train the model to adapt itself in a real-time environment.

Reinforcement learning is concerned with how an intelligent machine performs in any unknown or dynamic environment and learns from feedback. With the feedback loop, the agent improves its learning where a reward policy is implemented for correct decisions and penalized for incorrect decisions. Figure 6 shows the block diagram of the reinforcement learning process where the agent makes a decision based on the current state and changes its state based on the feedback from the environment. It continues three operations (action, change state, and get feedback) to learn and explore the environment. The applications of reinforcement learning are autonomous vehicles, gaming, virtual assistant, search engines, and chatbots, etc. Reinforcement learning is explored under different methods, such as Deep Q-learning Network (DQRN), Q-learning Network (QRN), temporal-difference (TD) learning, and deep reinforcement learning (DRL).

Autonomous vehicles need to interact dynamically, exhibiting cognitive characteristics like reasoning, perception, and analysis. Therefore, reinforcement learning

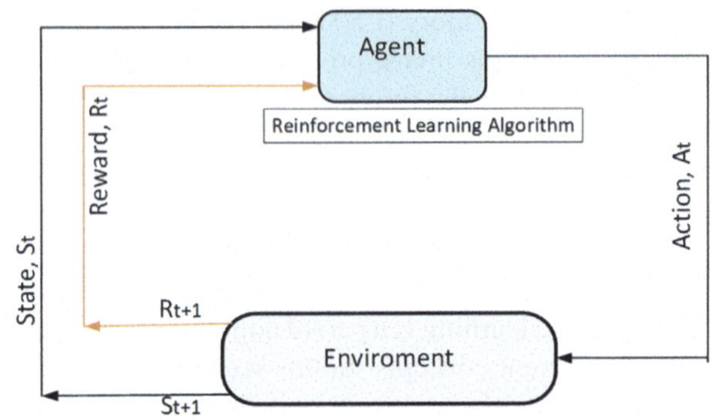

Fig. 6 Reinforcement learning mechanism

can be a good choice for autonomous training agents. However, this algorithm is more prevalent when the training dataset is small and the autonomous agents have the ability to learn through feedback from the environment. During the interaction, the agent considers achieving the goal of executing tasks upon specified context.

However, reinforcement learning does not always show the best cognitive characteristic as the learning agent may get stuck in a local optimum instead of finding global optima. The main reason behind this is that reinforcement learning stops exploring its environmental space after finding an optimal action with considerable results. There are different approaches like Soft Actor-Critic (SAC), Proximal Policy Optimization (PPO), and Asynchronous Advantage Actor-Critic (A3C) for reaching global optima. In A3C, the method has two components: Value and Policy function. The value function estimates the reward point and the policy function finds the action probability. In addition, the Proximal Policy Optimization (PPO) uses a policy function to control the autonomous vehicles' agent's actions. These existing approaches still not optimal as it requires further explorations.

To alleviate the problem, this paper [20] proposes to use the correlated time noise in the agent's policy function so that the learning agents move out for finding global output. To include the noise in the input space, the author uses Ornstein-Uhlenbeck process that generates random disturbance. This improves the agent's exploration method and the performance in learning new things from the environment. The experimental results state that the injection of noise expands the exploration space, and the agent improves its training and testing time. Moreover, the results show the improved performance for receiving awards in the different learning contexts. This shows the cognition power of reinforcement learning.

4 Wireless Communication Technologies in Mobile Computing

The critical component of cognitive mobile computing is wireless communication. It plays a vital role in addressing reliability and safety challenges in cyber-physical systems. In practice, we find different wireless technologies commonly used to transmit data in the CPS network. A few wireless technologies [21] are listed in Fig. 7 based on the data transmission range and the details are described below:

(a) *Wi-Fi*: Wi-Fi is a low-powered wireless technology that mainly helps to connect devices to the internet. It uses radio waves (RF) to connect with other devices and it follows the IEEE 802.11 standard. This standard describes different specifications, limits, values and algorithms to establish a Wi-Fi connection. The term Wi-Fi is often interchangeably used with other wireless connections: Wireless Local Area Network (WLAN) and Wireless Body Area Network (WBAN).
(b) *Bluetooth*: Bluetooth uses radio waves for communication between devices in the frequency range of 2.402–2.480 GHz. It covers a very short distance to wirelessly transfer data between devices (e.g., phones and tablets) without compromising heavily battery power. The advanced form of Bluetooth is Bluetooth Low Energy (BLE), and it requires less power than the standard Bluetooth.
(c) *ZigBee*: To implement Low-Power Wide-Area (LPWA) Networks, ZigBee has become very popular for wireless communication in the IoT environment. It operates on the IEEE 802.15 standard. It offers a better network in terms of lower cost, lower power consumption, and network topology (e.g., mesh) that helps distributed devices connect through multiple pathways. Unlike point-to-point

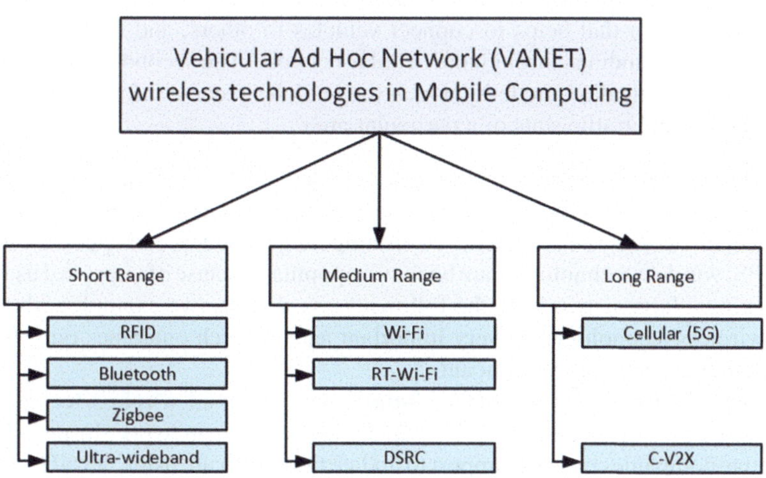

Fig. 7 Wireless technologies used in cognitive mobile computing

communication of Bluetooth, it uses the mesh network to provide connection stability among interconnected devices even if any intermediate device fails to connect in the network.

(d) *Cellular (3G/4G/5G)*: A cellular network is commonly known as a mobile network. It comprises many cells distributed over land areas. Each cell has the ability to transmit and receive data over a large area. When cells are joined together, they can cover a wide geographic area for wireless communication. Each cell migrates the data to the next cell when it requires to transfer the data at a longer distance. Over the years, cellular networks are evolved through 1G, 2G, 3G, 4G, and 5G networks. Nowadays, 4G is fully implemented in the ground. On the other hand, 5G is still immature, but it is already investigated for its effectiveness in providing high-speed data networking on large sites.

(e) *Dedicated short-range communications (DSRC)*: DSRC is a medium-range wireless communication channel designed for automobiles to facilitate an intelligent transportation system. It operates on the 5.9 GHz band of the radio frequency spectrum. It is an open-source protocol, and it provides low latency and high reliability in vehicle-to-vehicle (V2V) or vehicle-to-infrastructure (V2I) communications.

(f) *Real-time WiFi* [22]: Real-time WiFi is a high-speed wireless communication protocol. It uses the time-division multiple access (TDMA) method to provide a deterministic timing guarantee on data transmission. Moreover, it has a high sampling rate of up to 6kHz, low latency variance, and high reliability for wireless Cyber-Physical Control Applications..

(g) *Cellular Vehicle-to-Everything (C-V2X)*: C-V2X is a wide-area and short-range connectivity platform to enable vehicle-to-vehicle (V2V), vehicle-to-pedestrian (V2P), vehicle-to-infrastructure (V2I), and vehicle-to-cloud (V2C) communications. It offers transmission in two different modes. The first mode is direct communication that helps to connect vehicles to others, and it communicates with the surroundings through the cellular network. The second mode is the network communication mode by which the vehicles receive information on road conditions and traffic stats of a particular area.

To visualize the differences among the common wireless technologies, a comparison table is shown in Table 1. It shows the comparison based on different metrics such as standard, data rate, frequency, operating range, latency, and application area.

In CPS, wireless technologies are becoming popular because of improved usability in managing a large network. In the following, we discuss two examples where the use of wireless technologies is very important and research outcomes on solutions to the challenges are quite significant.

Wireless Technologies for OTA Update: The potential wireless technologies in the OTA update process are WiFi, DSRC, and cellular network. In most cases, automotive companies provide support to update their software using 3G, 4G, or Long Term Evolution (LTE) networks. Although the 5G network is already in a place for OTA updates on a large scale, it needs to be more investigated to handle different types of OTA update applications. For example, OTA updates can be categorized as

Table 1 Comparison of wireless technologies in mobile computing [22]

Technology	Standard	Max operating range	Data rate	Latency	Frequency	Potential use cases
Zigbee	IEEE 802.15	100 m	250 Kbps	<140 ms	2.4 GHz	Wireless sensors (monitoring)
Bluetooth	IEEE 802.15	200 m	1–3 Mbps	100 ms	2.4–2.5 GHz	Wireless sensors (monitoring, control)
WiFi	IEEE 802.11	70–250 m	Upto 600 Mbps	<1 ms	2.4–5 GHz	Mobile internet
4G	IEEE 802.16e	10 miles	Upto 1 Gbps	10 ms	<6 GHz	Streaming video
5G	3GPP	1500 feet	Upto 20 Gbps	<1 ms	30–300GHz	IoT, VR
DSRC	IEEE 802.11p	<100 m	27 Mbps	150 ms	<6 GHz	Vehicular networks
Real-time WiFi	IEEE 802.11 b/g/n	70–250 m	150 Mbps	<4.2 ms	2.4 GHz	Wireless control systems
C-V2X	C-V2X Rel. 16	107 m	4–50 Mbps	<50 ms	<6 GHz	Vehicular networks

critical and non-critical updates. A non-critical update (e.g., song playlist update) is always compromised with the system's safety, and it can be updated using cellular networks at any place. On the other hand, a critical update is considered a mandatory update by the vendor, and it needs to be completed instantly to avoid any catastrophic failure of the systems. In vehicular networks, the critical firmware OTA updates aim to fix bugs, improving the system's functionality and security to ensure safety. In such cases, the vehicles can update their firmware from any static position (e.g., parking area) to maintain stable wireless communication. The current 4G and LTE networks are evolved enough to handle static updates in vehicular networks. However, if the vehicles desire to update the critical software on the driving mode in a dynamic environment, the wireless network requires to satisfy a lot of constraints, including safety. The 5G/6G, C-V2X, and DSRC wireless technologies are still working to support critical updates in any mobile computing platform.

Wireless Technologies in Vehicle Platooning: Platooning refers to the formation of a group of vehicles that travel together closely in one direction, maintaining a safe speed [23]. In each platoon, a lead vehicle controls the speed and direction, and all the following vehicles respond to adjust their braking and acceleration. Modern wireless communication technologies such as Bluetooth, WiFi, GPS, cellular networks (e.g., 5G) have the potential to implement vehicle platooning. However, it requires a comprehensive investigation to evaluate the feasibility of these technologies. For example, if the lead vehicle does not respond on time to an obstacle, the following vehicles may get involved in the collision due to driving proximity. Therefore, it requires reliable wireless technology to ensure the safety of the systems and drivers. Recently, DSRC and C-V2X are suggested implementing vehicle platooning to improve safe driving and fuel consumption. For efficient vehicle platooning, context-aware computation is also required because all the vehicles rely on their surroundings and dynamic environment. Volvo has presented platooning using three motorcars driven autonomously behind a lead truck driven at speeds up to 90 kph [24]. In this scenario, if the platoons have more than three vehicles, the communication channel should be implemented with a low latency cellular network (e.g., 5G) or other wireless technologies.

5 Challenges in Cognitive Mobile Computing

(a) *Computational complexity*: Due to the high volume of training data, the computing platform requires parallel execution to provide faster results. Finding available resources and assigning different computational tasks to remote resources make the automation process challenging.
(b) *Risk of inaccurate results*: As the computing resources are mobile, synchronizing different computational tasks is challenging. Moreover, cognitive computing includes different machine learning algorithms to make decisions. Thus, the incorrect choice of algorithms and failure to satisfy system constraints may lead to an unstable system state.
(c) *Dynamic interaction*: The dynamic characteristics of cyber-physical systems make the overall cognitive computing process challenging as the computing process needs to satisfy both the application requirements and the environmental constraints.
(d) *Wireless connectivity*: In cognitive mobile computing, physical and software components are connected through the wireless network. Therefore, the communication medium plays an essential role in different aspects such as faster communication, data security, accessibility, constant connectivity, and user experiences. To ensure an efficient cognitive mobile computing, it introduces different research challenges. For example,

    - Communication delay: Due to dynamic characteristics of cognitive computing, longer network communication delay may cause data to be corrupted and lost in transit. Therefore, it shows the necessity to optimize the communication delay.

- Data security and privacy: The broadcasting character of wireless networks causes the conveyed data to be vulnerable to eavesdropping. Therefore, recent studies focus on different communication protocols that are used to secure the wireless connection and user data.
- Resource and inference management: With the increased number of emerging applications, wireless network requires to support required resources and interference management schemes.
- Network architecture and constant accessibility: An efficient network architecture is required to handle different application requirements. For example, integrated functionalities of a network should have the ability to determine where to compute a task to ensure safety and better quality of services. Therefore, the network architecture design and performance analysis are the key elements of any cognitive mobile computing.

6 Research Challenges in Cognitive Mobile Computing for CPS

Cognitive mobile computing can assist in different decision-making activities by leveraging the benefits of using machine learning. In the following, some research challenges are presented from the different perspectives discussed in [25].

(a) *Dynamic interaction*: One of the major components of cognitive mobile computing is the dynamic engagement or interactions with different system entities. The modern cyber-physical systems produce large structured and non-structured data. After analyzing these data, the system should interact to provide insight to assist humans in making decisions and hypotheses. Online chatbots and object detection for autonomous vehicles are good examples of cognitive mobile computing. An autonomous vehicle interacts with different physical components and environments to help in making decisions.

(b) *Intelligent decision making at runtime*: The next scope after the dynamic interaction is intelligent decision-making at system runtime. Due to dynamic interaction, cyber-physical systems require to make a decision within a short period of time. These decision-making capabilities are learned through model training and experience. For example, reinforcement learning can continuously teach a system based on new information, outcomes, and actions. After that, the system can help in making the right decision to select the best solutions from multiple options. A popular use case of cognitive computing is IBM Watson Virtual Assistant [25], which provides fast, consistent, and accurate answers for different queries. Cognitive computing will also have the ability to analyze patterns, history, and traces to make future decisions.

(c) *Discovering knowledge from data*: Finding knowledge is the potential scope of cognitive mobile computing where the system will find insights and understand vast sensor data. The computing systems will have developed human skills to analyze different situations (e.g., environment) to make decisions. The system model will be built with unsupervised machine learning to develop itself continuously. Moreover, future cognitive computing is considered the collection of distributed intelligent agents that extract raw streaming data and analyze them to create interactive systems that provide real-time monitoring and analysis.

7 Conclusion

Cognitive mobile computing is a requirement for any intelligence cyber-physical system today. The constraints of cyber-physical systems make cognitive mobile computing more challenging for ensuring resource allocation, self-adaptation, and efficient application execution. Therefore, it requires to use new technologies such as fog computing, transfer learning, context-aware computing, and runtime monitoring because components are connected with various sensors through different wireless communication networks. In the future, we see a potential use of cognitive mobile computing in the IoT environment. The extensive sensor data produced by sensing devices can provide important knowledge, but it is often considered a big data challenge. Thus, sensor data analytics with visualization in any mobile computing platform is a growing field of endeavor.

References

1. S. Deb, What is cognitive AI? Is it the future? Edureka. https://www.edureka.co/blog/cognitive-ai/. Last Accessed 16 Nov 2021 (2019)
2. J.S. Hurwitz, M. Kaufman, A. Bowles, *Cognitive Computing and Big Data Analytics* (Wiley, 2015)
3. M. Al Maruf, A. Azim, Extending resources for avoiding overloads of mixed-criticality tasks in cyber-physical systems. IET Cyber-Phys. Syst. Theory Appl. **5**(1), 60–70 (2019)
4. E. Christopoulou, Context as a necessity in mobile applications, in *Mobile computing: concepts, methodologies, tools, and applications* (IGI Global, 2009), pp. 65–83
5. A. Larkin, Disadvantages of cloud computing. Cloud Adoption, Cloud Academy. https://cloudacademy.com/blog/disadvantages-of-cloud-computing/. Last Accessed 16 Nov 2021 (2019)
6. M.A. Maruf, A. Azim, Requirements-preserving design automation for multiprocessor embedded system applications. J. Ambient Intell. Humanized Comput. **12**, 821–833 (2021)
7. K. Fizza, N. Auluck, A. Azim, M.A. Maruf, A. Singh, Faster ota updates in smart vehicles using fog computing, in *Proceedings of the 12th IEEE/ACM International Conference on Utility and Cloud Computing Companion* (2019), pp. 59–64
8. M. Al Maruf, A. Singh, A. Azim, N. Auluck, Resource efficient allocation of fog nodes for faster vehicular ota updates, in *2020 International Symposium on Networks, Computers and Communications (ISNCC)* (IEEE, 2020), pp. 1–6
9. M. Al Maruf, A. Singh, A. Azim, N. Auluck, Faster fog computing based over-the-air vehicular updates: a transfer learning approach. IEEE Trans. Serv. Comput. (2021)
10. S.J. Pan, Q. Yang, A survey on transfer learning. IEEE Trans. Know. Data Eng. **22**(10), 1345–1359 (2009)

11. A. Dosovitskiy, G. Ros, F. Codevilla, A. Lopez, V. Koltun, Carla: an open urban driving simulator, in *Conference on Robot Learning* (PMLR, 2017), pp. 1–16
12. P. Chawdhry, G. Folloni, S. Luzardi, S. Lumachi, European Wifi Hotspot signal strength coverage. European Commission, Joint Research Centre (JRC). https://data.europa.eu/89h/jrc-netbravo-netbravo-od-eu-wifi. Last Accessed 29 Dec 2021 (2016)
13. D. Raca, D. Leahy, C.J. Sreenan, J.J. Quinlan, Beyond throughput, the next generation: a 5g dataset with channel and context metrics, in *Proceedings of the 11th ACM Multimedia Systems Conference* (2020), pp. 303–308
14. N. Islam, A. Azim, A situation-aware adaptation framework for intelligent transportation systems, in *2020 IEEE 23rd International Symposium on Real-Time Distributed Computing (ISORC)* (IEEE, 2020), pp. 106–115
15. N. Islam, A. Azim, Assuring the runtime behavior of self-adaptive cyber-physical systems using feature modeling, in *Proceedings of the 28th Annual International Conference on Computer Science and Software Engineering* (2018), pp. 48–59
16. N. Islam, A. Azim, A multi-mode real-time system verification model using efficient event-driven dataset. J. Ambient Intell. Humanized Comput. 1–14 (2018)
17. R. Quan, L. Zhu, Y. Wu, Y. Yang, Holistic lstm for pedestrian trajectory prediction. IEEE Trans. Image Process. **30**, 3229–3239 (2021)
18. Y. Ma, X. Zhu, S. Zhang, R. Yang, W. Wang, D. Manocha, Trafficpredict: trajectory prediction for heterogeneous traffic-agents, in *Proceedings of the AAAI Conference on Artificial Intelligence*, vol. 33 (2019), pp. 6120–6127
19. J.M. Peixoto, A. Azim, Context-based learning for autonomous vehicles, in *2020 IEEE 23rd International Symposium on Real-Time Distributed Computing (ISORC)* (IEEE, 2020), pp. 150–151
20. M.J. Peixoto, A. Azim, Using time-correlated noise to encourage exploration and improve autonomous agents performance in reinforcement learning. Procedia Comp. Sci. **191**, 85–92 (2021)
21. M.N. Ahangar, Q.Z. Ahmed, F.A. Khan, M. Hafeez, A survey of autonomous vehicles: enabling communication technologies and challenges. Sensors **21**(3), 706 (2021)
22. Y.H. Wei, Q. Leng, S. Han, A.K. Mok, W. Zhang, M. Tomizuka, Rt-wifi: Real-time high-speed communication protocol for wireless cyber-physical control applications, in *2013 IEEE 34th Real-Time Systems Symposium* (IEEE, 2013), pp. 140–149
23. H. Zhou, R. Saigal, F. Dion, L. Yang, Vehicle platoon control in high-latency wireless communications environment: model predictive control method. Transp. Res. Record **2324**(1), 81–90 (2012)
24. S.E. Shladover, X. Yun, L. Yang, H. Ramezani, J. Spring, C.V. Nowakowski, D. Nelson, D. Thompson, A. Kailas et al., Cooperative adaptive cruise control (cacc) for partially automated truck platooning. Tech. rep., California. Dept. of Transportation. Division of Research and Innovation (2018)
25. What is cognitive computing? Features, scope & limitations. Maruti Techlabs. https://marutitech.com/cognitive-computing-features-scope-limitations/. Accessed 1 Jan 2021 (2021)

Akramul Azim is an Associate Professor in Software Engineering at the Ontario Tech University, Canada. His research interests include real-time embedded software, safety-critical software, machine learning, cognitive computing, software quality, and intelligent transportation systems. Dr. Azim is a senior member of IEEE and a licensed professional engineer (PEng) of Ontario, Canada.

Md. Al Maruf is a Ph.D. student at the Electrical, Computer and Software Engineering department of Ontario Tech University, Canada. His main research focuses on real-time embedded systems, task scheduling, cloud/fog computing, and machine learning.

Edge Intelligence for Autonomous Driving Cars

Latif U. Khan, Anselme Ndikumana, Nguyen H. Tran, and Choong Seon Hong

Abstract Self-driving cars have shown an immense interest from both academia and industry due to their wide range of features. These features are infotainment, collision avoidance alerts, driving with minimum possible user intervention, lane changing guidance for minimizing congestion, and accident reporting, among others. To enable these features, there is a need to efficiently deploy secure, secure, fault-tolerant, robust, scalable, and interoperable technologies. Additionally, there is a need for on-demand computing resources at the network edge to assist autonomous cars in performing complex computing tasks. To further improve the performance at the network edge, one can use machine learning. In this chapter, we investigate the key design aspects and technologies required for autonomous driving cars. A case study of using deep learning for enabling infotainment in autonomous cars is also presented. Finally, simulations results are presented for validation of the case study.

1 Introduction

1.1 Background and Motivations

Autonomous driving cars have witnessed a significant increase in interest from the research community as well as industry. The automobile industries have focused

---

L. U. Khan · C. S. Hong (✉)
kyung hee university, seoul, South Korea
e-mail: cshong@khu.ac.kr

L. U. Khan
e-mail: latif@khu.ac.kr

A. Ndikumana
département de génie des systèmes, école de technologie supérieure, montréal, québec h3c 1k3, Canada

N. H. Tran
university of sydney, sydney, Australia
e-mail: nguyen.tran@sydney.edu.au

A. K. Pathan (ed.), *Towards a Wireless Connected World: Achievements and New Technologies*, https://doi.org/10.1007/978-3-031-04321-5_10

on the next stage of autonomous driving, called self-driving, where cars will drive themselves without human driver intervention [1–3]. Self-driving cars are equipped with the smart sensors that generate significant amount of data. Such a data can be used to train various machine learning models [4–6]. These machine learning models will be then used for various applications (e.g., autonomous lane changing, accident reporting, and collision avoidance). In this work, we choose self-driving cars over human-driven cars because self-driving cars already have On-Board Units (OBUs) with Graphics Processing Units (GPUs), Field Programmable Gate Array (FPGA), and Application Specific Integrated Chip (ASIC) to handle in-car AI. This gives the self-driving cars the capability to observe, think, learn, and navigate in real driving environments [1]. Also, according to a study on the incremental time and what activities people will perform if everyone uses self-driving cars, it is estimated that there will be 22 billions of hours for extra media consummation in the US [7]. Therefore, with AI and OBUs that can handle Computation, Communication, Caching, and Control (4C) in self-driving cars, passengers will spend more time on infotainment services such watching media, playing games, and utilizing social networks. Overview of AI in enabling autonomous driving cars is shown in Fig. 1. To support this, self-driving cars should be equipped with recent emerging technologies for infotainment services such as AI-based games, Virtual, Augmented, and Mixed Reality [8]. However, retrieving infotainment contents from Data Centers (DCs) can worsen infotainment content delivery services due to the associated end-to-end delay and consumed backhaul bandwidth resource. As an example, watching a video in a car requires three components, namely a video source, screen, and sound system. Therefore, if the source of the video is not in the car, the car needs to download it from DC. Assuming the DC is distantly located, then the infotainment content delivery services will incur a high delay. Therefore, caching in self-driving cars will play an important role in enhancing infotainment services. Furthermore, for retrieving infotainment contents that need to be cached in self-driving cars, we consider Multi-access Edge Computing (MEC) [9, 10] as a suitable technology to support self-driving cars through caching infotainment contents near self-driving cars. In this work, MEC servers are deployed at RoadSide Units (RSUs).

1.2 Challenges for Infotainment Caching

- In human-driven cars, drivers choose the infotainment contents to display or play. However, in the absence of the driver, the self-driving car should determine itself the infotainment contents to cache and play that are likely to entertain its passengers.
- Some infotainment contents may not be appropriate for consumption by passengers depending on their age and area. Therefore, the self-driving car should determine itself the infotainment contents to cache that do not violate prohibited and restricted content access policies.

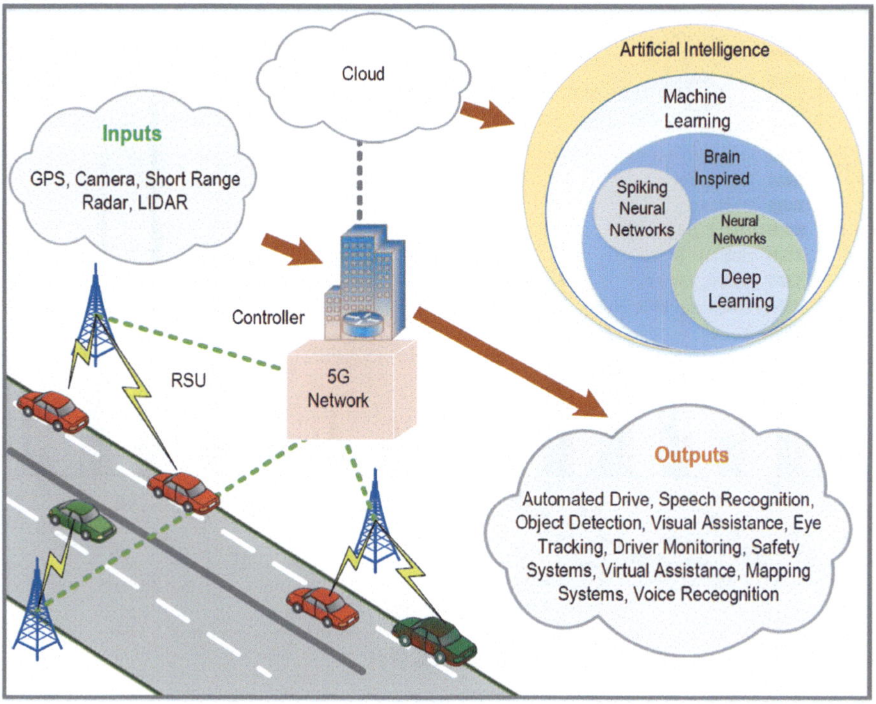

Fig. 1 Role of AI in autonomous driving cars [5]

- As shown in Fig. 2 generated from YouTube demographics dataset for one month available in [11], people have different content preferences, in which their choices depend on their features such as age and gender. Therefore, in the self-driving driving cars, caching decisions for the infotainment contents should depend on passengers' features.
- Self-driving cars will eventually deliver more heterogeneous infotainment contents such as movies, TV, music, and games as well as recent emerging technologies such as Virtual, Augmented, and Mixed Reality [8]. However, obtaining infotainment contents from DC can induce high car-DC delay. Therefore, self-driving cars need to be supported by MEC servers by caching infotainment content in close proximity to self-driving cars at RSUs.
- Self-driving cars are sensitive to delay due to their high mobility and connection in-motion. Therefore, to achieve less variation in transmission delay for downloading contents need to be cached, at the beginning of the journey, the self-driving car should select available MEC servers en-route that will be used to download infotainment contents.

To address the aforementioned challenges, we propose a deep learning based caching for self-driving cars, where caching decisions depend on passengers' features obtained using deep learning approaches and available communication, caching, and

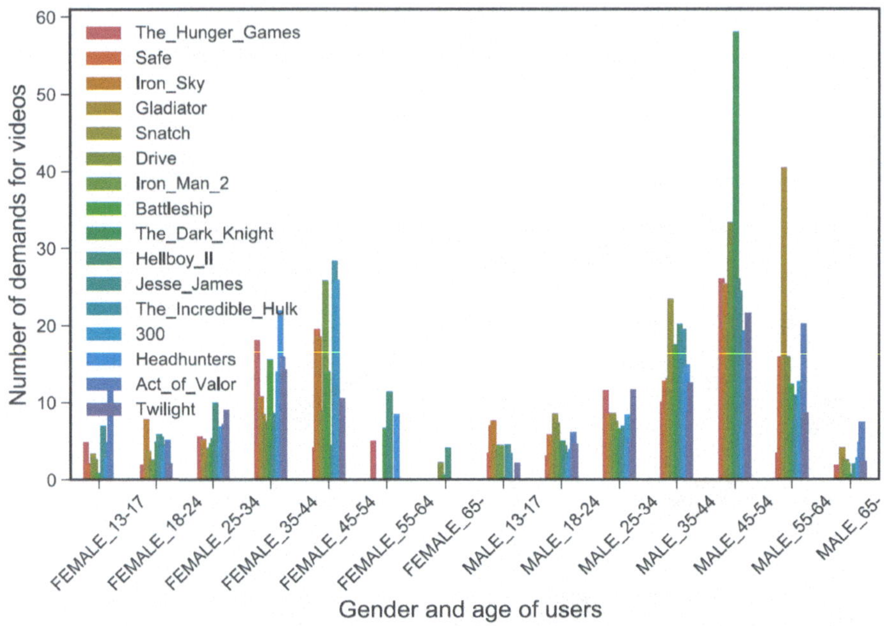

Fig. 2 Content preferences based on users' features [11]

computation (3C) resources. As an extended version of our earlier work published in [12], the main contributions of this chapter are summarized as follows:

- We present deep learning based caching for self-driving cars as a new application of Convolutional Neural Network (CNN), where caching decisions depend on passengers' features obtained using CNN model and facial images of the passengers. Here, we assume the CNN model is trained and tested at DC using dataset. Then, the CNN model is deployed at MEC servers attached to the RSUs in close proximity to the self-driving cars, where the self-driving cars can retrieve model with minimized delay.
- We present a Multi-Layer Perceptron (MLP) framework at DC to predict the probability of infotainment contents to be requested in specific edge areas of MEC servers. Then, the MLP prediction output is deployed at MEC servers. During off-peak hours, each MEC server uses MLP output to identify the infotainment contents that have high predicted probability values of being requested in its area, downloads and caches them. To identify the infotainment contents that are likely to entertain its passengers and need to be cached in the self-driving car, each self-driving car downloads and stores the CNN model and MLP output from the MEC server. The self-driving car uses the CNN model for predicting passengers' features via facial images captured by its camera. Then, the self-driving car compares the CNN output with the MLP output using classification [13, 14] for identifying the contents that meet passengers' features.

- We use a communication model that helps the self-driving car select available RSUs en-route. Then, the self-driving car uses these RSUs for retrieving identified infotainment contents that meet passengers' features and need to be cached.
- We use a computation model for cached infotainment contents, where the cached contents can be served in different formats and qualities depending on demands. Therefore, we consider that MEC servers and self-driving cars have computation resources, which can be used to compute or process cached contents in different formats and qualities.
- We formulate an optimization problem that links the formulated models (deep learning-based caching, communication, and computation models) into one optimization problem whose goal is to minimize the content downloading delay. However, the formulated problem is shown to be non-convex. Therefore, to make it convex, we proposed a convex surrogate problem, which is an upper-bound of the formulated problem. Then, we apply the Block Successive Majorization-Minimization (BS-MM) technique [15] for solving it. We chose BS-MM over other optimization techniques because BS-MM is a new technique that can decompose the original problem into small subproblems, where each subproblem can be solved separately.

2 System Model

The system model of deep learning based caching is depicted in Fig. 3.

Data Center (DC): We assume that DC has higher computation resources than the self-driving car and RSU. Therefore, to minimize computation time, we use DC and dataset to make, train, and test deep learning models (CNN and MLP models) that will be used for predicting passengers features and infotainment contents need to be cached at the RSUs and in self-driving cars. To reduce the communication delay between the self-driving cars and the DC, the trained and tested CNN model and MLP output are deployed at MEC servers attached to the RSUs.

RoadSide Unit (RSU): As defined in 3GPP TS 22.185 V15.0.0 [16], we consider eNB-type RSU as an entity that supports both evolved NodeB (eNB) functionalities and V2X applications. We assume that each RSU $r \in \mathcal{R}$ has access to the DC via a wired backhaul of capacity $\omega_{r,DC}$, where $\mathcal{R}$ is the set of RSUs. Also, each RSU $r \in \mathcal{R}$ has an MEC server. Therefore, unless stated otherwise, we use the terms "RSU" and "MEC server" interchangeably. Furthermore, as defined in 3GPP specifications in [16], we consider an MEC server as locally application server that serves a certain particular geographic area $n \in \mathcal{N}$, where $\mathcal{N} = \{1, 2, \ldots, N\}$ is a set of geographic areas. Furthermore, each MEC server $r \in \mathcal{R}$ has a cache storage of capacity c_r and computational resource of capacity p_r. Furthermore, during off-peak hours, by using backhaul communication resources, each RSU $r \in \mathcal{R}$ downloads CNN model and MLP output. Then, based on the MLP output, each MEC server downloads and cache infotainment contents that have high predicted probabilities of being requested in its

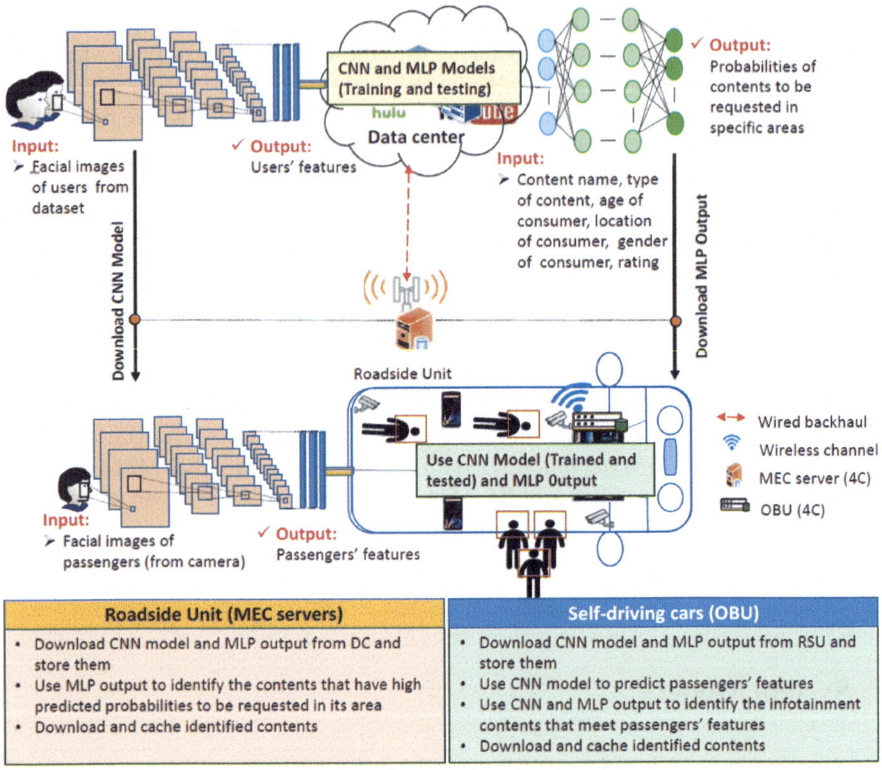

Fig. 3 Illustration of our system model

area. We use $\mathcal{I}$ to denote a set of infotainment contents, where each content $i \in \mathcal{I}$ has a size of $S(i)$ Mb. Also, we consider that people from different areas may need different infotainment contents [17]. Therefore, it is more reasonable to cache infotainment contents at RSUs based on probabilities of being requested in particular areas.

Self-driving car: We consider $\mathcal{V}$ as a set of self-driving cars, where each self-driving car $v \in \mathcal{V}$ has OBU that can handle 4C to support caching and computation of infotainment contents for passengers. Furthermore, each self-driving $v \in \mathcal{V}$ can get broadband Internet service from RSU $r \in \mathcal{R}$ through a wireless link of capacity $\omega_{v,r}$. Each self-driving car $v \in \mathcal{V}$ has a cache storage capacity of c_v and computation capability of p_v. Furthermore, to predict the passengers' features, we use the CNN model. This helps in deciding which infotainment contents to request and cache in the self-driving car that meet passengers' features. During off-peak hours, each self-driving car $v \in \mathcal{V}$ downloads CNN model and MLP output from MEC server. By using the k-means and binary classification, the self-driving car compares its CNN prediction with the predicted output from MLP. This helps the self-driving car identify the infotainment contents that are appropriate to the passengers' features.

Finally, the self-driving car downloads and caches the identified contents that meet passengers' features.

To avoid repetitive delivery of the same contents that require to use backhaul bandwidth, depending on demands, we consider that the computation resources of RSU and the self-driving car can be used to compute cached infotainment contents. As an example, content i' with the H.264 format may not be available in the cache storage. Instead, the cache storage may have content i with the MP4 format of the same content. Therefore, to satisfy the demand, by using the computational resource, cached infotainment content i can be converted to content i' (MP4 to H.264).

3 Deep Learning Based Caching in Self-driving Cars

In this section, to identify the infotainment contents need to be cached, we discuss the deep learning and recommendation model in Sect. 3.1. For retrieving the recommended contents requires communication resources. Therefore, in Sect. 3.2, we discuss the communication model. For caching downloaded contents, we present the caching model in Sect. 3.3. Furthermore, Based on the demands, cached contents can be converted or transcoded to different formats by using computational resources, where the computation model is described in Sect. 3.4 (Table 1).

3.1 Deep Learning and Recommendation Model

In this subsection, we discuss MLP for predicting infotainment contents need to be cached at RSUs nearby the self-driving cars, CNN model for predicting passengers' features, and recommendation model for identifying the contents that meet passengers' features and need to be cache in the self-driving cars.

3.1.1 Multi-layer Perceptron (MLP)

We propose MLP for predicting probabilities of contents to be requested in particular areas of RSUs. We choose MLP over other prediction methods such as AutoRegressive (AR) and AutoRegressive Moving Average (ARMA) models because MLP can cope with both linear and non-linear prediction problems [18]. We use a demographical dataset that will be described in Sect. 5. The input and output are described as follows:

- *Input*: In the dataset, we have infotainment content names, rating, viewer's age, gender, and location as the input of MLP. Furthermore, for predicting the probabilities of contents to be requested in specific areas, we use $\vec{x} = (x_1, x_2, \dots x_M)^T$

Table 1 Summary of key notations

Notation	Definition
$\mathcal{R}$	Set of RSUs, $\|\mathcal{R}\| = R$
$\mathcal{V}$	Set of self-driving cars, $\|\mathcal{V}\| = V$
$\mathcal{I}$	Set of contents, $\|\mathcal{I}\| = I$
$\mathcal{I}_r(n)$	Set of contents that need to be cached in area n of RSU r, $\|\mathcal{I}_r(n)\| = I_r(n)$
$\mathcal{U}$	Set of consumers of contents, $\|\mathcal{U}\| = U$
$\vec{x}$	Input of MLP
$\vec{\tilde{y}}$	Output of MLP
$\vec{y}$	Ground truth for MLP
M	The number of input features
N	The number of geographic areas
c_r	Caching capacity of each RSU $r \in \mathcal{R}$
p_r	Computation capability of RSU $r \in \mathcal{R}$
c_υ	Caching capacity of each car $\upsilon \in \mathcal{V}$
p_υ	Computation capability of car $\upsilon \in \mathcal{V}$
$\tau_u^{\text{Tot}}(\vec{q}, \vec{h}, \vec{\varrho})$	Total delay experienced by each passenger $u \in \mathcal{U}_\upsilon$
ψ_u^υ	Data rate for each passenger u via WiFi of self-driving car υ

to denote the input vector, where the subscripts are used to denote the features such as content names, rating, viewer's age, gender, and location.

- *Output*: From the input, MLP tries to predict $\vec{\tilde{y}} = (\tilde{y}_1, \tilde{y}_2, \ldots \tilde{y}_N)^T$ as the output vector and the subscripts are used to denote the geographic areas. Also, in the output layer, each area $n \in \mathcal{N}$ corresponds to one neuron, where the output layer predicts the probabilities of contents to be cached in each specific area $n \in \mathcal{N}$.

For MLP, we use l to denote the number of hidden layers, $\vec{x}$ for the input vector, $\vec{b}^{(1)}, \ldots, \vec{b}^{(l)}$ for the bias vectors, $\vec{W}^{(1)}, \ldots, \vec{W}^{(l)}$ for the weight matrices at each hidden layer, and $\vec{\tilde{y}}$ for the output vector. $\vec{\tilde{y}}$ can be expressed as follows:

$$\vec{\tilde{y}} = f(\vec{W}^{(l)} \ldots f(\vec{W}^{(2)} f(\vec{W}^{(1)}\vec{x} + \vec{b}^{(1)}) + \vec{b}^{(2)}) \cdots + \vec{b}^{(l)}). \tag{1}$$

where $f(.)$ is the activation function.

In our MLP, we use the Rectified Linear Unit (ReLU) as the activation function in all the layers except at the output layer. We chose ReLU over other activation functions, because it mitigates the vanishing gradient problem experienced by MLP during the training process [19]. Furthermore, in the output layer l, we use the softmax function as an activation function. The purpose of the softmax function is to squeeze the output vector $\vec{\tilde{y}}$ into a set of probability values, where softmax function is defined as:

$$softmax(\tilde{\vec{y}})^{(l)} = \frac{e^{\tilde{y}_l}}{\sum_{n=1}^{N} e^{\tilde{y}_n}}, \text{ for } l = 1, \ldots, N. \tag{2}$$

The output layer has N neurons that correspond to N areas of RSUs. Furthermore, for the error function, we chose the cross-entropy error function over other error functions since our MLP classifies the contents needs to be cached in N geographic areas of RSUs. This problem can be considered as a classification problem, where we interpret the output as probabilities of the contents to be requested in each specific area $n \in \mathcal{N}$. The cross-entropy error function $A(\vec{y}, \tilde{\vec{y}})$ can be expressed as follows:

$$A(\vec{y}, \tilde{\vec{y}}) = -\sum_{n=1}^{N} y_n \log \tilde{y}_n. \tag{3}$$

$A(\vec{y}, \tilde{\vec{y}})$ calculates the cross-entropy between the estimated class probabilities $\tilde{\vec{y}}$ and the ground truth $\vec{y}$.

Finally, to reduce the communication delay between the self-driving car and DC, as the DC may be located far from the self-driving cars, the output of the MLP are downloaded and stored to the RSUs based on their areas.

3.1.2 Convolutional Neural Network (CNN)

In our proposal, we do not focus on proposing new CNN model. Conversely, we focus on a new application of existing CNN model for automatic age, emotion, and gender prediction from facial images [20] in caching decision. We describe the CNN workflow for automatic age, emotion, and gender extraction as follows:

- *Input*: We consider $\vec{k}_0$ as the input image with three-dimensional space: height, width, and the number of color channels (red, green, and blue).
- *Convolution layer*: The convolution layer applies filters to the input regions and computes the output of each neuron. Each neuron is connected to local regions of the input, and using dot products between the weight and local regions, the convolution layer produces a feature map $\vec{k}_j$. We use $\vec{k}_j$ to denote the feature map produced after convolution layer j.
- *RELU layer*: In this layer, we apply the ReLU $(\max(0, \vec{k}_j))$ as an elementwise activation function. The ReLU keeps the size of its associated convolution layer j unchanged.
- *Max pooling layer*: After the convolution and RELU layers, we have a high-dimensional matrix. Therefore, for dimension reduction, we apply a max-pooling layer as a downsampling operation.
- *Fully-connected layer*: This layer is fully connected to all previous neurons and is used to compute the class scores that a face could potentially belong to. Here, we have two classes for gender (male and female), 101 classes for age (from 0 to 101), and 8 classes for emotion (anger, anticipation, disgust, fear, joy, sad, surprise, and trust). In other words, we use three fully-connected layers for age, gender, and emotion classification.

- *Softmax layer and output*: In this layer, for each facial image, we need to interpret the output as the probability values of classes for gender, emotion, and age that a facial image could potentially belong to. To achieve this, the softmax activation function is applied to the output of the fully-connected layers.

To reduce the communication delay between the self-driving cars and DC, the trained and tested CNN model is deployed to the RSUs. Then, each self-driving car $v \in \mathcal{V}$ downloads CNN model and uses it for predicting age, gender, and emotion of passengers from facial images. Once the facial image of a passenger is captured via a camera. The self-driving car can extract features such as eyes, nose, mouth, and chin and use them for classifying the passengers' faces into different age, emotion, and gender classes. As describe in below recommendation model, this helps the self-driving car identify the infotainment contents that meet passengers' features as recommended contents to cache. Here, we assume that the passengers are aware of the presence of the camera. In other words, the self-driving cars have warning signs that inform passengers on the presence of the cameras. The same techniques were used in the deployment of public video surveillance at streets or public places [21].

3.1.3 Recommendation Model

The workflow of the recommendation model for self-driving cars is illustrated in Fig. 4 and described as follows:

- *Step 1*: Each self-driving car $v \in \mathcal{V}$ downloads the MLP output and CNN model from MEC server attached to RSU.
- *Step 2*: By using the k-means algorithm for age and emotion-based grouping and binary classification for gender-based grouping on the MLP output, each self-driving car $v \in \mathcal{V}$ creates age, gender, and emotion-based clusters of content consumers and generates an initial recommendation for the contents that need to be cached and have high predicted probability values for being requested.
- *Step 3*: For each new passenger $u \in \mathcal{U}$, the self-driving car uses the CNN model for predicting its age, gender, and emotion from facial image.
- *Step 4*: The self-driving car uses these passenger's features to calculate the similarity of passenger $u \in \mathcal{U}$ with the existing users (i.e., content consumers) in age, gender, and emotion-based clusters. Then, based on the similarity calculation, each passenger $u \in \mathcal{U}$ will be assigned to a cluster.
- *Step 5*: After clustering the passenger(s), self-driving car $v \in \mathcal{V}$ selects top contents that have high predicted probability values for being requested as recommended contents to cache.
- *Step 6*: Finally, self-driving car $v \in \mathcal{V}$ downloads the recommended contents via RSU and caches them in its cache storage c_v.

For the k-means algorithm, first, we use age as numerical data. We denote $\vec{\tilde{y}}_n$ as the MLP output at each area $n \in \mathcal{N}$ and $\mathcal{X} = \vec{\tilde{y}}_n$ as the input of the k-means algorithm. The k-means partitions the consumer of the contents $\mathcal{X} = \{x_1, \ldots, x_U\}$

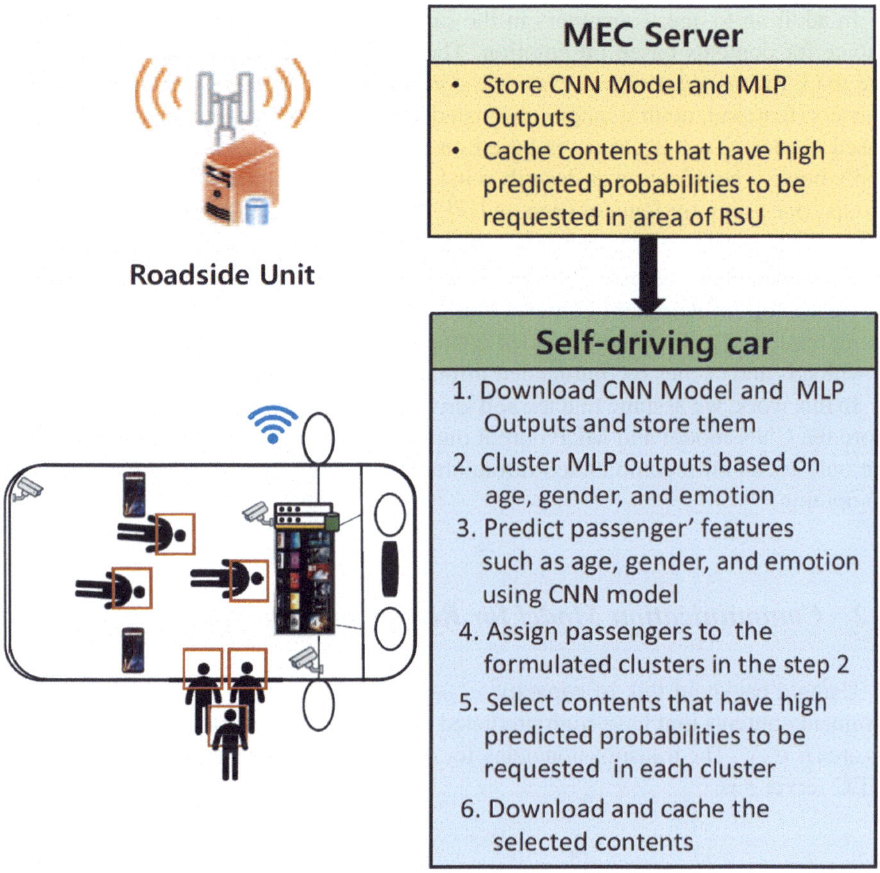

Fig. 4 Recommendation model for self-driving car

into K age-based clusters $\mathcal{X}_1, \ldots, \mathcal{X}_K$ such that $\mathcal{X}_1 \cup \mathcal{X}_2 \cup \cdots \cup \mathcal{X}_K = \mathcal{X}$. In k-means, consumers are grouped into clusters based on their age. In addition, the clusters are disjoint $\mathcal{X}_i \cap \mathcal{X}_j = \emptyset,\ i \neq j$. The goal of k-means is to assign users to age-based clusters such that the objective function below is minimized:

$$\min_{\{\mathcal{X}_j\}_{j=1}^{K}} \sum_{j=1}^{K} \sum_{x_u \in \mathcal{X}_j} \|x_u - \tilde{x}_j\|^2, \tag{4}$$

where $\tilde{x}_j$ is the centroid of cluster $\mathcal{X}_j$, which is defined as

$$\tilde{x}_j = \frac{\sum_{x_u \in \mathcal{X}_j} x_u}{|\mathcal{X}_j|}. \tag{5}$$

In addition to age, consumers in the same age-based cluster can have different choice for contents based on emotion. Therefore, in each age-based cluster j, we use the k-means algorithm to class the consumers of contents in E emotion-based clusters (fear, sad, neutral, angry, disgusted, surprised). Therefore, in each emotion-based cluster e, we group users based on gender. For gender-based grouping, we apply binary classification as described in [14], which results in the formation of two groups, one group for females (denoted $\mathcal{G}_{je}^{\text{female}}$) and another group for males (denoted $\mathcal{G}_{je}^{\text{male}}$) such that $\mathcal{G}_{je}^{\text{female}} \cap \mathcal{G}_{je}^{\text{male}} = \emptyset$. Then, inside $\mathcal{G}_{je}^{\text{female}}$ and $\mathcal{G}_{je}^{\text{male}}$ clusters, which are sub-clusters of age-based cluster j and emotion-based cluster e, the self-driving car select top infotainment contents that have high predicted probability values of being requested as the recommended contents to cache. Finally, the self-driving car downloads and caches recommended infotainment contents.

In this work, we assume that the self-driving cars and MEC servers download and store the CNN model and MLP output during off-peak hours. Therefore, hereafter, we only focus on recommended infotainment contents downloading, caching, and computing.

3.2 Communication Model for Retrieving Contents

Using a backhaul link of capacity $\omega_{r,DC}$, each MEC server downloads the infotainment contents that have high predicted probability values for being requested in its area $n \in \mathcal{N}$. The transmission delay for downloading contents from the DC to the MEC server r is:

$$\tau_r^{\text{DC}} = \frac{q^{\text{DC}\rightarrow r} \sum_{i \in \mathcal{I}_r(n)} S(i)}{\omega_{r,DC}}, \tag{6}$$

where $\mathcal{I}_r(n)$ for $n \in \mathcal{N}$ denotes the set of predicted contents that have high probability values for being requested in area n of RSU, and $q^{\text{DC}\rightarrow r}$ is a decision variable that indicates whether or not MEC server r is connected to the the DC, such that:

$$q^{\text{DC}\rightarrow r} = \begin{cases} 1, & \text{if MEC server } r \text{ is connected to the DC,} \\ 0, & \text{otherwise.} \end{cases} \tag{7}$$

As illustrated in Fig. 5, to have less variation in the transmission delay and hand-off before the self-driving car starts its journey, it can select RSUs that will be used to download the top-recommended contents. To discover RSUs located in a route of the self-driving car, Access Network Discovery and Selection Function (ANDSF) implemented in the cellular network and described in 3GPP TS 24.312 V15.0.0 [22] can be utilized. We assume each self-driving car $v \in \mathcal{V}$ moves in an area covered by macro Base Stations (BSs) and RSUs. Therefore, to obtain RSU information such as coordinate and coverage, the self-driving car sends a request to the ANDSF server

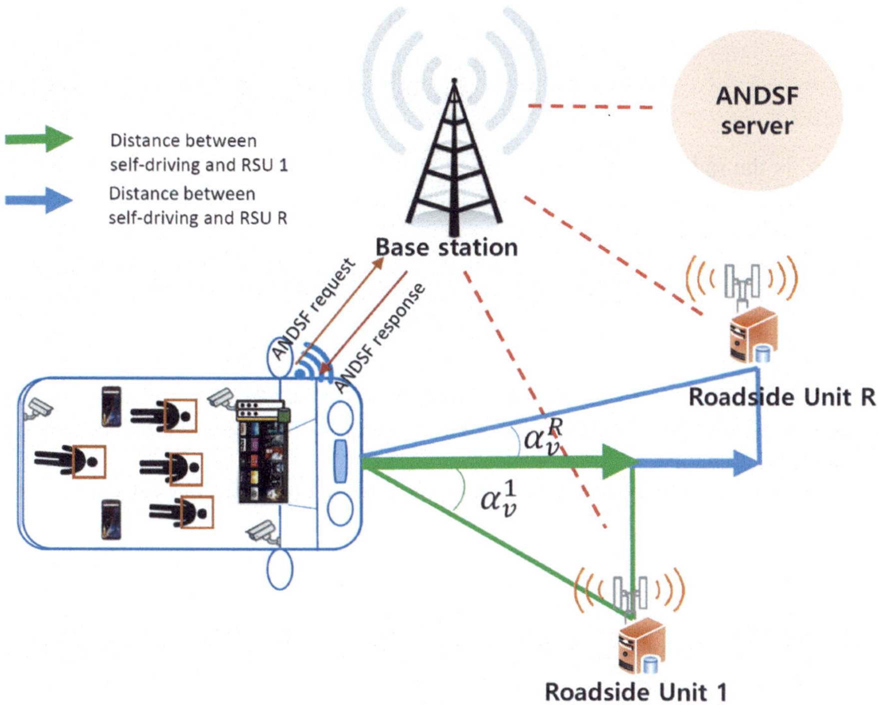

Fig. 5 Communication planning for self-driving car

via a BS [23]. The request includes a geographic location of the self-driving car, speed, and direction. On the other hand, in the ANDSF server's feedback includes the coordinates and coverage of all RSUs available in the direction of the self-driving car.

Each self-driving car v computes the following distance $\tilde{d}_v^r$ between each RSU r and its route:

$$\tilde{d}_v^r = g_v^r sin\alpha_v^r, \tag{8}$$

where α_v^r is the angle between the trajectory of movement of self-driving car v and the straight line from RSU $r \in \mathcal{R}$, and g_v^r is the geographical distance between self-driving car v and cache-enabled RSU r. In addition, each self-driving car v computes the following distance d_r^v remaining to reach each area covered by cache-enabled RSU $r \in \mathcal{R}$:

$$d_r^v = g_v^r cos\alpha_v^r. \tag{9}$$

We defined ρ_v^r as the probability that RSU $r \in \mathcal{R}$ will be selected as a source of infotainment contents to be cached in self-driving car v as follows:

$$\rho_v^r = \begin{cases} 1, & \text{if } \tilde{d}_v^r = 0, \\ \frac{\tilde{d}_v^r}{\gamma_r} & \text{if } 0 < \tilde{d}_v^r < \gamma_r, \\ 0, & \text{otherwise}, \end{cases} \quad (10)$$

where γ_r is the radius of the area covered by RSU $r \in \mathcal{R}$. Therefore, we define q_v^r as a decision variable that indicates whether or not the self-driving car is connected to RSU $r \in \mathcal{R}$ as follows:

$$q_v^r = \begin{cases} 1, & \text{if } \rho_v^r > 0 \text{ and } d_r^v = 0, \\ 0, & \text{otherwise}. \end{cases} \quad (11)$$

Equations (10) and (11) ensure that once the self-driving car v reaches an area covered by cache-enabled RSU $r \in \mathcal{R}$, it immediately starts downloading the recommended infotainment contents.

We assume each RSU r has a wireless channel of capacity $\omega_{v,r}$, where $\omega_{v,r}$ can be expressed as follows:

$$\omega_{v,r} = q_v^r B_r \log_2 \left(1 + \varphi_r |G_v^r|^2\right), \; \forall v \in \mathcal{V}, \; r \in \mathcal{R}, \quad (12)$$

where B_r is the bandwidth for the car to RSU communications, G_v^r is the channel gain between RSU r and self-driving car v, and φ_r is the transmission power of RSU r. Therefore, based on the channel capacity, the transmission delay for downloading contents that meet passengers' features from the MEC server to self-driving car v is expressed as:

$$\tau_v^r = \frac{\sum_{\tilde{i}_f, \tilde{i}_m \in \mathcal{I}_r(n)} q_v^r \left(S(\tilde{i}_f)) + S(\tilde{i}_m)\right)}{\omega_{v,r}}, \quad (13)$$

where $\tilde{i}_f \in \mathcal{G}_{je}^{\text{female}}$ is the recommended infotainment content for female passengers and $\tilde{i}_m \in \mathcal{G}_{je}^{\text{male}}$ is the recommended infotainment content for male passengers in each age and emotion-based cluster in area n, where $\tilde{i}_f, \tilde{i}_m \in \mathcal{I}_r(n)$.

Based on self-driving car's speed, we consider t_v^r as the time required by self-driving car $v \in \mathcal{V}$ to leave an area covered by RSU r. We can calculate t_v^r as follows:

$$t_v^r = \frac{2 q_v^r \gamma_r}{\mu_v}, \quad (14)$$

where μ_v is the speed of self-driving car v. When $\tau_v^r < t_v^r$, the self-driving can easily download the recommended infotainment content in the area coverage by RSU r. However, when $\tau_v^r \geq t_v^r$, the self-driving car can select the next RSU to use for downloading recommended infotainment contents.

Each self-driving car v has a WiFi Router on board that can be used to provide WiFi connectivity to its passengers. However, in the self-driving car, passengers are free to choose their appropriate connections. Here, we aim to minimize delay experienced by the passengers that are inside of the self-driving car and use WiFi connectivity of the self-driving car for getting infotainment contents. Therefore, the instantaneous data rate for each passenger u via the WiFi of self-driving car v is given by:

$$\psi_u^v = \frac{q_u^v \varphi_v \tilde{\psi}_u^v \xi_u^v(|\mathcal{U}_v|)}{|\mathcal{U}_v|}, \forall u \in \mathcal{U}_v,\ v \in \mathcal{V}_v, \tag{15}$$

where φ_v is the WiFi throughput efficiency factor and $|\mathcal{U}_v|$ is the number of passengers that are connected simultaneously to the WiFi of self-driving car v, where $\mathcal{U}_v \subset \mathcal{U}$. We use φ_v to denote the overhead related to the MAC protocol layering. Furthermore, $\tilde{\psi}_u^v$ is the maximum theoretical data rate that the WiFi can handle. Furthermore, $\xi_u^v(|\mathcal{U}_v|)$ is a channel utilization function, which is a function of the number of passengers connected simultaneously to the WiFi [24]. $\xi_u^v(|\mathcal{U}_v|)$ is used to determine the impact of contention over the WiFi throughput. Also, we use q_u^v as a decision variable that indicates whether or not passenger u is connected to the WiFi of self-driving v, specifically:

$$q_u^v = \begin{cases} 1, & \text{if the passenger } u \text{ is connected to the} \\ & \text{WiFi of the self-driving car } v, \\ 0, & \text{otherwise.} \end{cases} \tag{16}$$

For each passenger $u \in \mathcal{U}_v$, based on its instantaneous data rate ψ_u^v, the transmission delay τ_u^v for downloading content i from self-driving car v is given by:

$$\tau_u^v = \frac{\sum_{i \in \mathcal{I}_r(n)} q_u^v \left(S(\tilde{i}_f)) + S(\tilde{i}_m)\right)}{\psi_u^v}. \tag{17}$$

3.3 Caching Model for Retrieved Contents

We assume that the cache storage c_v of each self-driving car v is limited. Therefore, the sizes of the recommended infotainment contents that need to be downloaded from the MEC server and cached in the self-driving car must satisfy the cache resource constraint, which is expressed as follows:

$$q_v^r \sum_{j=1}^{K} \left(\sum_{\tilde{i}_f \in \mathcal{G}_{je}^{\text{female}}} o_v^{\tilde{i}_f} S(\tilde{i}_f)) + \sum_{\tilde{i}_m \in \mathcal{G}_{je}^{\text{male}}} o_v^{\tilde{i}_m} S(\tilde{i}_m) \right) \le c_v, \tag{18}$$

where $o_v^{\tilde{i}_f} \in \{0, 1\}$ is the decision variable that indicates whether or not self-driving car v has to cache infotainment content $\tilde{i}_f \in \mathcal{G}_{je}^{\text{female}}$, where $o_v^{\tilde{i}_f}$ is given by:

$$o_v^{\tilde{i}_f} = \begin{cases} 1, & \text{if self-driving car } v \text{ caches the content } \tilde{i}_f, \\ 0, & \text{otherwise.} \end{cases} \quad (19)$$

On the other hand, we let $o_v^{\tilde{i}_m} \in \{0, 1\}$ be the decision variable that indicates whether or not self-driving car v has to cache infotainment content $\tilde{i}_m \in \mathcal{G}_{je}^{\text{male}}$, where $o_v^{\tilde{i}_m}$ is given by:

$$o_v^{\tilde{i}_m} = \begin{cases} 1, & \text{if self-driving car } v \text{ caches the content } \tilde{i}_m, \\ 0, & \text{otherwise.} \end{cases} \quad (20)$$

Furthermore, for analyzing cache storage utilization, which is based on cache hit and cache miss, we assume that $\tilde{i}_f$ and $\tilde{i}_m$ are cached in the same cache storage c_v. Therefore, we omit the subscript and superscript on content, and use i to denote either content $\tilde{i}_f$ or $\tilde{i}_m$.

We use $h_i^{u \to v} \in \{0, 1\}$ to denote the cache hit indicator at self-driving car v for content $i \in \mathcal{I}_r(n)$ requested by customer $u \in \mathcal{U}$:

$$h_i^{u \to v} = \begin{cases} 1, & \text{if content } i \text{ requested by consumer } u \\ & \text{is returned from self-driving car } v, \\ 0, & \text{otherwise.} \end{cases} \quad (21)$$

In the case of a cache miss ($h_i^{u \to v} = 0$), the self-driving car needs to forward the demand for content i to its associated MEC server. Based on the MLP output at the RSU, we assume that the MEC server caches the contents that have high probabilities of being requested in area n, where cache allocation has to satisfy the following constraint:

$$q^{\text{DC} \to r} \sum_{i \in \mathcal{I}_r(n)} o_r^i S(i) \leq c_r, \quad (22)$$

where o_r^i is a decision variable that indicates whether or not MEC server r has to cache content $i \in \mathcal{I}_r(n)$, defined as follows:

$$o_r^i = \begin{cases} 1, & \text{if MEC server } r \text{ caches content } i \in \mathcal{I}_r(n), \\ 0, & \text{otherwise.} \end{cases} \quad (23)$$

Furthermore, we use $h_i^{r \to v} \in \{0, 1\}$ to denote the cache hit indicator at the MEC server for content $i \in \mathcal{I}_r(n)$ requested by self-driving $v \in \mathcal{V}$:

$$h_i^{r \to v} = \begin{cases} 1, & \text{if the content } i \text{ requested by self-dring} \\ & \text{car } v \text{ is cached at MEC server } r, \\ 0, & \text{otherwise.} \end{cases} \quad (24)$$

However, when the MEC server does not have content i in its cache storage, the MEC server forwards the demand for content i to the DC via a wired backhaul link.

3.4 Computation Model for Cached Contents

In self-driving cars, a passenger may request a content format (e.g., H.264) that is not available in the cache storage c_v. Instead, the cache storage may have other content formats (e.g., MP4) for the same content that can be transcoded to the desired format (H.264).

Therefore, to adopt this process of serving cached content after computation, we define the following decision variable:

$$h_{i'}^{v \to u} = \begin{cases} 1, & \text{if content } i' \text{ requested by consumer } u \\ & \text{is returned by car } v \text{ after computation,} \\ 0, & \text{otherwise.} \end{cases} \tag{25}$$

To ensure that self-driving car v returns only one format of the requested content, the following constraint should be satisfied:

$$h_i^{u \to v} + h_{i'}^{v \to u} \leq 1. \tag{26}$$

We assume that converting content i to content i' requires computation resource $p_v^{i \to i'}$ of self-driving car v, where the computational resource allocation $p_v^{i \to i'}$ is given by:

$$p_v^{i \to i'} = p_v \frac{h_i^{u \to v} \varrho_v^{i \to i'} z^{i \to i'}}{\sum_{u \in \mathcal{U}} \sum_{i \in \mathcal{I}} h_i^{u \to v} \varrho_v^{i \to i'} z^{i \to i'}}, \quad \forall v \in \mathcal{V}, \tag{27}$$

where $z^{i \to i'}$ is the computation workload or intensity in terms of CPU cycles per bit required for converting cached content i to i', while $\varrho_v^{i \to i'}$ is the computation decision variable, which is expressed as:

$$\varrho_v^{i \to i'} = \begin{cases} 1, & \text{if the cached content } i \text{ is converted to the} \\ & \text{desired format } i' \text{ in self-driving car } v. \\ 0, & \text{otherwise.} \end{cases} \tag{28}$$

In (27), for computational resources allocation, we use weighted proportional allocation [25] because it is simple to implement in practical communication systems such Vehicular Ad-hoc Networks (VANETs) and 4G & 5G cellular networks [10]. Furthermore, computation resource allocation should satisfy the following constraint:

$$\sum_{u=1}^{U}\sum_{i=1}^{I_r(n)} q_u^v h_i^{u\to v} \varrho_v^{i\to i'} p_v^{i\to i'} \leq p_v. \tag{29}$$

In addition, converting content i to content i' requires executing time. Therefore, in self-driving car v, the execution time $\tau_v^{i\to i'}$ is given by:

$$\tau_v^{i\to i'} = \frac{q_u^v h_i^{u\to v} \varrho_v^{i\to i'} z^{i\to i'} S(i)}{p_v^{i\to i'}}. \tag{30}$$

When constraint (29) cannot be satisfied due to insufficient computational resource for converting content i into the requested content i', the self-driving car forwards the demand for content i' to the MEC server.

At the MEC server, to convert cached content i into content i', it requires an execution time of $\tau_r^{i\to i'}$. Thus, the execution time at the MEC server is given by:

$$\tau_r^{i\to i'} = (1 - \varrho_v^{i\to i'}) \left(\frac{q_v^r h_i^{r\to v} \varrho_r^{i\to i'} z^{i\to i'} S(i)}{p_r^{i\to i'}} \right), \tag{31}$$

where $p_r^{i\to i'}$ is the required computation resource of MEC server r for converting content i to content i'. $p_r^{i\to i'}$ can be calculated in the same manner used in (27). We define a $\varrho_r^{i\to i'}$ computation decision variable, where $\varrho_r^{i\to i'}$ is expressed as follows:

$$\varrho_r^{i\to i'} = \begin{cases} 1, & \text{if the cached content } i \text{ is converted to} \\ & \text{desired format } i' \text{ at MEC server,} \\ 0, & \text{otherwise,} \end{cases} \tag{32}$$

We assume that the computation resources at the MEC server are limited, where computation allocation has to satisfy the following constraint:

$$\sum_{v=1}^{V}\sum_{i=1}^{I_r(n)} q_v^r h_i^{r\to v} \varrho_r^{i\to i'} p_r^{i\to i'} \leq P_r. \tag{33}$$

In addition, we define $h_{i'}^{r\to v}$ as a decision variable that indicates whether or not the MEC server returns the requested content i' to self-driving car v after computation, where $h_{i'}^{r\to v}$ is given by:

$$h_{i'}^{r\to v} = \begin{cases} 1, & \text{if content } i' \text{ requested by car } v \text{ is returned} \\ & \text{by MEC server } r \text{ after computation,} \\ 0, & \text{otherwise.} \end{cases} \tag{34}$$

To ensure that converting cached content i to the requested content i' is performed exactly at one location, either at the self-driving car or at MEC server, and self-driving car or MEC server sends exactly one format of content, we formulate the following constraints:

$$q_u^v(h_i^{u\to v} + h_{i'}^{v\to u}) + q_v^r \eta_v (h_i^{r\to v} + h_{i'}^{r\to v}) \leq 1, \tag{35}$$

$$\varrho_v^{i\to i'} + q_v^r(1 - \varrho_v^{i\to i'}) \leq 1. \tag{36}$$

Here, we use $\eta_v = 1 - (h_i^{u\to v} + h_{i'}^{v\to u})$. However, if the above constraints cannot be satisfied due to limited computation and caching resources, MEC server submits the request for content i' to the DC.

4 Problem Formulation and Solution

In this section, we present our optimization problem for minimizing delay in downloading the infotainment contents in Sect. 4.1. Then, in Sect. 4.2, we present a solution of the formulated optimization problem.

4.1 Problem Formulation

In the self-driving car, to coordinate deep Learning and recommendation, communication, caching, and computation models, we formulate an optimization problem that links the formulated models into one problem whose goal is to minimize total delay $\tau_u^{\text{Tot}}(\mathbf{q}, \mathbf{h}, \varrho)$ for retrieving infotainment contents, where $\tau_u^{\text{Tot}}(\mathbf{q}, \mathbf{h}, \varrho)$ is given by:

$$\begin{aligned}\tau_u^{\text{Tot}}(\mathbf{q}, \mathbf{h}, \varrho) = \tau_u^v h_i^{u\to v} + h_{i'}^{v\to u} \varrho_v^{i\to i'} \tau_v^{i\to i'} + \\ \left(1 - (h_i^{u\to v} + \varrho_v^{i\to i'} h_{i'}^{v\to u})\right) (\tau_v^r h_i^{r\to v} + \tau_r^{i\to i'} \varrho_r^{i\to i'} h_{i'}^{r\to v}) + \\ (1 - (h_i^{r\to v} + \varrho_r^{i\to i'} h_{i'}^{r\to v}))\tau_r^{\text{DC}}.\end{aligned} \tag{37}$$

In the above Eq. (37), a requested infotainment content can be retrieved in the self-driving car. However, if the requested content can not be retrieved in a self-driving car, the self-driving car sends a request to RSU, where RSU can return the requested content. In the worst case, if the requested content can not be retrieved from self-driving car or RSU, DC can be used. Therefore, our optimization problem can be expressed as follows:

$$\underset{\vec{q},\vec{h},\vec{\varrho}}{\text{minimize}} \sum_{u=1}^{U} \tau_u^{\text{Tot}}(\vec{q},\vec{h},\vec{\varrho}) \tag{38}$$

subject to:

$$\sum_{v=1}^{V} q_v^r \leq 1, \; \forall r \in \mathcal{R}, \tag{38a}$$

$$q_v^r \sum_{j=1}^{k} \Big(\sum_{\tilde{i}_f \in \mathcal{G}_{je}^{\text{female}}} o_v^{\tilde{i}_f} S(\tilde{i}_f) \Big) + \sum_{\tilde{i}_m \in \mathcal{G}_{je}^{\text{male}}} o_v^{\tilde{i}_m} S(\tilde{i}_m)) \leq c_v, \tag{38b}$$

$$\sum_{u=1}^{U} \sum_{i=1}^{I_r(n)} q_u^v h_i^{u \to v} \varrho_v^{i \to i'} p_v^{i \to i'} \leq p_v, \; \forall v \in \mathcal{V}, \; \forall n \in \mathcal{N}, \tag{38c}$$

$$q_u^v (h_i^{u \to v} + h_{i'}^{v \to u}) + q_v^r \eta_v (h_i^{r \to v} + h_{i'}^{r \to v}) \leq 1, \tag{38d}$$

$$q_u^v \varrho_v^{i \to i'} + q_v^r (1 - \varrho_v^{i \to i'}) \leq 1. \tag{38e}$$

The constraint in (38a) ensures that the self-driving car has to be connected to RSU $r \in \mathcal{R}$ to download the contents. The constraints in (38b) and (38c) guarantee that the caching and computational resource allocations have to be less than or equal to the available caching and computational resources of the self-driving car. Furthermore, constraint in (38b) is based on CNN output, where the self-driving car caches the contents based on passengers' features such as age, emotion, and gender. The constraint in (38d) ensures that the self-driving car or MEC server returns only one format of the requested content (either cached or computed from the cached content). The constraint (38e) ensures that converting i to i' is only executed at one location, either in self-driving car v or at MEC server r.

The formulated optimization problem in (38) is non-convex problem which makes it complicated to solve. Therefore, in the next Sect. 4.2, we propose a proximal convex surrogate problem of the formulated problem in (38) and apply Block Successive Majorization-Minimization (BS-MM) [15] for solving proximal convex surrogate problem.

4.2 Proposed Solution: Distributed Algorithm for Deep Learning Based Caching

For solving our optimization problem, we use BS-MM described in [15, 26]. We chose BS-MM over other distributed algorithms such as DC (Difference of Convex) programming, concave-convex, and successive convex approximation because BS-

MM is a new approach that allows to partition the problem into blocks and applies MM to one block of variables while keeping the values of the other blocks fixed [15]. The BS-MM may have computation overhead due to the computation of the best solution at each iteration, especially when the size of the problem is very large. Also, when BS-MM is fast, it may skip the true local minimum. If BS-MM is too slow, it may never converge because it tries to find a local minimum at each iteration. Therefore, to overcome these BS-MM challenges and ensure that all blocks are utilized, as suggested in [27], we use different selection rules such as Cyclic, Gauss-Southwell, and Randomized described in [27]. To apply BS-MM in (38), we consider $\mathcal{Q} \triangleq \{\vec{q} : \sum_{u=1}^{U} q_u^v \leq 1,\ q_u^v \in [0,1]\}$, $\mathcal{H} \triangleq \{\vec{h} : \sum_{u=1}^{U} (h_i^{u\to v} + h_{i'}^{v\to u}) + (1 - (h_i^{u\to v} + h_{i'}^{v\to u}))\,(h_i^{r\to v} + h_{i'}^{r\to v}) \leq 1, h_i^{u\to v}, h_{i'}^{v\to u}, h_i^{r\to v}, h_{i'}^{r\to v} \in [0,1]\}$, and $\mathcal{P} \triangleq \{\vec{\varrho} : \sum_{i,i'\in\mathcal{I}} \varrho_v^{i\to i'} + (1 - \varrho_v^{i\to i'})\varrho_r^{i\to i'} \leq 1, \varrho_v^{i\to i'}, \varrho_r^{i\to i'} \in [0,1]\}$ as non-empty and closed sets of the relaxed variables $\vec{q}$, $\vec{h}$, and $\vec{\varrho}$, respectively. Therefore, to simplify our notation, we use $\mathcal{F}(\vec{q},\vec{h},\vec{\varrho})$ to denote (38), where $\mathcal{F}(\vec{q},\vec{h},\vec{\varrho})$ is expressed as follows:

$$\mathcal{F}(\vec{q},\vec{h},\vec{\varrho}) = \sum_{u=1}^{U} \tau_u^{\text{Tot}}(\vec{q},\vec{h},\vec{\varrho}). \tag{39}$$

Both (38) and (39) have the same constraints. Therefore, to solve (39), we use the following steps:

- In the first step, called majorization, we propose a proximal convex surrogate problem $\mathcal{F}_j(\vec{q},\vec{h},\vec{\varrho})$ (40) of the formulated problem in (39), which is an upper-bound of (39).
- In the second step, called minimization, instead of minimizing (39) which is intractable, we minimize its proximal convex surrogate function $\mathcal{F}_j(\vec{q},\vec{h},\vec{\varrho})$(40).

The success of BS-MM relies on the surrogate function. Therefore, a surrogate function that is easy to solve and upper-bound of of the formulated problem in (39) is preferable. To achieve this, in the majorization step, we use the proximal upper-bound minimization technique described in [15]. Then, we propose the following proximal convex surrogate problem $\mathcal{F}_j(\vec{q},\vec{h},\vec{\varrho})$ (40) of the formulated problem in (39) by adding the quadratic term $(\frac{\varrho_j}{2}\|(\vec{q}_j - \vec{q}^{(0)})\|^2)$ to (39):

$$\mathcal{F}_j(\vec{q}_j,\vec{q}^{(t)},\vec{h}^{(t)},\vec{\varrho}^{(t)})\mathcal{F}(\vec{q}_j,\vec{q}^{(0)},\vec{h}^{(0)},\vec{\varrho}^{(0)}) + \frac{\alpha_j}{2}\|(\vec{q}_j - \vec{q}^{(0)})\|^2, \tag{40}$$

where $\vec{q}^{(0)}$, $\vec{h}^{(0)}$, and $\vec{\varrho}^{(0)}$ are the initial feasible points. Furthermore, the surrogate function in (40) can be applied to other vectors $\vec{h}$ and $\vec{\varrho}$. In addition, the quadratic term $(\frac{\alpha_j}{2}\|(\vec{q}_j - \vec{q}^{(0)})\|^2)$ makes the problem (40) to be convex and upper-bound of (39). In the minimization step, we minimize the surrogate function $\mathcal{F}_j(\vec{q},\vec{h},\vec{\varrho})$ (40) by taking steps proportional to the negative of the gradient in the direction toward the formulated problem in (39), where $\mathcal{J}^t$ is a set of indexes at each iteration t and α_j is a positive penalty parameter for $j \in \mathcal{J}^t$. At each iteration $t+1$, the solution is updated by solving the following problems:

$$\vec{q}_j^{(t+1)} \in \min_{\vec{q}_j \in \mathcal{Q}} \mathcal{F}_j(\vec{q}_j, \vec{q}^{(t)}, \vec{h}^{(t)}, \vec{\varrho}^{(t)}), \tag{41}$$

$$\vec{h}_j^{(t+1)} \in \min_{\vec{h}_j \in \mathcal{H}} \mathcal{F}_j(\vec{h}_j, \vec{h}^{(t)}, \vec{q}_j^{(t+1)}, \vec{\varrho}^{(t)}), \tag{42}$$

$$\vec{\varrho}_j^{(t+1)} \in \min_{\vec{\varrho}_j \in \mathcal{P}} \mathcal{F}_j(\vec{\varrho}_j, \vec{\varrho}^{(t)}, \vec{q}_j^{(t+1)}, \vec{h}_j^{(t+1)}). \tag{43}$$

To solve our problems in (41), (42), and (43) we use vectors $\vec{q}_j$, $\vec{h}_j$ and $\vec{\varrho}_j$ of relaxed variables. Therefore, we need to enforce $\vec{q}_j$, $\vec{h}_j$ and $\vec{\varrho}_j$ to be vectors of binary variables. To achieve this, we apply the rounding techniques described in [28]. As an illustration example, for a solution $q_v^{r*} \in \vec{q}_j^{(t+1)}$, we define the rounding threshold $\varphi \in (0, 1)$, such that the enforced binary value of q_v^{r*} is given by:

$$q_v^{r*} = \begin{cases} 1, & \text{if } q_v^{r*} \geq \varphi, \\ 0, & \text{otherwise.} \end{cases} \tag{44}$$

As highlighted in [10, 29], the rounding technique may violate 3C resource constraints. Therefore, to overcome this issue, we solve $\mathcal{F}_j$ in the form $\mathcal{F}_j + \beta_v \Delta_v$ by updating the constrains in (38a), (38b), and (38c) as follows:

$$\sum_{v=1}^{V} q_v^r a_v^r \leq 1 + \Delta_{v_a}, \ \forall r \in \mathcal{R}, \tag{45}$$

$$\sum_{u=1}^{U} \sum_{i=1}^{I_r(n)} q_u^v h_i^{u \to v} \varrho_v^{i \to i'} p_v^{i \to i'} \leq p_v + \Delta_{v_p}, \forall v \in \mathcal{V}, \tag{46}$$

$$q_v^r \sum_{j=1}^{k} \Big(\sum_{\tilde{i}_f \in \mathcal{G}_{je}^{\text{female}}} o_v^{\tilde{i}_f} S(\tilde{i}_f)) + \sum_{\tilde{i}_m \in \mathcal{G}_{je}^{\text{male}}} o_v^{\tilde{i}_m} S(\tilde{i}_m)) \leq c_v + \Delta_{v_c}, \tag{47}$$

where $\Delta_v = \Delta_{v_a} + \Delta_{v_p} + \Delta_{v_c}$ is the maximum violation of the 3C resource constraints and β_v as the weight parameter of Δ_v. Furthermore, the values of Δ_{v_a}, Δ_{v_p}, and Δ_{v_c} are given by:

$$\Delta_{v_a} = \max\{0, \sum_{v=1}^{V} q_v^r a_v^r - 1\}, \ \forall r \in \mathcal{R}, \tag{48}$$

$$\Delta_{v_p} = \max\{0, \sum_{u=1}^{U} \sum_{i=1}^{I_r(n)} q_u^v h_i^{u \to v} \varrho_v^{i \to i'} p_v^{i \to i'} - p_v\}, \ \forall v \in \mathcal{V}, \tag{49}$$

Algorithm 1 : Distributed algorithm for deep learning based caching.

1: **Preconditions:** MLP output and CNN models are deployed to the RSUs and in self-driving car;

2: **Input:** $\vec{U}$: A vector of passengers, $\vec{\omega}_v^r$: wireless link capacities, $\vec{\mathcal{X}}$: Vector of recommended contents for $\mathcal{G}_{je}^{\text{female}}$ and $\mathcal{G}_{je}^{\text{male}}$ in self-driving car v, ψ_u^v, p_v, and c_v;

3: **Output:** $\vec{q}^*$, $\vec{h}^*$, $\vec{\varrho}^*$;

4: Initialize $t = 0$;

5: Find initial feasible points $\vec{q}^{(0)}, \vec{h}^{(0)}, \vec{\varrho}^{(0)}$;

6: **repeat**

7: Choose index set $\mathcal{J}^t$;

8: Let $\vec{q}_j^{(t+1)} \in \min_{\vec{q}_j \in \mathcal{Q}} \mathcal{F}_j(\vec{q}_j, \vec{q}^{(t)}, \vec{h}^{(t)}, \vec{\varrho}^{(t)})$ (41);

9: Set $\vec{q}_k^{t+1} = \vec{q}_k^t, \forall k \notin \mathcal{J}^t$ and solve $\min_{\vec{q}_j \in \mathcal{Q}} \mathcal{F}_j(\vec{q}_j, \vec{q}^{(t)}, \vec{h}^{(t)}, \vec{\varrho}^{(t)})$;

10: For $\vec{h}_j^{(t+1)}$ and $\vec{\varrho}_j^{(t+1)}$, restart from step 4, salve (42) and (43);

11: $t = t + 1$;

12: **until** $\lim_{t \to \infty} \inf_{\vec{q}, \vec{h}, \vec{\varrho}} \|\mathcal{F}_j^{(t+1)} - \mathcal{F}_j^{(t)}\|_2 = 0$;

13: By rounding technique, enforce $\vec{q}_j^{(t+1)}$, $\vec{h}_j^{(t+1)}$, and $\vec{\varrho}_j^{(t+1)}$ to be vectors of binary variables;

14: Solve $\mathcal{F}_j + \beta_v \Delta_v$ and compute ϕ_j until $\phi_j \leq 1$;

15: Then, consider $\vec{q}^* = \vec{q}_j^{(t+1)}$, $\vec{h}^* = \vec{h}_j^{(t+1)}$, and $\vec{\varrho}^* = \vec{\varrho}_j^{(t+1)}$ as a solution.

$$\Delta_{v_c} = \max\{0, q_v^r \sum_{j=1}^{k} ((\sum_{\tilde{i}_f \in \mathcal{G}_{je}^{\text{female}}} o_v^{\tilde{i}_f} S(\tilde{i}_f)) + \sum_{\tilde{i}_m \in \mathcal{G}_{je}^{\text{male}}} o_v^{\tilde{i}_m} S(\tilde{i}_m)) - c_v\}. \quad (50)$$

Therefore, to ensure that the best solution is achieved, we use the integrality gap described in [28].

Definition 1 (*Integrality gap*) For the problems $\mathcal{F}_j + \beta_v \Delta_v$ and $\mathcal{F}_j$, the integrality gap is expressed as follows:

$$\phi_j = \min_{\vec{q}, \vec{h}, \vec{\varrho}} \frac{\mathcal{F}_j}{\mathcal{F}_j + \beta_v \Delta_v}. \quad (51)$$

The best solutions of $\mathcal{F}_j$ and $\mathcal{F}_j + \beta_v \Delta_v$ are obtained when $\phi_j \leq 1$.

We propose a distributed algorithm (Algorithm 1), which is based on BS-MM [15]. We assume that the MLP output and CNN model are already deployed at RSUs and in self-driving car. We consider a vector of passengers, vector of RSUs, vector of wireless link capacities, vector of recommended contents that need to be cached in self-driving car v, ψ_u^v, p_v, and c_v as the input. First, Algorithm 1 finds the initial feasible points $\vec{q}^{(0)}$, $\vec{h}^{(0)}$, and $\vec{\varrho}^{(0)}$. Then, Algorithm 1 starts an iterative process by choosing an index set $\mathcal{J}^t$ at each iteration t. At each iteration

Table 2 The used route for the self-driving bus

Route	Distance (km)	Max. speed (km/h)	RSUs
1	54.62	109.016	1–2
2	53.82	107.34	2–3
3	54.02	108.17	3–4
4	52.83	105.38	4–5
5	55.66	111.33	5–6

$t+1$, the solution is updated by solving the problems (41), (42), and (43) until $\lim_{t\to\infty} \inf_{\vec{q},\vec{h},\vec{\varrho}} \|\mathcal{F}_j^{(t+1)} - \mathcal{F}_j^{(t)}\|_2 = 0$, where $\lim_{t\to\infty} \inf_{\vec{q},\vec{h},\vec{\varrho}} \|\mathcal{F}_j^{(t+1)} - \mathcal{F}_j^{(t)}\|_2 = 0$ is the convergence criteria. Therefore, when $\lim_{t\to\infty} \inf_{\vec{q},\vec{h},\vec{\varrho}} \|\mathcal{F}_j^{(t+1)} - \mathcal{F}_j^{(t)}\|_2 = 0$, Algorithm 1 considers $\vec{q}_j^{(t+1)}, \vec{h}_j^{(t+1)}$, and $\vec{\varrho}_j^{(t+1)}$ as a solution. Then, Algorithm 1 forces the solution $\vec{q}_j^{(t+1)}$, $\vec{h}_j^{(t+1)}$, and $\vec{\varrho}_j^{(t+1)}$ to be vectors of binary variables via the rounding technique, where Algorithm 1 solves $\mathcal{F}_j + \beta_v \Delta_v$ and computes ϕ_j. Finally, when $\phi_j \leq 1$, Algorithm 1 considers $\vec{q}^* = \vec{q}_j^{(t+1)}$, $\vec{h}^* = \vec{h}_j^{(t+1)}$, and $\vec{\varrho}^* = \vec{\varrho}_j^{(t+1)}$ as a solution which does not violate 3C resource constraints. Furthermore, for the convergence of the proposed algorithm, based on the convergence of MM defined and proved in [15], we make the following remark:

Remark 1 (*Convergence of the proposed algorithm*) Based on the MM algorithm [15], the proposed Algorithm 1, which is based on BS-MM, converges to coordinate-wise minimum point which is stationary point, when the vectors $\vec{q}^* = \vec{q}_j^{(t+1)}$, $\vec{h}^* = \vec{h}_j^{(t+1)}$, and $\vec{\varrho}^* = \vec{\varrho}_j^{(t+1)}$ cannot find a better minimum direction, i.e., $\lim_{t\to\infty} \inf_{\vec{q},\vec{h},\vec{\varrho}} \|\mathcal{F}_j^{(t+1)} - \mathcal{F}_j^{(t)}\|_2 = 0$.

For complexity analysis of the proposed Algorithm 1, based on complexity analysis described in [27], we make the following remark:

Remark 2 (*Complexity of the proposed Algorithm 1*) The Algorithm 1, which is based on BS-MM, uses proximal upper-bound minimization technique. This makes it fall under the BSUM framework [27]. Therefore, for the iteration index $j \in \mathcal{J}^t$, the Algorithm 1 has $\mathcal{O}(1/j)$ iteration complexity, which is sub-linear.

5 Simulation Results and Analysis

In this section, we present a performance evaluation of the proposed deep learning-based caching in self-driving cars. We use Google Maps Services [30] for the self-driving car mobility analysis, Keras with Tensorflow [31] for the deep learning simulation, and pandas [32] for data analysis.

5.1 Simulation Setup

To predict the probabilities of contents to be requested in specific areas of MEC servers, we use a well-known dataset called Movie-Lens Dataset [33]. In the dataset, we have movies with related information such as movie titles, release date, and genre of movies such as comedy, drama, and documentary. We associate the emotion with the genre of movies, where sad users recommended to watch drama movies, disgust users recommended to watch musical movies, anger users recommended to watch comedy movies, anticipate users recommended to watch thriller movies, fear users recommended to watch adventure movies, joy users recommended to watch thriller movies, trust users recommended to watch western movies, and surprise users recommended to watch fantasy movies. However, the dataset does not have movie sizes and formats. Since our deep learning-based caching scheme uses content size, we randomly generate size $S(i)$ for each movie i in the range from $S(i) = 317$ to $S(i) = 750$ Mb and randomly assign each movie i a format. Furthermore, we have user's information such as age (as shown in Fig. 6), gender, rating, and ZIP codes. To identify the areas of users, we convert the ZIP codes into longitude and latitude coordinates and deploy 6 RSUs to the specific areas based on the movie watching counts, rankings, and the locations of users. We use MLP with 2 layers (for input and output) and 2 hidden layers to predict the probabilities of contents to be requested in specific areas of RSUs. In MLP, each layer has 32 neurons except the output layer which has 6 neurons. In the output layer, 6 neurons correspond to the probabilities of contents to be cached in specific areas of 6 RSUs. We use 60% of the dataset for training and 40% for testing. Furthermore, the learning rate is set to be equal to 0.002, while the batch size equals to 32.

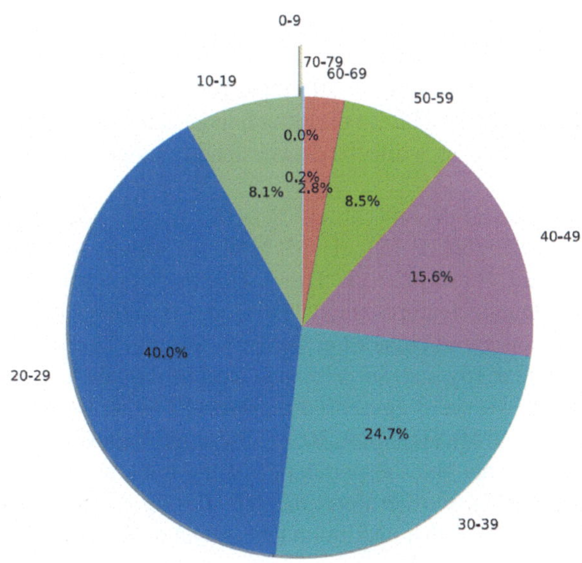

Fig. 6 Visualization of the used dataset [33] for movie watching based on age

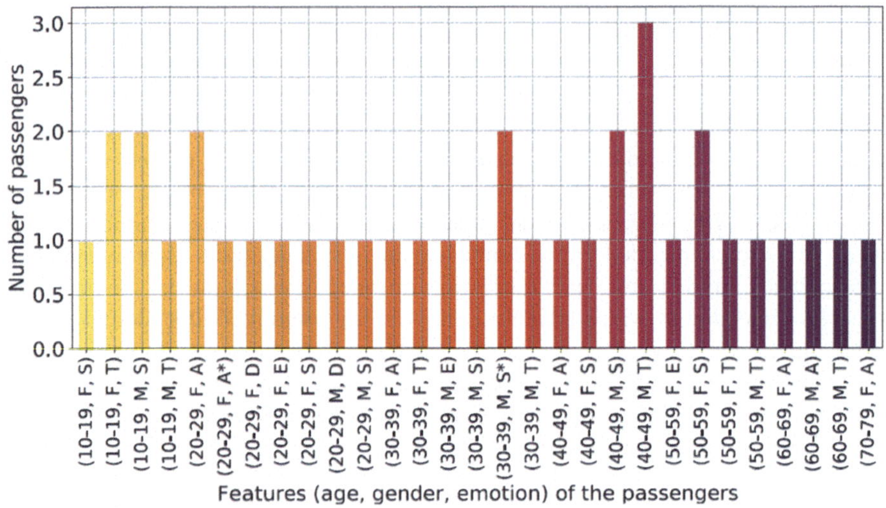

Fig. 7 Visualization of the used passengers' features for the self-driving bus

With the departure time and locations of the RSUs, the Google Maps service provides the distance and duration to reach each RSU $r \in \mathcal{R}$, where the duration is based on traffic conditions between the source and destination. Based on the distance (in terms of km) and duration (in terms of hours), we can calculate the speed (in terms of km/h) of the self-driving car and find the RSUs that the self-driving car can connect to for retrieving contents. However, based on Google Maps service [30], the distances between RSUs are very large. Therefore, to have realistic distances between RSUs, we update the RSU locations and create a routing table summarized in Table 2, where the self-driving car starts its journey at RSU 1 and ends at RSU 6. We set each RSU $r \in \mathcal{R}$ to be connected to the DC with a wired backhaul of capacity ranging from $\omega_{r,DC} = 60$ to $\omega_{r,DC} = 70$ Mbps. We assume that each RSU $r \in \mathcal{R}$ has a bandwidth of $\omega_{v,r} = 10$ MHz. On the other hand, each MEC server $r \in \mathcal{R}$ has a CPU of capacity $p_r = 3.6$ GHz, while the cache capacity ranges from $c_r = 100$ to $c_r = 110$ terabytes (TB).

For a self-driving car $v \in \mathcal{V}$, as shown in Fig. 7, we generated randomly features of 37 passengers (F: Female, M: Male, A: Anger, A*: Anticipation, D: Disgust, E: Joy, S*: Sad, S: Surprise, T: Trust). However, in a realistic implementation, for getting passengers' features, the CNN model described in Sect. 3.1.2 should be used. For emotion-based clustering, we use 8 emotion-based clusters: anger, anticipation, disgust, fear, joy, sad, surprise, and trust as the labels. Furthermore, for age-based clustering, we use 8 age-based clusters: $[0 \rightarrow 9, 10 \rightarrow 19, 20 \rightarrow 29, 30 \rightarrow 39, 40 \rightarrow 49, 50 \rightarrow 59, 60 \rightarrow 69, 70 \rightarrow 79]$ as the labels. We generated randomly demands for contents and the popularity of the contents follows Zipf distribution described in [34, 35]. Furthermore, the self-driving car has a WiFi bandwidth of 160 MHz (802.11ac) with a maximum theoretical data rate of $\tilde{\psi}_u^v = 3466.8$ Mbps. In addition,

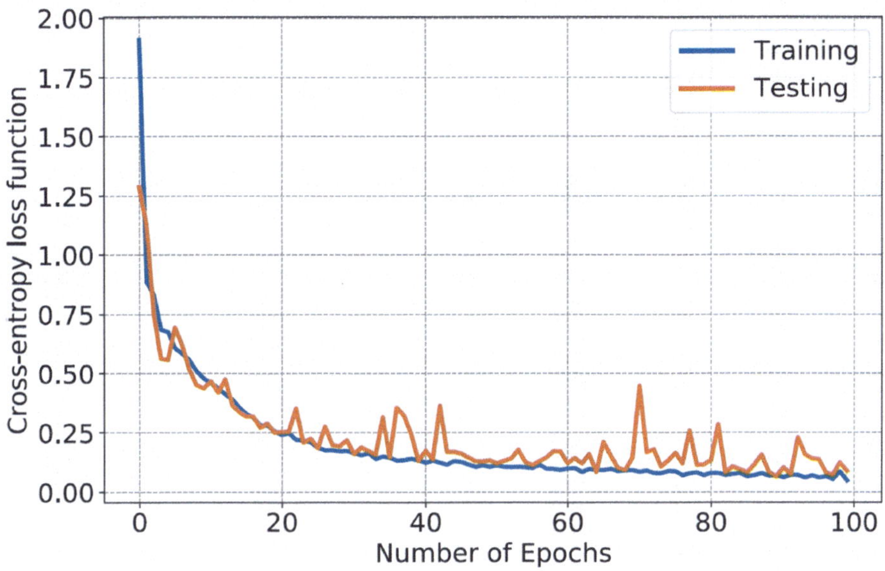

Fig. 8 Minimization of error function for predicting the probability of movies to be requested in the specific areas of RSUs (acc: 97.82%)

the computation capacity of the self-driving car is set to $p_v = 3.6$ GHz, while the cache capacity is set to $c_v = 100$ TB.

5.2 Evaluation Results

Based on video ratings and users' location information, we select six areas to deploy RSUs by using the k-means algorithm. In the selected six areas, we predict the probabilities of contents to be requested in these areas by using MLP. As shown in Fig. 8, in MLP, we minimize the cross-entropy loss function. An accuracy of 97.82% is achieved for predicting the probabilities of contents to be requested in 6 areas of RSUs. Each RSU $v \in \mathcal{V}$ caches movies by starting with the movies that have high ratings and predicted probabilities to be requested within the RSU area (in descending order) until the cache storage becomes full or there are no more movies to cache. As an example, Fig. 9 shows the top 8 movies that need to be cached at RSU 1 with their predicted probabilities using MLP.

Caching at the RSUs is based on location and movie ratings. However, in addition to location and movie ratings, caching in self-driving cars is based on passengers' features such as age, emotion, and gender. Therefore, when the self-driving car is connected to an RSU, it downloads the MLP output from the RSU. Then, it groups the MLP output based on age and emotion using the k-means algorithm and on

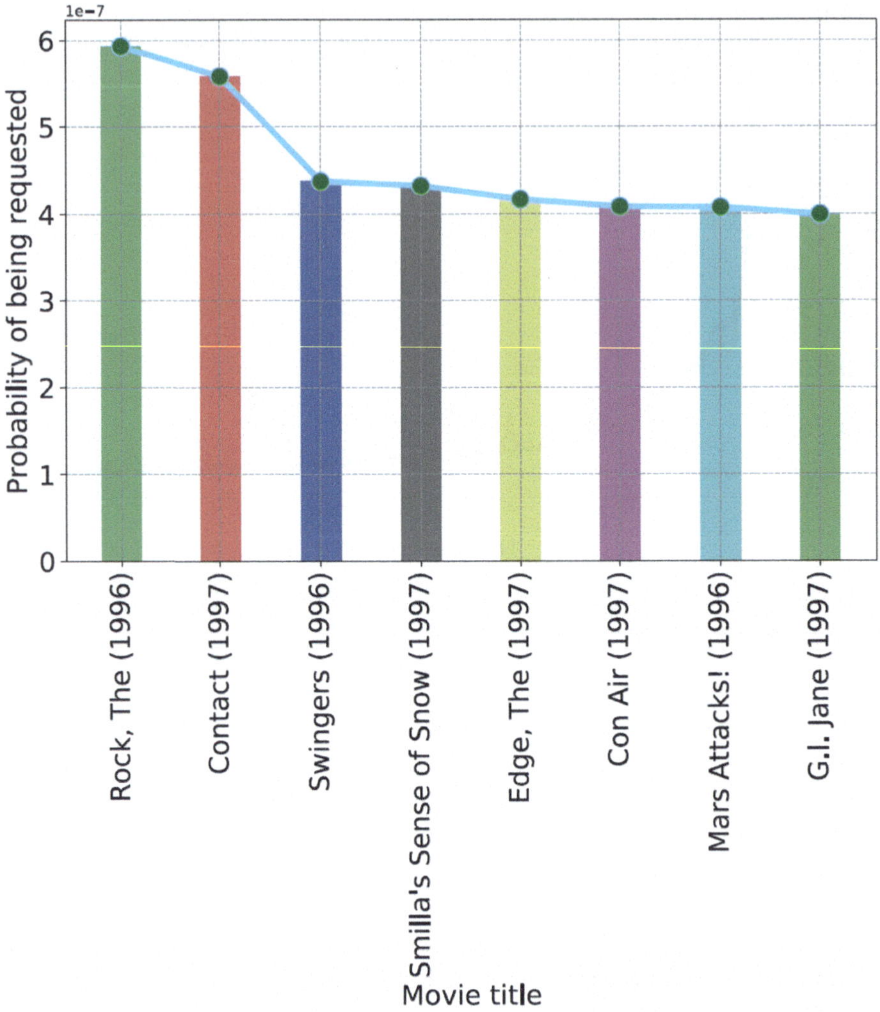

Fig. 9 Some high recommended movies to cache in close proximity of the self-driving cars at RSUs

gender using binary classification described in Sect. 3.1.3. Here, we use 8 age-based clusters, 8 emotion-based clusters, and 2 gender-based clusters. Furthermore, for the passengers, we use age, emotion, and gender features described in Fig. 7. However, CNN can be used to predict these features (age, emotion, and gender) using facial images of passengers captured by car's camera. The self-driving car uses k-means and binary classification to classify the passengers in different age, emotion, and gender-based clusters formed using MLP output. Then, inside the formed clusters, the self-driving car finds the movies that have high ratings and predicted probabilities to be requested as recommended movies for the passengers.

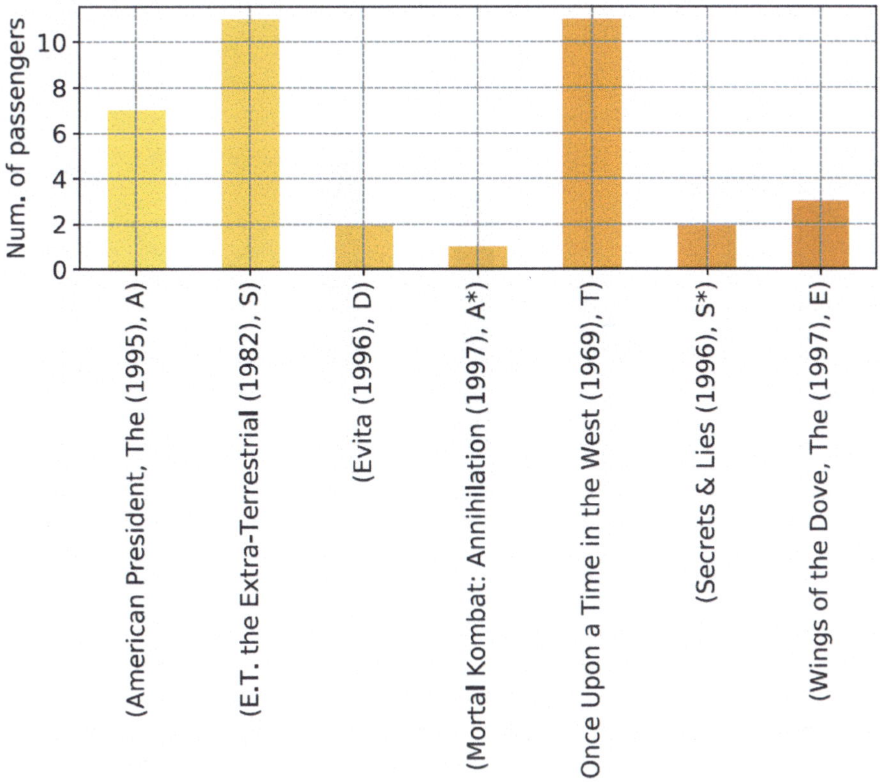

Fig. 10 Some high recommended movies to watch based on passengers' features (age, gender, and emotion)

Figure 10 shows recommended movies to watch depending on age, emotion, and gender of the passengers. As shown in this figure, based on these features, passengers may like similar movies (many passengers like Once Upon A Time and Secrets & Lies). Therefore, caching these recommended movies inside the car can prevent repetitive demands of the same movies that need to be sent to RSUs or DC. In other words, we can save bandwidth. Furthermore, we chose CNN and MLP-based recommendation for movies over collaborative filtering approaches because each passenger's features for infotainment contents are not a priori known by the self-driving car. The collaborative filtering approaches, which are described in [36], consist of establishing the relationship between prior known users' preferences and movies' features. However, after identifying passengers' features and movies' features, we compare our proposal denoted Deep3C with the well-known collaborative filtering approaches such as Singular Value Decomposition (SVD), Non-negative Matrix Factorization (NMF), K-Nearest Neighbors (KNN), and Co-clustering (Coclust).The simulation results in Fig. 11 show that our proposal (Deep3C) achieves better performance over existing collaborative filtering approaches.

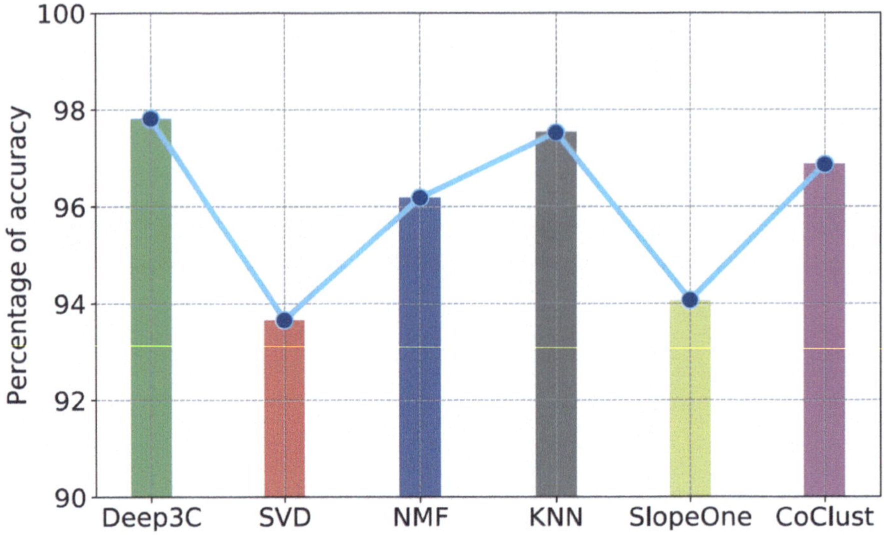

Fig. 11 Comparison of various collaborative filtering algorithms and our proposal (Deep3C)

We generated randomly demands of passengers for contents, where the popularity of the contents follows Zipf distribution [34]. We use Zipf parameter a with values from $a = 0.5$ to $a = 2.0$. The choice of a = 0.5 to a = 2.0 comes from the results presented in Fig. 12, where the difference in convergence is observed within a range of $a = 0.5$ to $a = 2.0$. Furthermore, based on the demands of the passengers, Fig. 13 shows the normalized cache hits for the cached movies. The movies that are not cached in the self-driving car (cache misses) need to be retrieved at the RSU or DC. In this figure, we present the cache hits for the contents cached at RSUs and self-driving car. In other words, the total cache hits at RSUs and the self-driving car equal to 61% of the whole demands, i.e., 39% of the demands need to be served by DC. Therefore, with edge caching at RSUs and self-driving cars, we can significantly save backhaul bandwidth. The results in this figure demonstrate that the cache hits increase with Zipf parameter, i.e., when $a = 2.0$ the small number of movies are very popular and requested by many passengers. In other words, the movies with high demands are characterized by high probabilities of being requested and caching these movies contribute to the high increase of cache hits.

Figure 14 shows the solution of the surrogate function (40), where (40) minimizes the total delays (transmission delay and computation delay). The surrogate function (40) converges to a coordinate-wise minimum point which is the stationary point through the use of different selection rules such as Cyclic, Gauss-Southwell, and Randomized. In other words, at a stationary point, the problem (40) cannot find a better minimum direction. Furthermore, in this figure, the self-driving car needs to download the recommended contents first, and then caches these recommended contents; this contributes to high latency at the first iterations. As described in Fig. 10,

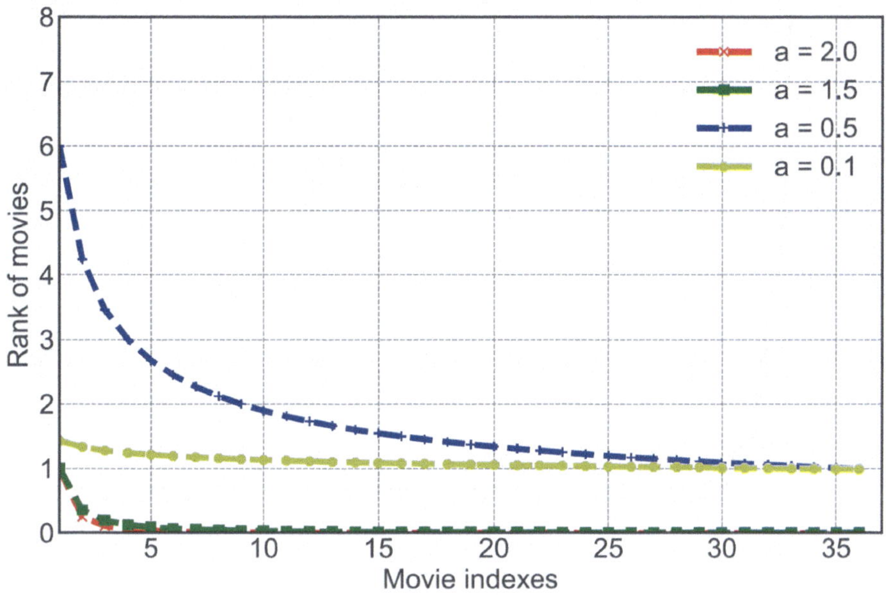

Fig. 12 Ranking of movie demands based on Zipf distribution

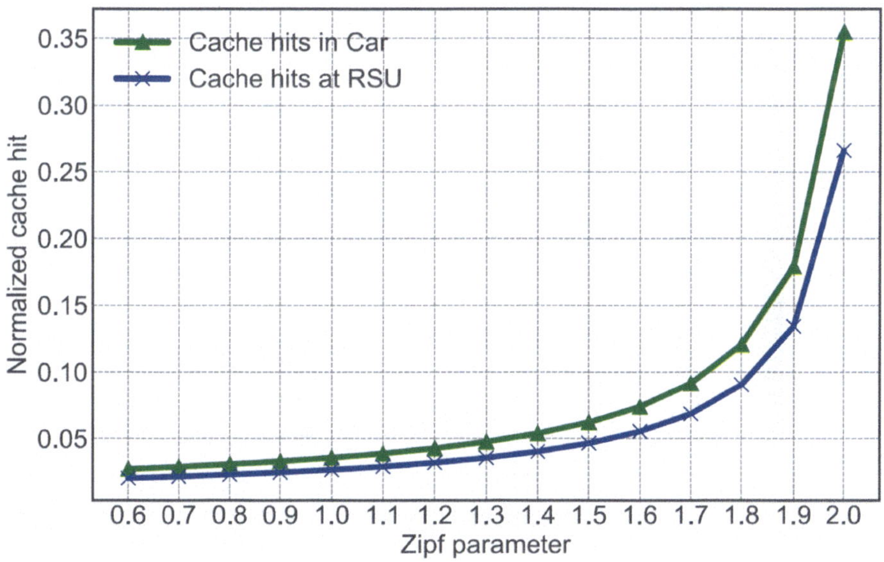

Fig. 13 Cache hits for the requested movies

some passengers may need to watch similar movies, i.e., many requests for movies can be satisfied from the cache storage.

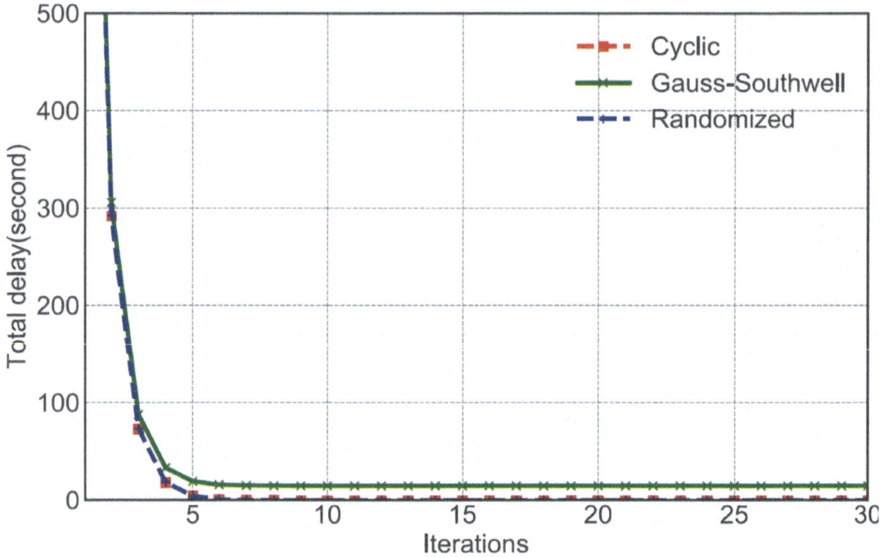

Fig. 14 The solution of total delay minimization problem (40)

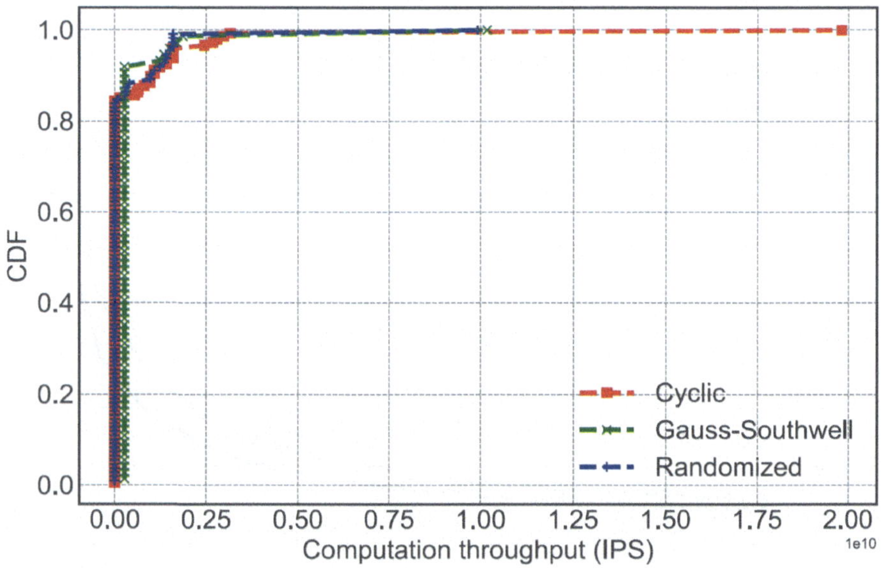

Fig. 15 Computation throughput for the cache contents

In Fig. 15, we present the Cumulative Distribution Function (CDF) of computational throughput in terms of the number of Instruction Per Second (IPS). Here, we define computation throughput as a measurement of how many units of tasks that can be computed by OBU for a given time. In this figure, the simulation results demonstrate that the Cyclic selection rule uses higher computational resource than Gauss-Southwell and Randomized selection rules. Cyclic selection rule has to choose index $j \in \mathcal{J}^t$ cyclically until all indexes in $\mathcal{J}^t$ are used.

6 Summary

In this chapter, we proposed a novel framework that uses deep learning for content caching in a self-driving car. In the proposed framework, at the DC, we proposed an MLP to predict the probabilities of contents being requested in specific areas. Then, the output is deployed in MEC servers (at the RSUs) close to the self-driving cars, where each MEC server downloads and caches the contents that have high probabilities of being requested in its coverage area. Furthermore, for a self-driving car, to cache infotainment contents that are appropriate regarding the age, emotion, and gender of the passengers, we proposed to use CNN approach for predicting the age, emotion, and gender. Then, the self-driving car downloads the MLP output from the MEC server and combines CNN output with the MLP output using k-means and binary classifications to identify the infotainment contents that meet passengers' features to be downloaded and cached. Therefore, we formulated the deep learning-based caching problem as an optimization problem that minimizes the content-downloading delay. The simulation results demonstrate that our caching approach can reduce 61% of the backhaul traffic, i.e., caching at RSUs and self-driving cars can serve 61% of the whole demands for infotainment contents. Furthermore, our prediction for the infotainment contents that need to be cached at the RSUs and the self-driving cars reaches 97.82% accuracy.

Acknowledgements This work was supported by the Institute of Information & Communications Technology Planning & Evaluation (IITP) grant funded by the Korea government (MSIT) (No. 2019-0-01287, Evolvable Deep Learning Model Generation Platform for Edge Computing) and by the Institute of Information and Communications Technology Planning and Evaluation (IITP) grant funded by the Korea Government (MSIT) (Artificial Intelligence Innovation Hub) under Grant 2021-0-02068. *Dr. CS Hong is the corresponding author.

References

1. M. Daily, S. Medasani, R. Behringer, M. Trivedi, Self-driving cars. Computer **50**(12), 18–23 (2017)
2. L.U. Khan, W. Saad, D. Niyato, Z. Han, C.S. Hong, Digital-twin-enabled 6g: vision, architectural trends, and future directions (2021), arXiv:2102.12169
3. L.U. Khan, Z. Han, W. Saad, E. Hossain, M. Guizani, C.S. Hong, Digital twin of wireless systems: overview, taxonomy, challenges, and opportunities (2021), arXiv:2102.12169
4. L.U. Khan, Z. Han, D. Niyato, C.S. Hong, Socially-aware-clustering-enabled federated learning for edge networks. IEEE Trans. Netw. Serv. Manag. (2021)

5. L.U. Khan, W. Saad, Z. Han, C.S. Hong, Dispersed federated learning: vision, taxonomy, and future directions. IEEE Wirel. Commun. **28**(5), 192–198 (2021)
6. L.U. Khan, W. Saad, Z. Han, E. Hossain, C.S. Hong, Federated learning for internet of things: recent advances, taxonomy, and open challenges. IEEE Commun. Surv. Tutor. (2021)
7. G. Jarvis, Keeping entertained in the autonomous vehicle, TU-Automotive Detroit, 6–7 June 2018
8. F. Fathi, N. Abghour, M. Ouzzif, From big data to better behavior in self-driving cars, in *Proceedings of the 2nd International Conference on Cloud and Big Data Computing* (ACM, 2018), pp. 42–46
9. Y.C. Hu, M. Patel, D. Sabella, N. Sprecher, V. Young, Mobile edge computing-a key technology towards 5G. ETSI White Paper **11**(11), 1–16 (2015)
10. A. Ndikumana, N.H. Tran, T.M. Ho, Z. Han, W. Saad, D. Niyato, C.S. Hong, Joint communication, computation, caching, and control in big data multi-access edge computing. IEEE Trans. Mob. Comput. (2019)
11. Next Analytics, YouTube video appeal demographics, https://www.nextanalytics.com/excel-youtube-analytic-insights-and-data-mining/page/4/ [Online; Accessed 22 June 2019]
12. A. Ndikumana, C.S. Hong, Self-driving car meets multi-access edge computing for deep learning-based caching, in *Proceedings of 2019 International Conference on Information Networking (ICOIN)*, 9–11 Jan 2019, Kuala Lumpur, Malaysia
13. J.J. Whang, I.S. Dhillon, D.F. Gleich, Non-exhaustive, overlapping k-means, in *Proceedings of the 2015 SIAM International Conference on Data Mining*, SIAM, 30 Apr–2 May 2015, British Columbia, Canada (2015), pp. 936–944
14. J. Martineau, T. Finin, A. Joshi, S. Patel, Improving binary classification on text problems using differential word features, in *Proceedings of the 18th ACM Conference on Information and Knowledge Management*, 02–06 Nov 2009, Hong Kong, China (2009), pp. 2019–2024
15. Y. Sun, P. Babu, D.P. Palomar, Majorization-minimization algorithms in signal processing, communications, and machine learning. IEEE Trans. Signal Process. **65**(3), 794–816 (2017)
16. 3GPP TS 22.185 V15.0.0, 3rd Generation Partnership Project; technical specification group services and system aspects; service requirements for V2X services; stage 1(release 15), June 2018
17. B. Blaszczyszyn, A. Giovanidis, Optimal geographic caching in cellular networks, in *Proceedings of 2015 IEEE International Conference on Communications (ICC)*, 8–12 June 2015, London, UK (2015), pp. 3358–3363
18. A. Azzouni, G. Pujolle, NeuTM: a neural network-based framework for traffic matrix prediction in SDN, in *Proceedings of IEEE/IFIP Network Operations and Management Symposium(NOMS)*, 23–27 Apr 2018, Taipei, Taiwan (2018), pp. 1–5
19. M.Z. Alom, T.M. Taha, C. Yakopcic, S. Westberg, M. Hasan, B.C. Van Esesn, A.A.S. Awwal, V.K. Asari, The history began from alexnet: a comprehensive survey on deep learning approaches (2018), arXiv:1803.01164
20. K. Simonyan, A. Zisserman, Dager: deep age, gender and emotion recognition using convolutional neural network (2017), arXiv:1702.04280
21. L. Van Zoonen, Privacy concerns in smart cities. Gov. Inf. Q. **33**(3), 472–480 (2016)
22. 3GPP TS 24.312 V15.0.0, 3rd Generation Partnership Project; technical specification group core network and terminals; access network discovery and selection function (ANDSF) management object (MO) (release 15), June 2018
23. E. Ndashimye, N.I. Sarkar, S.K. Ray, A novel network selection mechanism for vehicle-to-infrastructure communication, in *Proceedings of IEEE 14th International Conference on Pervasive Intelligence and Computing, 2nd International Conference on Big Data Intelligence and Computing and Cyber Science and Technology Congress (DASC/PiCom/DataCom/CyberSciTech)*, 8–12 Aug 2016, Auckland, New Zealand (2016), pp. 483–488
24. N. Cheng, N. Lu, N. Zhang, X. Zhang, X.S. Shen, J.W. Mark, Opportunistic wifi offloading in vehicular environment: a game-theory approach. IEEE Trans. Intell. Transp. Syst. **17**(7), 1944–1955 (2016)

25. S. Mosleh, L. Liu, J. Zhang, Proportional-fair resource allocation for coordinated multi-point transmission in LTE-advanced. IEEE Trans. Wirel. Commun. **15**(8), 5355–5367 (2016)
26. A. Ndikumana, N.H. Tran, C.S. Hong, Deep learning based caching for self-driving car in multi-access edge computing (2018), arXiv:1810.01548
27. M. Hong, X. Wang, M. Razaviyayn, Z.-Q. Luo, Iteration complexity analysis of block coordinate descent methods. Math. Progr. **163**(1–2), 85–114 (2017)
28. U. Feige, M. Feldman, I. Talgam-Cohen, Oblivious rounding and the integrality gap, in *Approximation, Randomization, and Combinatorial Optimization. Algorithms and Techniques (APPROX/RANDOM 2016)*, ser. Leibniz International Proceedings in Informatics (LIPIcs)vol. 60, ed. by K. Jansen, C. Mathieu, J.D.P. Rolim, C. Umans (Schloss Dagstuhl–Leibniz-Zentrum fuer Informatik, Dagstuhl, Germany, 2016), pp. 8:1–8:23 [Online], http://drops.dagstuhl.de/opus/volltexte/2016/6631
29. N. Zhang, Y.-F. Liu, H. Farmanbar, T.-H. Chang, M. Hong, Z.-Q. Luo, Network slicing for service-oriented networks under resource constraints. IEEE J. Sel. Areas Commun. **35**(11), 2512–2521 (2017)
30. Google, Python client library for google maps api web services, https://github.com/googlemaps/google-maps-services-python [Online; Accessed 12 Aug 2018]
31. Keras, Keras: The Python Deep Learning library, https://keras.io/ [Online; Accessed 22 June 2019]
32. W. McKinney, Pandas: a foundational python library for data analysis and statistics. Python High Perform. Sci. Comput. 1–9 (2011)
33. F.M. Harper, J.A. Konstan, The Movielens datasets: history and context. ACM Trans. Interact. Intell. Syst. **5**(4), 19 (2016)
34. M.E. Newman, Power laws, pareto distributions and zipf's law. Contemp. Phys. **46**(5), 323–351
35. A. Ndikumana, K. Thar, T.M. Ho, N.H. Tran, P.L. Vo, D. Niyato, C.S. Hong, In-network caching for paid contents in content centric networking, in *Proceedings of IEEE Global Communications Conference (GLOBECOM)*, 4–8 Dec 2017, Singapore (2017), pp. 1–6
36. V. Subramaniyaswamy, R. Logesh, M. Chandrashekhar, A. Challa, V. Vijayakumar, A personalised movie recommendation system based on collaborative filtering. Int. J. High Perform. Comput. Netw. **10**(1–2), 54–63 (2017)

Latif U. Khan received his Ph.D. (Computer Engineering) and M.Sc. (Electrical Engineering) with distinction from Kyung Hee University (KHU), South Korea in 2021 and University of Engineering & Technology, Peshawar, Pakistan in 2017, respectively. He worked as a leading researcher in the intelligent Networking Laboratory under a project jointly funded by the prestigious Brain Korea 21st Century Plus and Ministry of Science and ICT, South Korea. Prior to joining the KHU, he has served as a faculty member and research associate in the UET, Peshawar, Pakistan. He has published his works in highly reputable conferences and journals. He is the author/co-author of two conference best paper awards. He is also author of two books, such as "Network Slicing for 5G and Beyond Networks" and "Federated Learning for Wireless Networks". His research interests include analytical techniques of optimization and game theory to edge computing, end-to-end network slicing, and federated learning for wireless networks.

Anselme Ndikumana received B.S. degree in Computer Science from the National University of Rwanda in 2007 and Ph.D. degree in Computer Engineering from Kyung Hee University, South Korea in August 2019. Since 2020, he has been with the Synchromedia Lab, École de Technologie Supérieure, Université du Québec, Montréal, QC, Canada where he is currently a postdoctoral fellow. His professional experience includes Lecturer at the University of Lay Adventists of Kigali from 2019 to 2020, Chief Information System, a System Analyst, and a Database Administrator at Rwanda Utilities Regulatory Authority from 2008 to 2014. His research interest includes AI for wireless communication, multi-access edge computing, 5G networks, information-centric networking, and in-network caching.

Nguyen H. Tran received BS and Ph.D. degrees, from HCMC University of Technology and Kyung Hee University, in electrical and computer engineering, in 2005 and 2011, respectively. He was an Assistant Professor with Department of Computer Science and Engineering, Kyung Hee University, from 2012 to 2017. Since 2018, he has been with the School of Computer Science, The University of Sydney, where he is currently a Senior Lecturer. His research interests include distributed computing, machine learning, and networking. He received the best KHU thesis award in engineering in 2011 and several best paper awards, including IEEE ICC 2016 and ACM MSWiM 2019. He receives the Korea NRF Funding for Basic Science and Research 2016–2023 and ARC Discovery Project 2020–2023. He was the Editor of IEEE Transactions on Green Communications and Networking from 2016 to 2020, and the Associate Editor of IEEE Journal of Selected Areas in Communications 2020/2021 in the area of distributed machine learning/Federated Learning.

Choong Seon Hong received the B.S. and M.S. degrees in electronic engineering from Kyung Hee University, Seoul, South Korea, in 1983 and 1985, respectively, and the Ph.D. degree from Keio University, Tokyo, Japan, in 1997. In 1988, he joined KT, Gyeonggi-do, South Korea, where he was involved in broadband networks as a member of the Technical Staff. Since 1993, he has been with Keio University. He was with the Telecommunications Network Laboratory, KT, as a Senior Member of Technical Staff and as the Director of the Networking Research Team until 1999. Since 1999, he has been a Professor with the Department of Computer Science and Engineering, Kyung Hee University. His research interests include future Internet, intelligent edge computing, network management, and network security. Dr. Hong is a member of the Association for Computing Machinery (ACM), the Institute of Electronics, Information and Communication Engineers (IEICE), the Information Processing Society of Japan (IPSJ), the Korean Institute of Information Scientists and Engineers (KIISE), the Korean Institute of Communications and Information Sciences (KICS), the Korean Information Processing Society (KIPS), and the Open Standards and ICT Association (OSIA). He has served as the General Chair, the TPC Chair/Member, or an Organizing Committee Member of international conferences, such as the Network Operations and Management Symposium (NOMS), International Symposium on Integrated Network Management (IM), Asia-Pacific Network Operations and Management Symposium (APNOMS), End-to-End Monitoring Techniques and Services (E2EMON), IEEE Consumer Communications and Networking Conference (CCNC), Assurance in Distributed Systems and Networks (ADSN), International Conference on Parallel Processing (ICPP), Data Integration and Mining (DIM), World Conference on Information Security Applications (WISA), Broadband Convergence Network (BcN), Telecommunication Information Networking Architecture (TINA), International Symposium on Applications and the Internet (SAINT), and International Conference on Information Networking (ICOIN). He was an Associate Editor of the IEEE TRANSACTIONS ON NETWORK AND SERVICE MANAGEMENT and the IEEE JOURNAL OF COMMUNICATIONS AND NETWORKS. He currently serves as an Associate Editor for the International Journal of Network Management.

C-ITS Applications, Use Cases and Requirements for V2X Communication Systems—Threading Through IEEE 802.11p to 5G

Zeeshan Hameed Mir and Fethi Filali

Abstract The existence of multiple wireless technologies and standards spurs the development of a wide range of cooperative applications for vehicular networking environments. Vehicle-to-Everything (V2X) communication systems enable several vehicular applications, and each poses a different set of performance requirements. Implementing these applications promises to bring substantial improvements to the way we travel. Therefore, identifying and addressing these requirements is necessary for developing future V2X communication systems. This chapter first describes the main application classes and analyzes their needs in several system performance metrics. Next, we provide a detailed review of the existing and next-generation communication technologies for multiple relevant objectives, including IEEE 802.11p, IEEE 802.11bd, LTE-V2x, C-V2X, 5G NR, and heterogeneous V2X. Finally, we presented a perspective on different V2X communication systems in their capabilities to support envisioned vehicular applications and use cases. Based on the analysis, we determine the suitability of communication technologies to satisfy the requirements.

1 Introduction

The rapid development of digital and wireless technologies over the last decades has led to the widespread use of technological advancements in many fields, including transportation. The Intelligent Transportation System (ITS) combines Information and Communication Technology (ICT) and traffic/transportation engineering concepts by placing roadside infrastructure and in-vehicle or onboard communication systems. Cooperative-ITS (C-ITS) focuses on communication, where vehicles com-

Z. Hameed Mir (✉)
Faculty of Computer Information Science, Higher Colleges of Technology (HCT), PO Box 4114, Fujairah, UAE
e-mail: zhameed@hct.ac.ae

F. Filali
Qatar Mobility Innovations Center (QMIC), Qatar University, PO Box 210531, Doha, Qatar
e-mail: filali@qmic.com

A. K. Pathan (ed.), *Towards a Wireless Connected World: Achievements and New Technologies*, https://doi.org/10.1007/978-3-031-04321-5_11

municate with each other and roadside infrastructure to provide safer, efficient, and convenient traveling methods.

C-ITS enables data exchange among vehicles, field devices attached to the roadside infrastructure, e.g., Roadside Units (RSUs), road users, and traffic management entities operating their services in a data center/cloud. C-ITS supports four different communication modes, i.e., Vehicle-to-Vehicle (V2V), Vehicle-to-Infrastructure (V2I), Vehicle-to-Pedestrian (V2P), and Vehicle-to-Network (V2N), as illustrated in Fig. 1. The term Vehicle-to-Everything (V2X) constitutes all four communication modes. Examples of V2X communication are vehicles or pedestrians near a signalized intersection exchange messages with RSUs containing local sensor measurements, positioning, and path trajectory predictions data to provide road users an unprecedented, extensive field of view on the surroundings. V2X allows road users to react to hazardous events appropriately within a critical time window. Other potential benefits include improving traffic flow, thus reducing road congestion and enhancing road user comfort.

V2X communication supports various applications and growing use cases in vehicular environments. Each application and its functional variants, i.e., use cases, have their own functional and performance requirements. These requirements have to be met simultaneously to achieve the benefits described above. For example, an advanced use case such as autonomous driving demands high throughput, lower latency, and ultra-high reliability. These challenges hinder the ability of a single

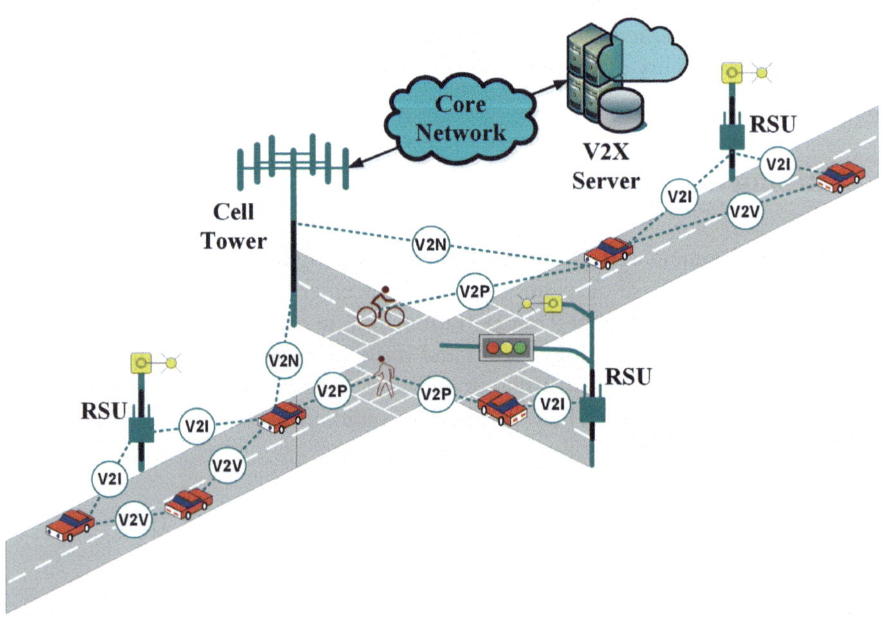

Fig. 1 Vehicular network scenario with four V2X communication modes, i.e., V2V, V2I, V2P, and V2N

wireless system to support the effective implementation of vehicular applications while catering to a diverse set of performance requirements. Several analytical, simulation and test-bed-based studies have been conducted on the feasibility of wireless technologies and standards for providing V2X communication. Among them, the most prominent ones managed by the Institute of Electrical and Electronics Engineers (IEEE) are IEEE 802.11p [1], IEEE 802.11bd [2]. Other alternatives include 4th Generation (4G)/Long Term Evolution (LTE) enabled LTE-V2x by 3rd Generation Partnership Project (3GPP) [3], and its continued evolutions Cellular-V2X (C-V2X) [4], and 5th Generation (5G) New Radio (NR) C-V2X [5]. In addition, many researchers argue the potential of using multi-tiered, heterogeneous networks [6, 7] for V2X communication.

The wireless technologies must meet the performance requirements to enable diverse applications through V2X communication and allow reliable functionality. Identifying these requirements will be a significant step in designing a multi-technology V2X communication system. For this purpose, we presented an intensive study on different types of V2X communication applications and use cases, along with a detailed analysis of their performance requirements. The primary emphasis is on analyzing system capabilities requirements such as radio and network communication and data requirements such as vehicle positioning capabilities. The study quantified each application's performance and functional requirements in metrics like beaconing frequency, minimum End-to-End (E2E) latency, communication range, reliability, data rate, and communication pattern. The main requirements are defined in European Telecommunications Standards Institute (ETSI) [8], United States (US) Department of Transportation (DOT) [9], and Car 2 Car-Communication Consortium (C2C-CC) [10] documents. Moreover, data available in most recently published review and survey literature such as [7, 11–15] is recapitulated and used as a central source of information to derive the requirements.

Next, we provide a comprehensive review of the existing and emerging wireless technologies and standards for supporting V2X communications. It includes reviewing both leading contending standards, i.e., IEEE 802.11p and LTE-V2x. We also have a detailed outlook of the enhancements incorporated in each emerging alternative, such as IEEE 802.11bd, C-V2X, 5G NR C-V2X, and the heterogeneous V2X networks. Finally, a qualitative analysis of different communication technologies is given based on their capabilities and the extent to which they meet the performance requirements of the various V2X communication applications and use cases.

The rest of the chapter is organized as follows. Section 2 summarizes the V2X applications use cases and analyzes their performance requirements. Section 3 describes the two existing communication technologies, IEEE 802.11p and LTE-V2x, supporting V2X communication. Section 4 presents the emerging V2X communication technologies and their capabilities to support vehicular application requirements. Section 5 includes a detailed analysis of to what extent the communication technologies described in previous sections can meet the performance requirement posed by different V2X application classes. Section 6 concludes the chapter.

2 V2X Applications, Use Cases, and Requirements

This section groups C-ITS applications into four main classes. The main classes are further divided into more specific sub-classes. Each sub-class represents a group of related use cases which are analyzed in terms of their communication pattern, V2X mode, and system performance requirements such as beaconing frequency (Hz), latency (ms), communication range (m), and message reception reliability (%). Figure 2 depicts some of the selected C-ITS applications and use cases.

Road Safety-Cooperative Awareness: The cooperative awareness applications aim at supporting road safety using inter-vehicle communications. For most road safety applications, vehicles periodically broadcast or geocast messages [10] within a limited time window, typically less than 100 ms [8]. The expected minimum requirement for the messaging frequency 10 Hz [8] which should be provided at a range of at least 300 m, and up to 1000 m, [10]. The required data rate has to be between 10 kbps to a maximum of 100 kbps, considering a typical message size between 60 and 1500 bytes [8] with needed reliability range from 90% to 95% [7].

Road Safety-Cooperative Maneuver: Cooperatively exchanging information about the surrounding traffic and movement trajectory information can reduce the risk of collisions while performing traffic mergers and lane-change maneuvers. The vehicles either conduct periodic broadcasts or establish unicast sessions [9], sending messages at the beaconing frequency 2 Hz–10 Hz [8] with very high reliability, i.e., over 99% [7]. The maximum tolerable latency has to be less than 100 ms [8]. The required range of communication has to be at least 150 m [9], and up to 500 m [12]

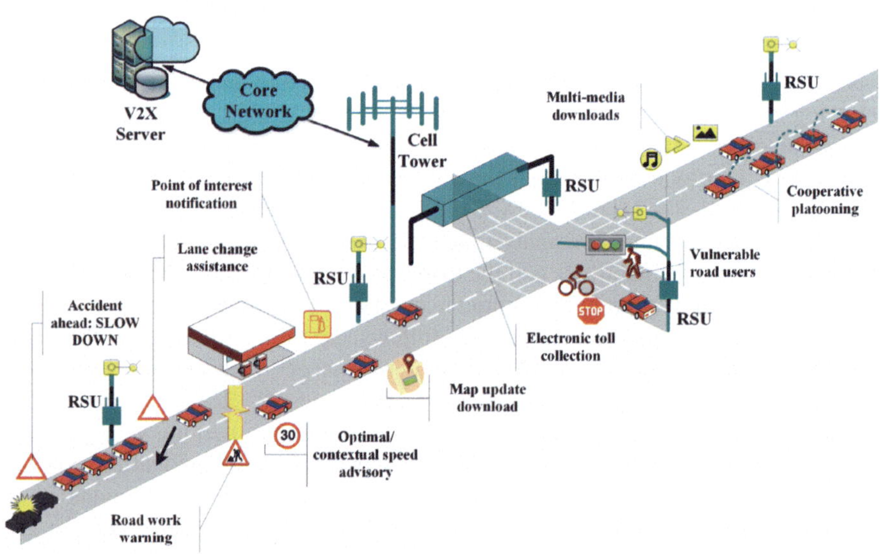

Fig. 2 Selected C-ITS applications and use cases in the domain of active road safety, traffic efficiency, infotainment, and eV2X

in high-speed scenarios. Since the use cases rely on information such as positioning and path trajectory prediction with lane-level accuracy ($\leq$2 m) [8], velocity, direction, and acceleration, the estimated data rate varies between 10 and 5000 kbps [7].

Road Safety-Road Hazard Warning: The road hazard warnings aim to indicate hazardous road situations such as a stationary vehicle or roadwork caused by or involving other road users. For most use cases, the periodic warning alerts are sent at the frequency 1 Hz–10 Hz [8], with the expectation that 90%–95% [7] of messages will be delivered successfully in 100 ms [8], and up to 1000 ms [9]. The indented warnings are generated either for road safety or traffic management purposes, each requiring a communication range over 200 m [9] and 2000 m [10], respectively. The data rate requirements vary between 1 and 10 kbps [11].

Road Safety-Vulnerable Road User Warning: The vulnerable road user warnings notify vulnerable road users (e.g., pedestrian, cyclist, motorcyclist) equipped with hand-held devices (e.g., smartphone) of an approaching vehicle and vice versa. The timely prevention of collision requires messages to be generated at a frequency 1 Hz [8] or 10 Hz [9], with the data rate of 10 kbps [7]. The expected latency has to be smaller than 100 ms [8] with 95% [7] messages successfully delivered. A radio link range of 200 m [9] is required to provide sufficient time for approaching road users or vehicles to react according to the event.

Road Safety-Cooperative Sensing: If a collision is unavoidable and imminent, the involved vehicles exchange relevant sensor and safety actuator, e.g., Anti-lock Braking System (ABS) data, to reduce the crash severity, and warning messages are broadcasted to all vehicles in the surrounding area. The vehicles may optionally share the messages with RSUs, which further relay them to more distant vehicles [8]. The required communication range is 50 m [9], but more significant distances i.e., up to 100 m [10] is beneficial to increase reaction time. Mainly these messages are considered to be highly time-critical with an expected maximum latency of 50 ms [8], or 20 ms [9], and a minimum generation rate 10 Hz [8], 50 Hz [9]. Since these use cases require a very high update frequency and reliability, the data rate has to be at least 50 kbps, and success rate of more than 95%, respectively [7]. The cooperative sensing and perception applications employ additional sources of information such as High-Definition (HD) streaming in see-through and Light Detection And Ranging (LiDAR) sensor use cases, each requiring data rates of 10 Mbps and 25 Mbps, respectively [7].

Traffic Efficiency-Cooperative Speed Management/Navigation: Traffic efficiency applications are intended to improve traffic flow, enhance traffic coordination/management and reduce the environmental impact of traffic congestion. The traffic management authorities initiate a periodic message broadcast via roadside infrastructure in Infrastructure-to-Vehicle (I2V) mode, and vehicles send and receive updates using V2N mode. The subclass of cooperative speed management enables the adaptation of traffic speed and routing control by cooperatively exchanging information about local speed or general traffic regulation constraints. Similarly, the cooperative navigation applications enhance possibilities for traffic management by periodical information exchanged between interconnected road users and adaptively reacting to on-road events. Traffic authorities will typically initiate itinerary recommendation messages. Depending on the use case type, the message generation frequency is required to be 1 Hz–10 Hz [8] with a tolerable end-to-end latency between 100 and 500 ms [8]. Typically, the data rate required is between 10 and 3000 kbps with reliability as high as 90% [7]. The required transmission range depends on the speed limit and will be between 50 m [9] and 500 m [8] for road safety and direct cooperation.

Infotainment-Local and Internet-based Services: Infotainment applications provide the user with information to enhance passenger convenience and enable global internet services. The subclass of cooperative local services application facilitates the use of different services offered by external providers where vehicular communication allows Location-based Services (LBS) such as Point of Interest (POI) notifications and parking management. Similarly, the subclass of Global Internet-based services includes use cases such as electronic toll collection, fleet management, and media downloading. To enable such services, either temporary or full internet access for Machine-to-Machine (M2M) communication must be provided. For most use cases, the beaconing frequency needed for RSUs to transmit information messages 1 Hz [8]. The minimum required communication range varies between 50 and 200 m [9] for the cooperative-local services when the vehicle is parked/moving slowly within close vicinity of a dedicated RSU. However, internet-based services with a fully mobile service provisioning might require link ranges of up to 2000 m. For nearly all use cases in the infotainment category, the average allowable latency is 500 ms [8]. The minimum expected data rate is 1 kbps for the local services to a maximum of several hundreds of kbps or Mbps for multimedia downloading/streaming (depends on content and quality) [12, 13]. The use cases like map or media download have medium reliability requirements as compared with a use case like local e-commerce, which requires a higher message reception rate [13].

Enhanced V2X (eV2X) applications and use cases: As part of the Release 15 [14] and 16 [15], 3GPP defines a set of enhanced V2X (eV2X) use cases which include advanced scenarios such as cooperative platooning and remote and autonomous driving, each with much more stringent requirements. The envisioned eV2X applications such as Advanced Driver Assistance Systems (ADAS) and Connected and Autonomous Driving (CAD) and evolving existing use cases such as cooperative platooning push the envelope of requirements by demanding even further performance improvements. For example, new situational awareness use cases involving autonomous vehicles require processing and exchanging more vehicle data from sensors such as cameras, Radio Direction And Ranging (RaDAR), and LiDAR with positioning and velocity data. The received information is then summarized to provide an extensive field of view on the driving environment. Compared to the other general categories of vehicular networking applications and use cases, ADAS and CAD require a high level of real-time data acquisition and processing to allow autonomous vehicles to react to a potentially hazardous event appropriately within a critical time window. With a maximum payload of 6500 bytes, a data rate of up to 1000 Mbps is needed to support use cases like fully automated driving. The eV2X use cases require message transmission with a frequency between 2 Hz to 50 Hz, over a distance between 50 and 1000 m. These use cases demand ultra-reliability up to 99.999% of successful message reception rate with ultra-lower end-to-end latencies, which varies between 3 and 100 ms.

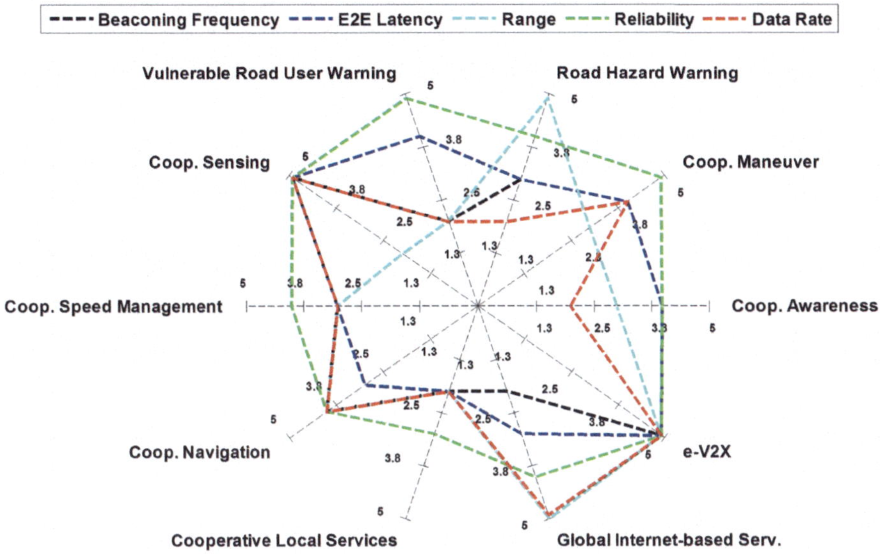

Fig. 3 The leading performance requirements, on the stringency scale for each of the C-ITS application categories, in terms of beaconing frequency, latency, coverage, reliability, and data rate

Table 1 summarizes the V2X applications and use cases with their performance requirements. Figure 3 illustrates the major performance requirements, along with their stringency levels for each of the C-ITS application categories. The higher level on the stringency scale represents more stringent requirements, i.e., higher beaconing frequency, lower latency, more extensive coverage range, etc. Here, it is noteworthy that all these applications pose a diverse set of performance requirements from the underlying V2X communication systems. Although except the eV2X advanced use cases, none of the application categories is high on all the dimensions, V2X communication systems capable of supporting long-range, reliable communication with lower end-to-end latencies and high data rates are highly sought after.

3 Existing V2X Communications Systems

3.1 Low Cost, Low Latency, and Proven IEEE 802.11p

The Dedicated Short Range Communications (DSRC) is a V2X technology based on IEEE 802.11p standard, which operates in the 5.9 GHz ITS spectrum. The physical (PHY) and medium access control (MAC) layers defined in IEEE 802.11p are derived from the IEEE 802.11a standard. The motivation was to support safety-critical, short-distance communications between the vehicles and between vehicles

Table 1 V2X communication applications and use cases with performance requirements

Application class	Application sub-class	Use case examples	Comm. pattern	V2X mode	Beacon freq. (Hz)	E2E latency (ms)	Comm. range (m)	Reliability (%)	Data rate (kbps)
Active road safety	Co-operative awareness	Emergency or slow vehicle, traffic turn, or intersection collision warnings	Event-triggered periodic broadcast [10]	V2X	10 [8]	100 [8]	≥300 ≤1000 [10]	90–95 [7]	10–100 [7]
	Co-operative maneuver	Merging traffic turn assistance, lane change assistance	Event triggered periodic broadcast; unicast for direct cooperation [9]	V2X	≥2≤10 [8]	100 [8]	≥150 [9] ≤500 [12]	≥99 [7]	10–5000 [7]
	Road hazard warning	Emergency electronic brake lights, Wrong-way driving, Stationary vehicle, Traffic condition, Signal violation, or roadwork warnings	Event-triggered periodic broadcast [8]	V2X/I2V	1–10 [8]	100 [8]–1000 [9]	200 [9]–2000 [10]	90–95 [7]	2–10 [11]
	Vulnerable road user warning	Vulnerable Pedestrian, motorcycle, bicyclist alert	Periodic [9], broadcast by RSU [8]	V2P / I2V	1 [8], 10 [9]	100 [8]	≤200 [9]	95 [7]	10 [7]
	Co-operative sensing	Pre-crash sensing, pre-collision brake assist	Unicast (high priority), broadcast [8]	V2X/I2V	10 [8], 50 [9]	20 [9], 50 [8]	50 [9], 100 [10]	≥95 [7]	20–25000 [7]
Traffic information & efficiency	Co-operative Speed Management	Regulatory/contextual speed limits notification, Traffic light optimal speed advisory	Periodic broadcast by RSU connected to cloud service [7, 8]	V2N/I2V	1–10 [8]	100 [8], 1000 [9]	≥50 ≤250 [9]	≤90 [7]	10–2000 [7]
	Co-operative navigation	Traffic information and recommended itinerary, Map download and updates	Periodic broadcast by RSU, Unicast session with backend server [7, 8]	V2N/I2V	1–10 [8]	≥500 [8]	≥300 [9]		
Infotainment	Co-operative Local Services	Point of interest notification, Automatic access control and parking management	Machine-to-Machine (periodic broadcast and unicast) [12]	V2N/I2V	1 [8]	≤500 [8]	≥50 ≤200 [9]	≈ 80 [13]	1–10 [13]
	Global Internet-based Services	Insurance/Financial service, Fleet/Loading zone management, Browsing, streaming, Media/software download	User access to Internet [12]	V2N/I2V	1 [8]	≤500 [8]	≥50 ≤2000 [9]	≈ 80–90 [13]	1- tens of Mbps [12, 13]
Enhanced V2X (eV2X)	Tele-operated and Remote Automated Driving	Vehicle platooning, extended sensors, advanced and remote driving	Periodic or event triggered groupcast, broadcast, unicast [15]	V2N/V2X	2–50 [15]	3–100 [15]	50–1000 [15]	90–99.999 [15]	1–1000 Mbps [15]

and the roadside infrastructure (e.g., RSU). Since most warning and awareness messages are transmitted locally, the design objectives were to support the transmission frequency 1 Hz–10 Hz over a communication range between 100 and 800 m while maintaining a communication latency of less than 10 ms. The physical layer of IEEE 802.11p uses Orthogonal Frequency Division Multiple Access (OFDMA). It includes reduced bitrate and subcarrier spacing by half of IEEE 802.11a and increased OFDM symbol and preamble durations by twofold to support V2V and V2I communication in a rapidly changing vehicular environment. Operating in the ad-hoc mode, i.e., in the absence of network infrastructure, the IEEE 802.11p uses Carrier Sense Multiple Access with Collision Avoidance (CSMA/CA) protocol as the multiple access mechanism. The IEEE 802.11p also implements the Enhanced Distributed Channel Access (EDCA) to prioritize channel access, allowing safety and time-critical messages to be sent with relatively lower contention delays. In low vehicle density and congestion scenarios, the IEEE 802.11p provides stable performance in terms of reliability and latency. However, the performance degrades significantly in high traffic, speed, and vehicle density environments.

3.2 *Scalable, Reliable, and Ubiquitous LTE-V2x*

3GPP initiated the V2x study item in 2015 with the initial mandate to explore the service aspects of LTE-based V2X communication, i.e., LTE-V2x [16]. Building upon the classic cellular 4G/LTE uplink (UL)/downlink (DL) connectivity platform, the basic idea is to leverage the LTE standard for delivering services like traffic information and efficiency and infotainment for automotive. The LTE-V2x establishes the foundation for the C-V2X and continues evolution to the future 5G NR C-V2X that will incorporate the radio interfaces to support V2X communication through direct modes and network mode paradigms. Based on the LTE direct device-to-device (D2D) design, it operates both in-coverage and out-of-coverage modes. The direct method allows vehicles to communicate directly with or without infrastructure assistance. The in-coverage (also cellular-assisted) relies on an infrastructure node such as a base station to coordinate communications between the vehicles. The infrastructure node performs link scheduling and resource allocation to support extremely low latency and high-reliability operations. Vehicles perform distributed scheduling without involving the infrastructure node in the out-of-coverage (also cellular unassisted) mode. The indirect model leverages existing UL/DL LTE communication where vehicles send a unicast message to the base station (UL) that further relays to an ITS server (edge or backend). The ITS server reaches other vehicles using LTE broadcast (DL).

4 Outlook to Emerging V2X Communication Systems

The emerging vehicular applications and use cases require the continued evolution of V2X technology. This section provides an outlook on the evolutionary standards such as IEEE 802.11bd for DSRC-based IEEE 802.11p, C-V2X, and 5G NR C-V2X for LTE-V2x and multiple radio access technologies (RAT) (also multi-RAT) heterogeneous communication systems.

4.1 Next-Generation IEEE 802.11bd V2X

The current IEEE V2X specification IEEE 802.11p is primarily based on IEEE 802.11a for the PHY and MAC layers. With lower reliability and higher delays in high-density, high-speed, and high-transmission scenarios, it becomes increasingly difficult to guarantee the current and future requirements of the vehicular application and use cases. Formed in January 2019, the IEEE 802.11bd Task Group is mandated to improve the IEEE 802.11 MAC and PHY layers for Next Generation V2X (NGV) communications. The IEEE 802.11bd is dubbed as the evolution of IEEE 802.11p for V2X communications leverages the MAC, and PHY layer enhancements from IEEE 802.11n, IEEE 802.11ac, IEEE 802.11ax [17] to provide next-generation V2X protocol. The goals are to double the throughput, relative speed support, and communication range of IEEE 802.11p while reducing the E2E latency. The other key features include coexistence, backward compatibility, fairness between IEEE 802.11p and IEEE 802.11bd devices, and cooperative positioning techniques [18].

- Support for higher throughput and reliability: The following technology enhancements are being investigated at the PHY and MAC layers to support the higher throughput and reliability requirements.
  - Higher bandwidth: By enabling 20 MHz bandwidth in the 5.9 GHz ITS band, the IEEE 802.11bd specification can significantly increase bandwidth, thus meeting the 2 times more than IEEE 802.11p throughput goal. The OFDMA numerology from IEEE 802.11ac and ax can be leveraged to use the 20MHz bandwidth further efficiently [19]. Furthermore, mmWave frequencies (60 GHz and above) can enhance V2X by supporting 1 Gbps or more data rates while maintaining a lower latency [20].
  - Reliable operation: To improve reliability, IEEE 802.11bd can include an adaptive retransmission mechanism [21] and associated PHY signaling [22]. The fully interoperable schemes exploit channel load information to control the number of retransmissions. The performance improvement in reliability is achieved for both legacy IEEE 802.11p and NGV stations without creating congestion. In addition, the spatial and frequency diversity gain through Low-Density Parity Check (LDPC) coding can also help improve the throughput and reliability performance. Moreover, already supported by IEEE 802.11p devices, IEEE

802.11bd devices can ensure fair channel access by using EDCA categories at the MAC layer.

- Support for longer range and higher relative speed: The IEEE 802.11p standard suffered from performance degradation due to shorter communication range and high-speed scenarios. The following technology guidelines are being investigated towards meeting the requirements of a more extended range and double the relative speed, i.e., up to 500 km/h.
  - Longer communication range: The IEEE 802.11bd complements an extra-range mode using Dual-Carrier Modulation (DCM), where performance gain (frequency diversity) of 3 dB or higher is achieved by replicating the same symbols twice on different subcarriers [19]. Similarly, another proposal suggested using Space-Time Block Coding (STBC) from IEEE 802.11n, which takes two antennas to transmit the same information, combined to achieve gains (spatial diversity) [23].
  - Higher speed robustness: The midamble-based channel estimation is proposed to address fast-varying channels' high Doppler frequency issues during a high-speed environment. With carefully selected mid-amble insertion periodicity between the OFDM data symbols, it is reported that an accurate channel estimation can reduce the Doppler Effect and improve the performance under high mobility scenarios [24].
- Support for positioning: Many vehicular applications require positioning, and path trajectory prediction with accuracy between 1m to 10m [8]. Several candidate solutions for faster ranging and positioning exist, such as Fine Timing Measurement (FTM) protocol defined in IEEE 802.11mc [25] or its enhancement such as EDCA FTM. Other proposals include Passive non-trigger-based (nTB) and multi-channel ranging [26, 27] as defined in IEEE 802.11az [28], also referred to as Next Generation Positioning (NGP).
- Support for interoperability, backward compatibility, and coexistence: IEEE 802.11p and IEEE 802.11bd systems must interoperate and coexist without changing the higher layers operations and segmenting the applications into legacy and NGV. The interoperability and backward compatibility require IEEE 802.11p devices to decode messages sent by IEEE 802.11bd devices and vice versa and IEEE 802.11bd devices to operate in a fully interoperable mode with the IEEE 802.11p devices, respectively. On the other hand, the coexistence only necessitates both IEEE 802.11bd and IEEE 802.11p devices to be aware of each other's presence and defer channel access [29]. Several interoperable and backward compatible PHY layer enhancements [30–34] have been investigated while ensuring fairness and performance improvements in reliability, higher throughput, and more extended communication range.

4.2 Evolvable and Expandable C-V2X

Motivated by the potential of developing a single adaptable technology and offering versatile performance in the V2X context, 3GPP in Release 14 introduced C-V2X. Initially defined as LTE-V2x, it leverages the existing 4G/LTE networks and describes two complementary transmission methods, i.e., LTE Direct and LTE Broadcast network communications [35].

4.2.1 PC5 Interface Enabled V2X

Similar to the concepts of LTE D2D communications, the objective is to provide direct communication capabilities and support for latency-sensitive V2X use cases native to cellular networks operating in the ITS 5.9GHz band. Direct communications work either with or without infrastructure assistance, i.e., Evolved Node B (eNodeB) to allocate resources among the communicating vehicles, also referred to as in-coverage and out-of-coverage modes. Transmission Mode 3 is defined as the network-assisted mode where the network, i.e., eNodeB, handles the allocation of resources. Mode 3 utilizes many scheduling mechanisms such as semi-persistent, cross-carrier, and UE report-based. Transmission Mode 4 is characterized by distributed scheduling and congestion control without network assistance. Unlike CSMA/CA-based resource selection in IEEE 802.11p, the vehicles autonomously perform the resource selection based on the channel sensing with the semi-persistent transmission.

In contrast to asynchronous IEEE 802.11p, the focus is on Global Navigation Satellite System (GNSS) based synchronization. With scalable access control mechanism, extended network coverage, and Frequency Division Multiplexing (FDM) usage in C-V2X provides an advantage compared to IEEE 802.11p. The research finding suggests that LTE-V2V is capable of attaining superior performance as compared with IEEE 802.11p in terms of Packet Reception Ratio (PRR) and range improvements [36, 37]. However, higher complexity and cost associated with the solutions might lead to performance problems and must be addressed before it is considered viable end-to-end solutions for providing V2V communications.

The direct communication transmission method (over PC5 interface) builds upon several enhancements to overcome some of the technical limitations of the existing LTE networks in a V2X communication context. However, the direct method has to overcome several challenges such as high speed/high Doppler scenarios, near-far problems, resource allocation, half-duplex, and synchronization issues. Towards this end, many solutions have been proposed, such as enhancing the signal design and transmission structure, geo-zoning, efficient resource allocation with the energy-based selection, transmission repetition, and the usage of satellites for time synchronization. These solutions have higher spectral efficiency, reliability with increased communication ranges, and lower channel access overhead.

4.2.2 LTE-Uu Interface Enabled V2X

The primary focus is to enhance the existing LTE cellular uplink/downlink communications (over Uu interface) to support latency tolerant V2V use cases and access cloud services via V2N in traditional mobile operators' licensed spectrum. The V2X communication via network involves a V2X server, which receives unicast messages from vehicles using LTE unicast (uplink transmissions). At the same time, the base stations (eNodeB/eNodeB-type RSU [35]) broadcast messages from a V2X back-end server to vehicles using LTE Broadcast (downlink transmissions). In addition to enhancing network range/coverage through RSU deployment, the combination of macro and small cells potentially improves the network capacity. It also allows enhanced functionalities such as multicast/broadcast in the downlink transmission to reduce the latency and improve radio efficiency via the Single-Cell Point To Multipoint (SC-PTM) mechanism, which uses Evolved Multimedia Broadcast Multicast Services (eMBMS) architecture.

4.3 Futureproof 5G NR Based C-V2X

C-V2X offers an *evolutionary path* towards 5G NR C-V2X. For advanced and eV2X use cases, 5G NR by 3GPP (5G NR C-V2X) has emerged as a natural successor to LTE-based C-V2X for vehicular communications. The 5G NR design is built on four pillars: NR design, massive Multiple-Input and Multiple-Output (MIMO) and beamforming, multi-connectivity, and network visualization.

- 5G New Radio Design: 5G NR will use a wide array of spectrum bands and types. In addition to low bands (below 1 GHz) and mid bands (between 1 and 6 GHz), it is proposed to utilize high bands ranging between 24 and 100 GHz (also mmWave), resulting in three main advantages, (1) mmWave frequency band is less utilized, (2) Higher frequency carries much more capacity and higher data rates than lower frequency bands, and (3) Make possible to use Massive MIMO techniques. The choice of radio waveform and multiple access techniques profoundly impact the 5G NR design. Unlike LTE, the 5G NR allows flexible sub-carrier frequency spacing, enabling faster delivery of low-latency payloads. The scalable/flexible OFDM numerology leads to an increased number of slots per subframe (or short symbols), thus increasing the number of symbols transmitted in a given time. Another critical 5G NR design component is multiple access schemes such as Resource Spread Multiple Access (RSMA) and Sparse code multiple access (SCMA). The non-orthogonal multiple access schemes are well suited for use cases requiring asynchronous and grant-free access with full mobility support.
- Massive MIMO and beamforming: The higher frequencies work well with smaller-sized antennas, thus making it possible to install many antenna elements in the base station and user devices. The 3GPP 5G NR supports robustness, extended coverage, and sustained high traffic capacity at a higher frequency by combining Massive

Multi-user (MU) MIMO with hybrid beamforming and beam tracking techniques, especially in Non-line-of-sight (NLOS) environments and user mobility.

- Multi-Connectivity Multi-RAT 5G NR: Multi-connectivity (MC) in 5G NR extends the dual connectivity (DC) concept in LTE to aggregate the radio resources of multiple base stations from different RATs like LTE, Wi-Fi, etc. The benefits include throughput and data rate enhancements via splitting data from various sources. Simultaneous connectivity to other network layers and RAT also provides seamless mobility with data duplication boosting reliability. To this aim, efficient MC mode configuration [38] and multi-cell scheduling algorithms [39] are required to obtain an optimal tradeoff between reliability and latency due to buffering delay, especially in the presence of high cell traffic load.
- Network Softwarization and virtualization: Network slicing provides the elasticity required to run multiple and diverse C-ITS applications over a unified 5G network platform. The end-to-end approach to network slicing spans 5G Core (5GC) and Next-Generation Radio Access Network (NG-RAN). It is implemented utilizing softwarization of the control plan and user plan network functionalities (NFs) using Network Function Virtualization (NFV) and Software-Defined Networking (SDN) [40]. NFV allows deployment of related NFs as Virtual NFs (VNFs) containers or virtual machines running either in the remote cloud or edge through the Multi-access Edge Computing (MEC) paradigm. SDN enables flexible interconnection and reconfiguration of VNFs based on service requirements and network infrastructure dynamics. MEC brings cloud computing and services closer to the users or edge, thus reducing latency.

With much higher data rates and Ultra-Reliability and Low Latency Communications (URLLC), 5G NR promises to support stringent performance requirements posed by remote and automated driving. 5G NR C-V2X Release 16 and onwards introduces complementary capabilities to support advanced use cases.

- Support for high data rates and increased capacity: The 5G NR C-V2X includes scalable OFDM and flexible Demodulation reference signal (DMRS) to achieve higher spectral efficiency at high-speed scenarios. The wide-band carrier support provides higher data rates and system capacity. Moreover, increased capacity is achieved through link-level gain, Hybrid automatic repeat request (HARQ) feedback, and resource allocation enhancements [5]. In addition, the use of physical layer numerology and different sub-carrier spacing support configuration flexibility.
- Support for URLLC operations: The 5G NR C-V2X utilizes advanced channel (LDPC/polar) coding for increased reliability. The specification support ultra-reliable and low latency communication due to self-contained slot structure (mini-slots) and flexible resource allocation. Semi-persistent scheduling for periodic traffic with similar packet sizes and per-packet scheduling with variable traffic and packet sizes support low latency V2X communications. In addition, groupcasting is where vehicles within a certain distance can form groups and reliably connect based on similar interests in services. The efficient Single Frequency Network

(SFN) feedback of Negative Acknowledgement (NAK) from the receivers also provides reliable unicast and multicast support.

4.4 Unified Multi-RAT Heterogeneous V2X

The multi-RAT, heterogeneous vehicular advocates argue that a single technology cannot satisfy all requirements. Therefore, the higher bandwidth requirements posed by emerging cooperative sensing and perception use cases can only be met by various solutions which are not limited to DSRC and C-V2X coexistence and integration but include a broad spectrum of communication technologies such as Visible light communication (VLC) [41], mmWave [42], and TV white space [6].

In heterogeneous solutions for V2X communication systems, vehicles are equipped with multiple RATs to leverage the DSRC for providing short-range direct communications and the suitability of C-V2X for delivering long-range network communications [43]. The US DOT equally supports this idea, and European Union (EU) [44]. In a report [45] by National Highway Traffic Safety Administration (NHTSA), while proposing a mandate to use IEEE 802.11p, it also allows provisions for other alternative wireless communication technologies that are interoperable with IEEE 802.11p. Similarly, the EU, in their report [46] on C-ITS strategy, fully endorsed the hybrid communication approach, which combines ETSI ITS-G5 compliant DSRC and existing 4G/LTE, as well as 5G, enabled C-V2X for providing C-ITS applications and services. The envisioned approach has at least two realizations.

4.4.1 Coexistence of Multi-RAT V2X

The main idea is that the two prominent radio access technologies coexist on two channels in the same 5.9 GHz ITS spectrum with a guard band separating both channels [47]. Different coexistence strategies have been investigated, including DSRC and Wi-Fi [48] and Wi-Fi and C-V2X technologies coexistence [49]. One alternative is frequency sharing between DSRC and C-V2X and their respective evolutionary successors, i.e., IEEE 802.11bd and 5G NR C-V2X [50]. The proponents of these ideas like to point out the advantages like interoperability and leveraging the complete ITS band to support cohesive end-to-end solutions for V2X communications. The opponents argue that extreme caution must be exercised against the negative impacts of sharing the ITS spectrum of safety networks. The implications include higher interference, bandwidth reduction for safety applications, and non-interoperable technologies.

4.4.2 DSRC and C-V2X Integrated Hybrid V2X

The other alternative is multi-tiered [51], multi-technology-enabled hybrid vehicular network architecture [52] which assumes that vehicles with dual interfaces are capable of alternating between them using RAT selection algorithms and by performing Vertical Handover (VHO). In [53], Mohamed et al. have successfully implemented and demonstrated IEEE 802.11p, and 4G/LTE enabled the hybrid vehicular network to support multimedia content delivery. They reported significant performance improvements in terms of reliability and packet delivery metrics. Similar concepts have been extensively explored earlier in the context of combining different Wireless Wide Area Network (WWAN) technologies such as Code Division Multiple Access (CDMA) 2000, Universal Mobile Telecommunications System (UMTS), CDMA 2000 1x-EV, and 3rd Generation (3G) with collocated IEEE 802.11 based Wireless Local Area Network (WLAN). The Vehicular Ad hoc Network (VANET) VANET-LTE integration gained widespread attention where the cellular network is utilized either as a backup or for offloading data [54]. The former cases follow the Always Best Connectivity (ABC) paradigm [55] which allows vehicles to communicate over the best available communication channel via VHO. The definition of the best varies depending on the communication system capabilities, application requirements, policies, security, network coverage, and resource utilization. In the latter cases, primarily the DSRC interface is used for message transmissions among the vehicles and a set of specialized gateway entities. The mobile or static gateways are equipped with dual interfaces and relay the messages towards vehicles directly or via roadside or cellular infrastructure.

Figure 4 shows a comparison among different communication technologies in their ability to meet multiple functional requirements of V2X applications such as (a) Operation in the absence of a network, (b) Support of V2V, (c) Support of safety-critical use cases, (d) Support of V2I/I2V, (e) Support of multimedia services, (f) Network coverage, and (g) Advanced PHY [56, 57].

5 Radio Technology Implications for C-ITS Applications Requirements

This section provides a review of the capabilities of the existing communication technologies to support the application requirements identified in Table 1. Figure 5 illustrates the capabilities of various communication technologies to support different application performance requirements using a radar graph. The suitability of a particular technology to support a use case and its requirements can only be guaranteed under specific circumstances and subjected to adjustment of operating conditions and system parameters and their configuration.

Since most road safety applications and use cases can be characterized as time-critical with higher messaging frequencies and reliable communication among vehi-

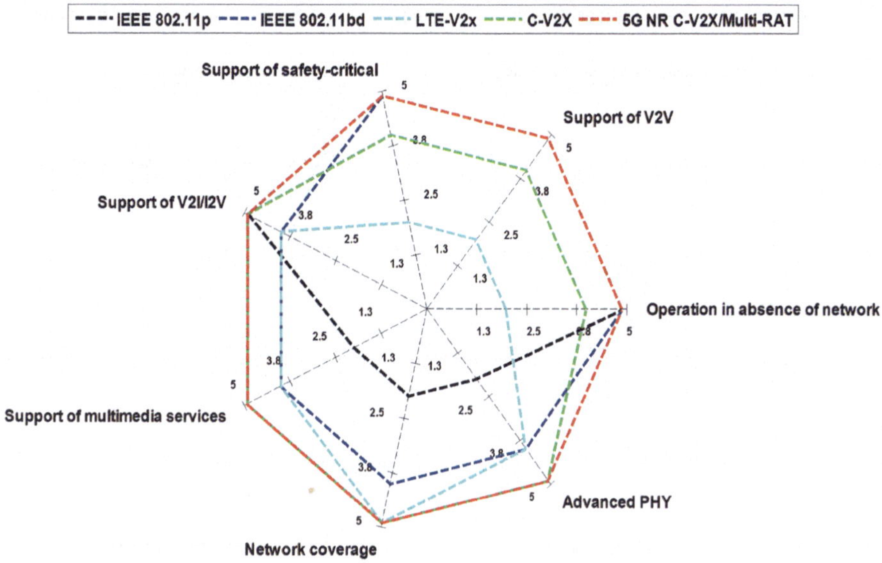

Fig. 4 The ability of communication technologies to meet seven leading functional requirements of V2X applications

cles (V2V), the IEEE 802.11p and IEEE 802.11bd seem to be a perfect fit. However, in high densities and high load scenarios, it becomes challenging to continue to support these requirements [58]. In such situations, the use of priority/Quality-of-Service (QoS) mechanisms available in IEEE 802.11p, such as Decentralized Congestion Control (DCC) and EDCA, might be helpful to minimize the expected delays [59], which has to be smaller than 100 ms. With the PHY/MAC enhancements and the 2x performance improvements promised, the IEEE 802.11bd supports all defined DSRC use cases and the more demanding applications such as cooperative maneuver and sensing.

On the other hand, without native support, functions, and processes explicitly designed to meet stringent delay requirements of V2V communications, 4G/LTE has limited applicability in use cases where the direct unicast operation is necessary. Especially in the presence of higher background cellular traffic load, the delay increases significantly [58]. For example, use cases such as cooperative sensing characterized by excessive-high messaging frequency and stringent lower latencies requirements, a direct DSRC link, or unified connectivity with C-V2X (PC5 interface based) will be necessary for the vehicles. The 4G/LTE as primary radio access technology will not be sufficient for such use cases. For reliable message dissemination over a longer distance, it may be advantageous to simultaneously use cellular technologies (LTE-V2x/C-V2X, or 5G) or exploit multi-RAT connectivity. The application of cellular network as a primary radio access technology is recommended in all use cases where:

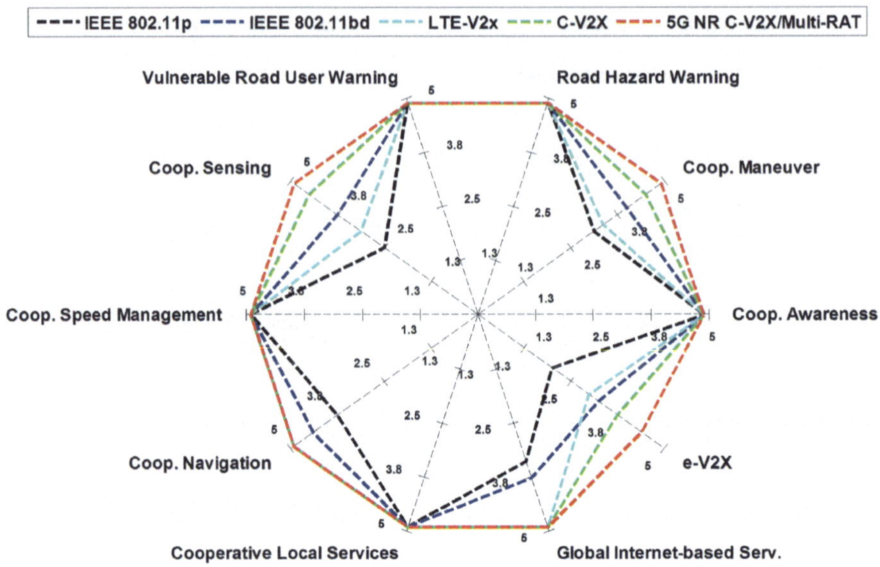

Fig. 5 The ability of the V2X communication technologies to support major categories of C-ITS applications and use cases

- No or only a lossy IEEE 802.11p/bd links are available. These situations often arise due to propagation conditions, channel congestion, or at intersections where the Line-of-Sight (LOS) might be obstructed by buildings or other objects such as foliage and vehicles.
- A more substantial area needs to be covered quickly (e.g., in highway scenarios) to increase the probability of message delivery.
- Communication with infrastructure, i.e., a base station (eNodeB/eNodeB-type RSU) or interactions with the background servers for accessing the cloud service is required.

Since traffic information data cannot be considered time-critical unless applied to the emergency scenarios, a maximum tolerable end-to-end latency of 500ms will be sufficient to run most of the use cases. Therefore, whenever a reliable message transmission via DSRC based communication link cannot be provided, the expandable C-V2X systems can be considered for these use cases. C-V2X exhibits superior performance thus has strongly emerged as an alternative to DSRC-based V2V communication. The performance gains in extended coverage, scalability, reliability, latency, and mobility support make the C-V2X and its successor the most suitable candidate for supporting diverse and continuously evolving use cases and their requirements. Moreover, a broad communication range of several hundred meters will benefit traffic information and itinerary recommendation use case. The extended coverage can be achieved either by using multi-hop broadcasting, the interconnection of local RSU or by using cellular technologies for whole area radio coverage.

Mobile networks can provide more significant link capacity and distances (especially in suburban and rural macro-cells). When applied in traffic efficiency and management context, some use cases require the propagation of notification messages across distances up to 2 km and high data rates. For IEEE 802.11p and its successor IEEE 802.11bd, the local radio link range might provide enough time for approaching vehicles to react accordingly. The event forwarding/multi-hop message dissemination will be extensively used to allow warning notifications over distances of several kilometers. Therefore, applying cellular technologies can be beneficial to reduce the DSRC channel congestion and the delay. The C-V2X communication technologies can support use cases with long-range and high data rate requirements (such as map download and upload) within a wide area without depending on fast varying ad-hoc network topologies.

For supporting infotainment use cases, temporary or full internet access and M2M type communication must be provided using DSRC based communication between the vehicle and an RSU or over the cellular networks. If a DSRC link is used for internet access, RSUs will be required to work as an access point and be attached to a fixed network or wireless internet access. A direct cellular connection will preferably be used as internet access technology to support fully mobile interaction. Depending on the implementation, either RSU itself may be required to run the service, or a background application server will be necessary. This application class is particularly suited to be used over a cellular network.

The eV2X applications and use cases such as ADAS involve high throughput sensor data sharing, low latency real-time local updates, and ultra-reliable coordinated driving. The two emerging communication technologies that can meet these requirements coexisting and backward compatible with their predecessors are IEEE 802.11bd and 5G NR C-V2X. More recent research results indicate that these two technologies outperformed their respective precursor technologies [60]. The theoretical studies and link-level/system-level simulations show that 5G NR-V2X performed superior to IEEE 802.11bd in data rates, transmission latency, and communication range [61–63]. It has been reported that multi-tiered network architecture supported by multi-RAT communication systems tends to increase gains in terms of reliability and greater robustness against channel congestion and higher speeds. The multi-RAT communication systems bring further performance improvements [64].

For a particular application and use case, the stringency of performance requirements increases with automation level in driving [65]. For example, from limited driving assistance to conditional automated and fully-automated driving, the data rate, latency, spectral and energy efficiency, and reliability vary to sense and map the surrounding environment accurately [66]. In addition to this, higher vehicle mobility and density, massive connectivity, and heavy traffic load conditions pose further challenges far beyond the capabilities of the existing wireless technologies [67]. Many believe that the solution lies in the 6G wireless communication network for V2X or 6G-V2X that will fulfill the requirements of next-generation and beyond V2X applications. The critical technology enablers for 6G-V2X include THz communication, edge intelligence (EI) supported MEC, distributed Artificial Intelligence

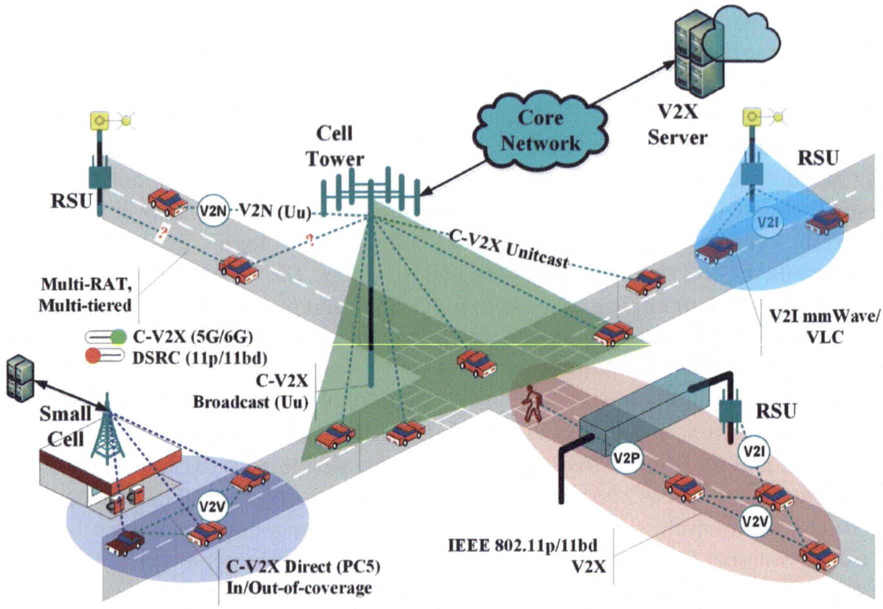

Fig. 6 Heterogeneous multiple-technology enabled vehicular network architecture with constituent components, such as IEEE 802.11p/bd, LTE-V2x, C-V2X, 5G NR V2X and other emerging technologies

(AI)-enabled wireless intelligence, and intelligent Unmanned Aerial Vehicle (UAV)-assisted extended communication infrastructure [68–70].

Figure 6 visualize the multi-tiered, multi-RAT, heterogeneous network architecture, where different RATs can not only coexist but aggregate resources to enable existing and future use cases.

6 Conclusion

This chapter identified essential performance and functional requirements for different C-ITS applications and use cases such as road safety, traffic efficiency, infotainment, and eV2X. These use cases are categorized, and their requirements are quantified in terms of V2X communication performance metrics like beaconing frequency, latency, communication range, reliability, and data rates. A detailed review of well-established and future emerging communication technologies is presented in multiple objectives relevant to V2X communication systems. Finally, the requirements are compared against the capabilities of different communication technologies such as IEEE 802.11p, IEEE 802.11bd, LTE-V2x, C-V2X, 5G NR, and multi-RAT V2X. The active road safety and advanced ADAS/CAD use cases pose the most stringent latency, throughput, and reliability requirements. The road safety applications requirements can be satisfied by direct V2V/V2P communication modes using DSRC-based technologies such as IEEE 802.11p/bd under low density and acceptable network load conditions and appropriate QoS/priority mechanisms in place. Most traffic efficiency and infotainment use cases are less time-critical but demand high reliability and broader coverage. To this end, IEEE 802.11bd and 5G NR C-V2X communication fulfill most requirements. Finally, combining the best from multiple wireless technologies in the form of spectrum sharing, multi-RAT heterogeneous network, and a unified 5G/6G C-V2X networking solutions hold the key to the successful implementation of future use cases with evolving performance requirements.

References

1. IEEE Std 802.11-2016, *IEEE Standard for Information technology—Telecommunications and information exchange between systems Local and metropolitan area networks–Specific requirements-Part 11: Wireless LAN Medium Access Control (MAC) and Physical Layer (PHY) Specifications*, Revision of IEEE Std 802.11-2012, 1-3534, (2016). https://doi.org/10.1109/IEEESTD.2016.7786995
2. *IEEE P802.11-TASK GROUP BD (NGV) Meeting Update*
3. LTE, *Overview of 3GPP Release 8* (2010)
4. 5G Automotive Association (5GAA), *The Case for Cellular V2X for Safety and Cooperative Driving*. White Paper (2016)
5. QualComm, *5G NR based C-V2X*, Presentation (2019)
6. J. Gozalvez, M. Sepulcre, R. Molina, O. Altintas, *Heterogeneous V2X Networks for Connected and Automated Vehicles*. Speaker Presentation at IEEE 5G Summit (2017)
7. M. Boban, A. Kousaridas, K. Manolakis, J. Eichinger, W. Xu, Connected roads of the future: use cases, requirements, and design considerations for vehicle-to-everything communications. IEEE Vehicular Technol. Mag. **13**(3), 110–123 (2018). https://doi.org/10.1109/MVT.2017.2777259
8. ETSI TR 102 638, *Intelligent Transport Systems (ITS); Vehicular Communications; Basic Set of Applications; Definitions*, Std. ETSI TR 102 638 V1.1.2, (2009)
9. DOT HS 809 859, *Vehicle Safety Communications Project Task 3 Final Report: Identify Intelligent Vehicle Safety Applications Enabled by DSRC*, (2005)
10. C2C-CC Manifesto, *CAR 2 CAR Communication Consortium Manifesto Overview of the C2C-CC System*, V 1.1 (2007)
11. D. Bowman, S. Baker, S. Stone, Z. Doerzaph, R. Hanowski, *Development of Performance Requirements for Commercial Vehicle Safety Applications*, (Report No. DOT HS 811 772) (National Highway Traffic Safety Administration, Washington, DC, 2013)
12. G. Pocovi, M. Lauridsen, B. Soret, K.I. Pedersen, P. Mogensen, Automation for on-road vehicles: use cases and requirements for radio design, in *2015 IEEE 82nd Vehicular Technology Conference (VTC2015-Fall)* (2015). https://doi.org/10.1109/VTCFall.2015.7390848
13. K. Zheng, Q. Zheng, P. Chatzimisios, W. Xiang, Y. Zhou, Heterogeneous vehicular networking: a survey on architecture, challenges, and solutions. IEEE Commun. Surv. Tutor. **17**(4), 2377–2396 (2015). https://doi.org/10.1109/COMST.2015.2440103

14. 3rd Generation Partnership Project; Technical Specification Group Services and System Aspects, *5G; Service requirements for enhanced V2X scenarios*, 3GPP TS 22.186 version 15.3.0 Release 15, 3GPP TR 22.186, 3GPP, (2018)
15. 3rd Generation Partnership Project; Technical Specification Group Services and System Aspects; *Study on Enhancement of 3GPP Support for 5G V2X Services* (v16.2.0, Release 16), 3GPP TR 22.886, 3GPP, (2018)
16. 3GPP TR 22.885, *Study on LTE Support for Vehicle to Everything (V2X) Services*, (Release 14), 3GPP Technical Specification Group Radio Access Network, v1.0.0, September 2015
17. IEEE P802.11ax/D8.0, *IEEE Draft Standard for Information technology– Telecommunications and information exchange between systems Local and metropolitan area networks–Specific requirements Part 11: Wireless LAN Medium Access Control (MAC) and Physical Layer (PHY) Specifications Amendment 1: Enhancements for High Efficiency WLAN,* 1-820, (2020)
18. G. Naik, B. Choudhury, J. Park, *IEEE 802.11bd & 5G NR V2X: Evolution of Radio Access Technologies for V2X Communications*, IEEE Access, 7 (2019), pp. 70169–70184. https://doi.org/10.1109/ACCESS.2019.2919489
19. L. Jianhan, A. Gary, P. Thomas,*IEEE 802.11-19/0774r1 Modulation Scheme for 11bd Range Extension Update*, IEEE NGV Meeting (2019)
20. M. Hiroyuki et al., *IEEE 802.11-18/1187r1 mmW for V2X use cases* (IEEE NGV Meetings, 2018)
21. F. Michael et al., *IEEE 802.11-19/784r0 Adaptive Repetition Scheme for NGV* (IEEE NGV Meeting, 2019)
22. Y. Rui et al., *IEEE 802.11-19/1596r0 PHY Signaling for Adaptive Repetition of 11p PPDU* (IEEE NGV Meeting, 2019)
23. S. Coffey, *IEEE 802.11-19/1299r0 Extended Range Modes in 11bd* (IEEE NGV Meeting, 2019)
24. L. Dongguk et al., *IEEE 802.11-19/332r2 PHY designs for 11bd* (IEEE NGV Meeting, 2019)
25. IEEE P802.11MC DRAFT, *Information Technology—Telecommunications and Information Exchange Between Systems Local And Metropolitan Area Networks—Specific Requirements PART 11: Wireless LAN Medium Access Control (MAC) AND Physical Layer (PHY) Specifications* (2016)
26. S. Stephan et al., *IEEE 802.11-19/0365r0 Consideration on Positioning with 802.11bd* (IEEE NGV Meeting, 2019)
27. S. Stephan et al., *IEEE 802.11-20/1728r1 802.11bd NGV Ranging Status and Types* (IEEE NGV Meeting, 2020)
28. IEEE P802.11az/D1.0, *IEEE Draft Standard for Information Technology—Telecommunications and Information Exchange Between Systems Local and Metropolitan Area Networks—Specific Requirements Part 11: Wireless LAN Medium Access Control (MAC) and Physical Layer (PHY) Specifications—Enhancements for Positioning* (2019), pp. 1–187
29. B. Sun et al., *IEEE 802.11-19/0202r0 TGbd agreed terminology and requirements* (IEEE NGV Meeting, 2019)
30. F. Michael et al., *IEEE 802.11-18/1577r0 Additional Details About Interoperable NGV PHY Improvements* (IEEE NGV Meeting, 2018)
31. F. Michael et al., *IEEE 802.11-18/1186r0 Interoperable NGV PHY Improvements* (IEEE NGV Meeting, 2018)
32. F. Michael et al., *IEEE 802.11-19/0082r3 Interoperable Approach for NGV New Modulations* (IEEE NGV Meeting, 2019)
33. H. Onn et al., *IEEE 802.11-18/XXXXr0 Backward Compatible PHY Feasibility* (IEEE NGV Meeting, 2018)
34. L. Dongguk et al., *IEEE 802.11-19/0009r0 Consideration on Features for 11bd* (IEEE NGV Meeting, 2019)
35. 3GPP TR 36.885, *Study on LTE-based V2X Services*, (Release 14), 3GPP Technical Specification Group Radio Access Network, v14.0.0 (2016)
36. A. Papathanassiou, A. Khoryaev, Cellular V2X as the Essential Enabler of Superior Global Connected Transportation Services. IEEE 5G Tech Focus **1**(2), (2017)

37. R. Molina-Masegosa, J. Gozalvez, *System Level Evaluation of LTE-V2V Mode 4 Communications and its Distributed Scheduling* (IEEE VTC2017-Spring, 2017)
38. N.H. Mahmood, M.M.L. Lechuga, D. Laselva, K.I. Pedersen, G. Berardinelli G, Reliability oriented dual connectivity for URLLC services in 5G New Radio, in *15th IEEE International Symposium on Wireless Communication Systems (ISWCS)* (2018)
39. A. Karimi, K.I. Pedersen, N.H. Mahmood, J. Steiner, P. Mogensen, 5G centralized multi-cell scheduling for URLLC: algorithms and system-level performance. IEEE Access **6**, 72253–72262 (2018). https://doi.org/10.1109/ACCESS.2018.2880289
40. C. Campolo, A. Molinaro, A. Iera, F. Menichella, 5G network slicing for vehicle-to-everything services. IEEE Wireless Commun. **24**(6), 38–45 (2017). https://doi.org/10.1109/MWC.2017.1600408
41. A. Memedi, F. Dressler, Vehicular visible light communications: a survey. IEEE Commun. Surv. Tutor. **23**(1), 161–181 (2021). https://doi.org/10.1109/COMST.2020.3034224
42. T. Zugno, M. Drago, M. Giordani, M. Polese, M. Zorzi, Toward standardization of millimeter-wave vehicle-to-vehicle networks: open challenges and performance evaluation. IEEE Commun. Mag. **58**(9), 79–85 (2020). https://doi.org/10.1109/MCOM.001.2000041
43. S. Zeadally, M.A. Javed, E.B. Hamida, Vehicular communications for ITS: standardization and challenges. IEEE Commun. Standards Mag. **4**(1), 11–17 (2020). https://doi.org/10.1109/MCOMSTD.001.1900044
44. K. Wevers, M. Lu, V2X Communication for ITS—from IEEE 802.11p Towards 5G. IEEE 5G Tech Focus **1**(2), (2017)
45. National Highway Traffic Safety Administration (NHTSA), Department of Transportation (DOT), Federal Motor Vehicle Safety Standards; V2V Communications, Notice of Proposed Rulemaking (NPRM), NHTSA-2016-0126, (2016)
46. European Commission, A European strategy on cooperative intelligent transport systems, a milestone towards cooperative, connected and automated mobility, in *Communication from the Commission to the European Parliament, the Council, the European Economic and Social Committee and the Committee of the Regions* (COM, 2016) 766 final
47. K.Z. Ghafoor, M. Guizani, L. Kong, H.S. Maghdid, K.F. Jasim, Enabling efficient coexistence of DSRC and C-V2X in vehicular networks. IEEE Wireless Commun. **27**(2), 134–140 (2019). https://doi.org/10.1109/MWC.001.1900219
48. G. Naik, J. Liu, J.-M. Park, *Coexistence of Dedicated Short Range Communications (DSRC) and Wi-Fi: Implications to Wi-Fi Performance* (Proc, IEEE INFOCOM, 2017)
49. G. Naik, J. Liu, J.-M. Park, Coexistence of wireless technologies in the 5 GHz bands: a survey of existing solutions and a roadmap for future research. IEEE Commun. Surv. Tutor. (2018)
50. 5G Automotive Association, *ITS spectrum utilization in the Asia Pacific Region*, White Paper (2018)
51. Z.H. Mir, F. Filali, Applications, requirements, and design guidelines for multi-tiered vehicular network architecture, 2018 Wireless Days (WD). Dubai (2018). https://doi.org/10.1109/WD.2018.8361686
52. Z.H. Mir, J. Toutouh, F. Filali, Y.-B. Ko, Enabling DSRC and C-V2X integrated hybrid vehicular networks: architecture and protocol. IEEE Access **8**, 180909–180927 (2020). https://doi.org/10.1109/ACCESS.2020.3027074
53. M.B. Brahim, Z.H. Mir, W. Znaidi, F. Filali, N. Hamdi, QoS-aware video transmission over hybrid wireless network for connected vehicles. IEEE Access **5**, 8313–8323 (2017). https://doi.org/10.1109/ACCESS.2017.2682278
54. N. Dreyer, A. Moller, Z.H. Mir, F. Filali, T. Kurner, A data traffic steering algorithm for IEEE 802.11p/LTE hybrid vehicular networks, in *2016 IEEE 84th Vehicular Technology Conference (VTC-Fall)* (2016). https://doi.org/10.1109/VTCFall.2016.7880850
55. J.A. Olivera, I. Cortazar, C. Pinart, A. Los Santos, I. Lequerica, VANBA: a simple handover mechanism for transparent, always-on V2V communications, in *IEEE 69th Vehicular Technology Conference (VTC2009-Spring)* (2009)
56. A. Filippi, Wireless connectivity in automotive, in *CWTe 2016 Research Retreat, Centre for Wireless Technology Eindhoven (CWTe)* (2016)

57. Z.H. Mir, F. Filali, Evaluation of DSRC and LTE-V2x: need for next-generation V2X communication systems, in *International Conference on Computing and Communication Networks (ICCCN-2022)* (2022)
58. Z.H. Mir, F. Filali, LTE and IEEE 802.11p for vehicular networking: a performance evaluation. EURASIP JWCN 2014(89) (2014)
59. D. Puthal, Z.H. Mir, F. Filali, H. Menouar, Cross-layer architecture for congestion control in Vehicular Ad-hoc Networks, in *2013 International Conference on Connected Vehicles and Expo (ICCVE)* (2013), pp. 887–892. https://doi.org/10.1109/ICCVE.2013.6799921.
60. W. Anwar, S. Dev, A. Kumar, N. Franchi, G. Fettweis, PHY abstraction techniques for V2X enabling technologies: modeling and analysis. IEEE Trans. Vehicular Technol. **70**(2), 1501–1517 (2021). https://doi.org/10.1109/TVT.2021.3053425
61. R. Jacob, W. Anwar, N. Schwarzenberg, N. Franchi, G. Fettweis, System-level performance comparison of IEEE 802.11p and 802.11bd draft in highway scenarios, in *2020 27th International Conference on Telecommunications (ICT)* (2020). https://doi.org/10.1109/ICT49546.2020.9239538
62. W. Anwar, A. Traßl, N. Franchi, G. Fettweis, On the reliability of NR-V2X and IEEE 802.11bd, in *2019 IEEE 30th Annual International Symposium on Personal, Indoor and Mobile Radio Communications (PIMRC)* (2019). https://doi.org/10.1109/PIMRC.2019.8904104
63. W. Anwar,N. Franchi , G. Fettweis, Physical layer evaluation of V2X communications technologies: 5G NR-V2X, LTE-V2X, IEEE 802.11bd, and IEEE 802.11p, in *2019 IEEE 90th Vehicular Technology Conference (VTC2019-Fall)* (2019). https://doi.org/10.1109/VTCFall.2019.8891313
64. R. Jacob, W. Anwar, G. Fettweis, J. Pohlmann, Exploiting multi-RAT diversity in vehicular ad-hoc networks to improve reliability of cooperative automated driving applications, *2019 IEEE 90th Vehicular Technology Conference (VTC2019-Fall)* (2019). https://doi.org/10.1109/VTCFall.2019.8891072
65. A. Qayyum, M. Usama, J. Qadir, A. Al-Fuqaha, Securing connected and autonomous vehicles: challenges posed by adversarial machine learning and the way forward. IEEE Commun. Surv. Tutor. **22**(2), 998–1026 (2020). https://doi.org/10.1109/COMST.2020.2975048
66. J. He, K.-H. Yang, H. Chen, 6G cellular networks and connected autonomous vehicles. IEEE Netw. **35**(4), 255–261 (2021). https://doi.org/10.1109/MNET.011.2000541
67. M. Noor-A-Rahim, Z. Liu, H. Lee, M.O. Khyam, J. He, D. Pesch, K. Moessner, W. Saad, H.V. Poor, *6G for Vehicle-to-Everything (V2X) Communications: Enabling Technologies, Challenges, and Opportunities* CoRR (2020). arXiv:2012.07753. Last accessed 16 Jan 2022
68. X. Zhou, W. Liang, J. She, Z. Yan, K.I.-K. Wang, Two-layer federated learning with heterogeneous model aggregation for 6G supported internet of vehicles. IEEE Trans. Vehicular Technol. **70**(6), 5308–5317 (2021). https://doi.org/10.1109/TVT.2021.3077893
69. J. Hu, C. Chen, L. Cai, M.R. Khosravi, Q. Pei, S. Wan, UAV-assisted vehicular edge computing for the 6G internet of vehicles: architecture. Intell. Challeng. IEEE Commun. Standards Mag. **5**(2), 12–18 (2021). https://doi.org/10.1109/MCOMSTD.001.2000017
70. J. Sanghvi, P. Bhattacharya, S. Tanwar, R. Gupta, N. Kumar, M. Guizani, Res6Edge: an edge-AI enabled resource sharing scheme for C-V2X communications towards 6G, in *2021 International Wireless Communications and Mobile Computing (IWCMC)* (2021), pp. 149–154. https://doi.org/10.1109/IWCMC51323.2021.9498593

Zeeshan Hameed Mir (Senior Member, IEEE, and Fellow HEA) is currently an Assistant Professor and Program Team Leader (PTL) in the CIS division at Higher Colleges of Technology (HCT), UAE. From 2013 to 2016, he was with QMIC, Qatar University, Qatar as a Research Scientist. From 2009 to 2012, he worked as a Member of Technical Staff (MTS) in Electronics and Telecommunication Research Institute (ETRI), South Korea. From 2001 to 2005, he was in the Faculty of CS department at the Institute of Business Administration (IBA), Pakistan. He received his Ph.D. from Ajou University, South Korea, in 2009, the M.S. from NUST, Pakistan, in 2004, and the B.S. from SSUET, Pakistan, in 1999. He has published his research work in major research publications worldwide and also served on the program/reviewer committees of several reputed conferences and journals. His research interests are mobile/ubiquitous computing, wireless networking/communications, smart mobility, and urban analytics.

Fethi Filali (Senior Member, IEEE) received a Ph.D. degree in computer science and a Habilitation degree from the University of Nice Sophia Antipolis, France, in 2002 and 2008, respectively. He was with the Mobile Communications Department, EURECOM, France, as an Assistant Professor and then an Associate Professor for eight years. He is currently the Director of technology and research with the Qatar Mobility Innovations Center (QMIC), Qatar University. He is also leading the technology development of QMIC's solutions in smart cities, the Internet of Things, intelligent transportation systems, and connected and automated vehicles solutions. He has invented technologies and developed algorithms that have been shipped in many QMIC products, including Masarak, Labeeb, and WaveTraf, creating commercial impact in the order of millions of dollars. His research grants include 15 competitive awards from several funding agencies, including the European Commission, the French National Research Agency, and the Qatar National Research Fund. He was a Ph.D. Director for over ten Ph.D. students in intelligent transportation, wireless sensor and mesh networks, vehicular communications, big data analytics, the Internet of Things, and mobility management. He has co-authored over 130 research papers in international peer-reviewed conferences and journals. He holds over ten patent applications.

Outage Performance Comparison of DF/AF Cooperative Relaying System with SC/MRC Diversity Techniques

Shailendra Singh and Matadeen Bansal

Abstract In future wireless communication systems, cooperative relaying (CR) has been considered as a promising technique for reliable communication between a source-destination pair, when the direct link between them fails to provide the desired throughput due to channel impairments. Employing a dedicated relay node between the source and destination may improve the performance and enhances the network coverage. In this chapter, a CR system is analyzed and compared when the dedicated relay works under decode-and-forward (DF) or amplify-and-forward (AF) mode and selection combining (SC) or maximal ratio combining (MRC) diversity technique is adopted at the destination to recover the symbol. The exact closed-form expressions for the outage probability of the CR system over Rayleigh fading channels are derived and analyzed. The accuracy of the analytical results is verified through the Monte Carlo simulations.

1 Introduction

Cooperative relaying (CR) has been considered as one of the promising techniques in fifth-generation (5G) and beyond 5G wireless communication systems [1]. CR is an intelligent way to achieve reliable communication between a source-destination pair when the direct link between them is either affected by shadowing, path loss, fading, etc., or the distance between them is larger than the communication range of the source [2, 3]. Adding a dedicated relay node between the source and destination improves the performance as well as enhances the coverage of the network. The basic architecture of a relay network comprises of a source, a destination, and one or more relay nodes, which provide two-hop communication between the source

S. Singh (✉) · M. Bansal
PDPM-Indian Institute of Information Technology, Design and Manufacturing, Jabalpur, India
e-mail: shailendra.ten@gmail.com

M. Bansal
e-mail: mbansal@iiitdmj.ac.in

A. K. Pathan (ed.), *Towards a Wireless Connected World: Achievements and New Technologies*, https://doi.org/10.1007/978-3-031-04321-5_12

and destination. First, the source transmits the signal towards the relay, which is re-transmitted to the destination after performing some processing at the relay.

Currently, amplify-and-forward (AF) and decode-and-forward (DF) modes of CR are in focus, and many pieces of research were carried out considering these modes [4–8]. For the AF mode, the received signal is directly amplified at the relay and then re-transmitted to the destination. Therefore, the additive noise is also amplified and transmitted by the relay, which is destructive to recover the signal successfully at the destination [9]. However, for the relay working in DF mode, the received signal is recovered, re-encoded and then re-transmitted to the destination. In the case of incorrect decoding at the relay, DF cooperative relay may suffer from error propagation [10]. The analysis in [9] and [10] reveals that the AF relay is more susceptible to noise, but it offers a good trade-off between overall performance and implementation complexity. However, the DF relays are less vulnerable to noise, but the encoding/decoding errors at the relay can be forwarded to the destination. Also, the complexity and the processing delay of the DF relays are more than that of the AF relays due to the additional decoding operations at the relays.

There are several works in the literature which analyzed the performance of CR systems. Particularly, in [11], the performance of distributed diversity was investigated in AF single relay cooperative communication system. In AF relaying mode, the performance of cognitive radio network was studied in [12]. A distributed AF scheme through message partitioning was proposed in [13], which makes cooperative communication more practical. However, in [14], optimal cooperative diversity in AF cooperative system under slow-fading channel was studied. In [15], an adaptive AF relay model was proposed, where the receiving and forwarding time duration of the signal depends on the channel conditions. However, in [16], authors analyzed the performance of opportunistic relaying in a dual-hop AF relay system. Moreover, in [17], authors investigated the optimal transmit power and optimal AF relay location between the source and destination in order to achieve better coverage and performance of the network.

On the other hand, in the DF CR mode, the performance of the CR network was investigated in a different manner, depending upon the decoding and encoding techniques at the relay. Particularly, in [18], the outage probability of DF cooperative relay multi-cast system over Rayleigh fading channels with best relay selection was presented. The error probability of multi-relay opportunistic DF was analyzed in [19] over Nakagami-m fading channels. In [20], an opportunistic DF relaying system was proposed with an adaptive modulation scheme to further improve the spectral efficiency. The performance of a spatial multiplexing system with a linear receiver with DF/AF CR modes over Rayleigh fading channels was analyzed in [21]. Furthermore, authors in [29] studied the optimal position of the DF dedicated relay between the source and destination to get the better outage performance of the overall system. These works [11–29] have demonstrated that the AF relaying protocol is simple to implement as it does not attempt to decode and encode the received signal. The only limitation of the AF relay is that the added noise along with the received signal from the source gets amplified when forwarding to the destination. In comparison to the AF relays, DF relays are less vulnerable to noise, because the noise is removed during the decoding of the symbol, which improves the outage performance of the network. In addition, DF relays provide more flexibility of adopting the different transmissions protocols at the source and destination [10]. Therefore, AF/DF CR expected to attain more efficient network coverage and reliability.

Moreover, there are many works on the performance improvement of the CR communication systems by employing selection-combining (SC) or maximal-ratio-combining (MRC) technique to decode the symbols at the destination. Particularly, in [23], authors proposed a relay-assisted cooperative non-orthogonal multiple access (NOMA) system, where the destination receives the same symbol in two-time slots and the destination employs the SC technique to decode the symbol. The work in [23] proposed to improve the outage performance of the symbol. In [24], MRC was employed to jointly decode the symbols received in two consecutive time-slots. A cooperative NOMA relaying network was presented in [25], where the destination jointly detects the superimposed symbols received in the first time-slot through the direct channel and the symbol received in the second time-slot via relay using MRC. A CR system based on NOMA with an AF relay has been investigated in [26], in which both successive interference cancellation (SIC) and MRC schemes are applied two times in order to detect both the symbols at the destination.

Most of the aforementioned works either considered the DF relay or AF relay in CR systems, however, they were not analyzed jointly. The performance comparisons for the DF and AF CR systems were presented only in [27] and [28]. In [27], the imperfect channel state information (CSI) was assumed to be known at the source, and the diversity gain was investigated for the DF and AF relaying system. However, the work in [27] focused on power allocation and optimal time duration. In [28], the DF and AF CR networks were compared under a cognitive radio scenario, and closed-form expressions for the network-wide throughput were evaluated. The work in [28] have shows that each relaying technique performs better in a certain application and parameter range, and for these two relaying strategies, there is no case of dominance.

Motivated by the advantages of CR systems and the aforementioned benefits and limitations of employing AF/DF relays from the existing literature, this chapter investigates and compares a CR system, consisting of a source, a destination, and a dedicated DF/AF relay with SC/MRC technique to decode the symbols. In the first time-slot, the source broadcasts a symbol to the destination and the relay. In the second time-slot, the relay re-transmits the previously received symbol either by forwarding the symbol after decoding and encoding it (as in the DF mode) or by simply forwarding the symbol after amplifying it (as in the AF mode). The same symbol received in two time slots from different channels is recovered either by using SC or MRC technique at the destination. Therefore, the aim of this chapter is to analyze and compare the outage performance of the CR system assuming DF/AF mode at the relay and SC/MRC diversity at the destination over Rayleigh fading channels. The exact closed-form expressions of the outage probability are derived and investigated. Simulations are performed to validate the analytical results. To the best of our knowledge, the outage performance comparison of the AF/DF CR system considered in this chapter has not been studied yet.

The rest of the chapter is organized as follows: The signal and system model of the CR-based system with dedicated AF/DF relay is explained in detail in Sect. 2. The outage probability is evaluated and analyzed in Sect. 3. Section 4 discusses the numerical and simulation results. Lastly, Sect. 5 concludes the chapter.

2 Signal and System Model

Consider a CR network consisting of a source S, a dedicated relay R (working in either AF or DF mode), and a destination D, as shown in Fig. 1. Here, it is considered that all the links, i.e., the links $S-D$, $S-R$, and $R-D$ are available and R works in a half-duplex (HD) mode. Let $|h_z|$, $z \in \{SD, SR, RD\}$ be the magnitude of the channel gain between the nodes, which is assumed to be independent and Rayleigh distributed with average power β_z. Since the distance between $S-D$ is assumed to be greater than that between $S-R$ and between $R-D$, we assume that $\beta_{SD} < \beta_{SR}$ and $\beta_{SD} < \beta_{RD}$.

Throughout this chapter, we assume perfect CSI at R and D, no CSI at S, frequency-flat fading, and perfect synchronization between the nodes. Perfect CSI at the destination implies that the channel condition of $S-R$ link is available at R, while the individual channel conditions of $S-D$, $R-D$ and $S-R$ links are available at D. Since, synchronization becomes increasingly challenging in larger systems, the assumption of synchronization is most critical.

Since, $|h_z|$ is considered to be Rayleigh distributed, $|h_z|^2$ is exponentially distributed. The probability density function (PDF) and cumulative distribution function (CDF) of $|h_z|^2$ are represented, respectively, as

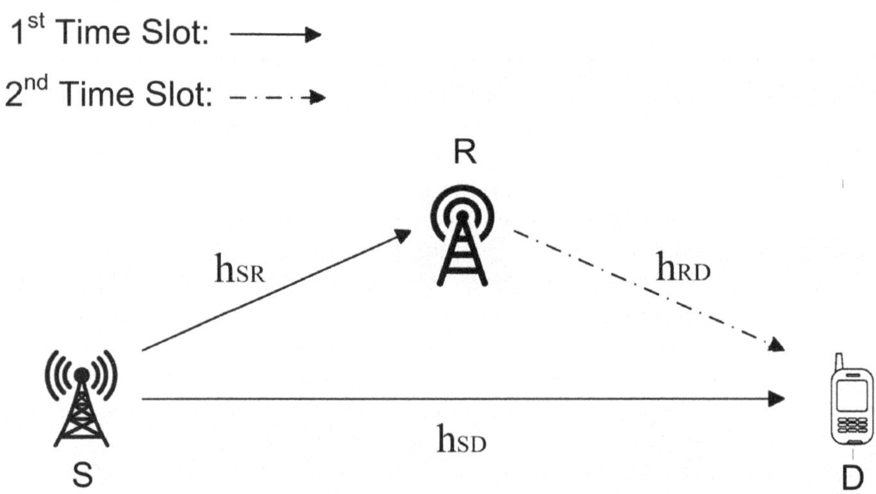

Fig. 1 CR system with half-duplex relay

$$f_{|h_z|^2}(x) = \frac{1}{\beta_z}\exp\left(-\frac{x}{\beta_z}\right) \text{ and} \tag{1}$$

$$F_{|h_z|^2}(x) = 1 - \exp\left(-\frac{x}{\beta_z}\right). \tag{2}$$

In the first time-slot, S broadcasts the symbol s as signal $\sqrt{P_S}s$, where $\mathbb{E}[|s|^2] = 1$ and P_S denotes the transmit power at S. The signal transmission in the first time-slot is the same for both the DF and AF modes of R. The received signals at R and D are, respectively, given as

$$y_{SR} = h_{SR}(\sqrt{P_S}s) + \eta_{SR} \text{ and} \tag{3}$$

$$y_{SD} = h_{SD}(\sqrt{P_S}s) + \eta_{SD}, \tag{4}$$

where η_{SR} and η_{SD} denote the additive white Gaussian noise (AWGN) with zero mean and variance σ^2. The received signal-to-noise ratio (SNR) for the symbol s at R using (3) is given by

$$\gamma_{SR} = \frac{P_S|h_{SR}|^2}{\sigma^2} = \delta|h_{SR}|^2, \tag{5}$$

where, $\delta \triangleq P_S/\sigma^2$ represents the transmit SNR. Similarly, the received SNR for the symbol s at D using (4) is given by

$$\gamma_{SD} = \delta|h_{SD}|^2. \tag{6}$$

2.1 *When DF Relay Is Employed*

During the first time-slot, R recovers the received symbol. Upon successful decoding at R, the symbol s is re-encoded and forwarded to the destination in the second time-slot. Therefore, the received signal in the second time-slot at D is given by

$$\hat{y}_{RD} = h_{RD}(\sqrt{P_R}s) + \eta_{RD}, \tag{7}$$

where P_R is the transmit power at R, η_{RD} is AWGN with mean zero and variance σ^2 at D. Here, for the sake of simplicity, we assume $P_R = P_S$.

2.1.1 When SC Technique Is Employed at *D*

In case of SC diversity technique, only the best received SNR signal from all the channels is selected for further processing at the destination. Therefore, if SC technique is employed at D, the received SNR for the symbol s in the second time-slot at D using (7) is given as

$$\hat{\gamma}_{RD,\ sc} = \frac{P_S|h_{RD}|^2}{\sigma^2} = \delta|h_{RD}|^2. \tag{8}$$

Now, the received SNR for the symbol s through R can be written as

$$\hat{\gamma}_{viaR,\ sc} = min\{\gamma_{SR},\ \hat{\gamma}_{RD,\ sc}\}. \tag{9}$$

2.1.2 When MRC Technique Is Employed at D

In MRC diversity technique, all the available channels are used to receive the symbol at the destination. At the destination node, MRC is employed to combine the received symbols. If MRC technique is used at D for decoding the symbol s, then the received signals in two consecutive time slots, i.e., y_{SD} and $\hat{y}_{RD}$, are weighted by the factors of w_{SD} and $\hat{w}_{RD}$, respectively. Thus, the combined received signal at D for detecting symbol s is given by

$$\hat{y}_{mrc} = w_{SD}y_{SD} + \hat{w}_{RD}\hat{y}_{RD}. \tag{10}$$

To optimize the received SNR for decoding the symbol s at D, the weighting factors are chosen as $w_{SD} = \frac{\sqrt{P_S}h_{SD}^H}{\sigma^2}$ and $\hat{w}_{RD} = \frac{\sqrt{P_S}h_{RD}^H}{\sigma^2}$. The denominator of these factors denotes the noise power at y_{SD} and $\hat{y}_{RD}$. Also, $(.)^H$ denotes the conjugate transpose operation.

After substituting the optimized values of the weighting factors in (10), the received SNR at D is given by

$$\hat{\gamma}_{mrc} = \delta|h_{SD}|^2 + \delta|h_{RD}|^2. \tag{11}$$

The end-to-end received SNR for the symbol s at D is given as

$$\hat{\gamma}_{e2e,\ mrc} = \min\left\{\gamma_{SR},\ \hat{\gamma}_{mrc}\right\}. \tag{12}$$

2.2 *When AF Relay Is Employed*

When R works in AF mode, it simply amplify and transmits the received symbol as $\sqrt{\tilde{P}_R}(h_{SR}\sqrt{P_S}s + \eta_{SR})$ in the second time-slot. The received signal at D through R in this case is given by

$$\tilde{y}_{viaR} = h_{RD}\sqrt{\tilde{P}_R}\left(h_{SR}\sqrt{P_S}s + \eta_{SR}\right) + \eta_{RD}, \tag{13}$$

where $\tilde{P}_R$ is the amplification factor at R. Here, for the sake of simplicity and without the loss of generality, we assume $\tilde{P}_R = \frac{P_S}{P_S|h_{SR}|^2+\sigma^2}$.

2.2.1 When SC Technique Is Employed at *D*

With SC technique at D, the received SNR through R is given as

$$\tilde{\gamma}_{viaR,\ sc} = \frac{P_S^2|h_{SR}|^2|h_{RD}|^2}{P_S|h_{SR}|^2\sigma^2 + P_S|h_{RD}|^2\sigma^2 + \sigma^4} = \frac{\delta^2|h_{SR}|^2|h_{RD}|^2}{\delta|h_{SR}|^2 + \delta|h_{RD}|^2 + 1}. \tag{14}$$

2.2.2 When MRC Technique Is Employed at *D*

In this case, the received signals in two consecutive time slots, i.e., y_{SD} and $\tilde{y}_{viaR}$, are weighted by the factors w_{SD} and $\tilde{w}_{viaR}$, respectively. Thus, the combined received signal at D for detecting the symbol s is given by

$$\tilde{y}_{mrc} = w_{SD}y_{SD} + \tilde{w}_{viaR}\tilde{y}_{viaR}. \tag{15}$$

To optimize the received SNR for detecting symbol s at D, the weighting factors are chosen as $w_{SD} = \frac{\sqrt{P_S}h_{SD}^H}{\sigma^2}$ and $\tilde{w}_{viaR} = \frac{\sqrt{P_S}h_{SR}^H h_{RD}^H\sqrt{\tilde{P}_R}}{|h_{RD}|^2\tilde{P}_R\sigma^2+\sigma^2}$. The denominator of these factors represents the the total interference and noise power at y_{SD} and $\tilde{y}_{viaR}$. After substituting the optimized values of weighting factors, the end-to-end received SNR for the symbol s at D is given as

$$\tilde{\gamma}_{e2e,\ mrc} = \left\{\delta|h_{SD}|^2 + \frac{\delta^2|h_{SR}|^2|h_{RD}|^2}{\delta|h_{SR}|^2 + \delta|h_{RD}|^2 + 1}\right\}. \tag{16}$$

3 Performance Analysis

In this section of the chapter, analytical expressions for the outage probabilities are evaluated at D.

Definition: The outage probability is defined as the probability of the event that the received SNR (γ) is less than some preset detection threshold SNR (γ_{th}), i.e.,

$$P^{out} = Prob(\gamma \leq \gamma_{th}).$$

3.1 Outage Probability with DF Relay and SC Technique at D

Theorem 1 *The outage probability for the symbol s at D for the direct link S − D is represented as*

$$\hat{P}^{out}_{SD,\,sc} = 1 - \exp\left(-\frac{\gamma_{th}}{\delta\beta_{SD}}\right). \tag{17}$$

Proof

$$\begin{aligned}\hat{P}^{out}_{SD,\,sc} &= Prob(\gamma_{SD} \leq \gamma_{th})\\ &= Prob\left(\delta|h_{SD}|^2 \leq \gamma_{th}\right)\\ &= Prob\left(|h_{SD}|^2 \leq \frac{\gamma_{th}}{\delta}\right)\\ &= \int_0^{\frac{\gamma_{th}}{\delta}} \frac{1}{\beta_{SD}} \exp\left(-\frac{y}{\beta_{SD}}\right) dy.\end{aligned}$$

After evaluating the integral, we get (17). □

Theorem 2 *The outage probability for the symbol s at D for the end-to-end link through R is represented as*

$$\hat{P}^{out}_{viaR,\,sc} = 1 - \exp\left\{-\frac{\gamma_{th}}{\delta}\left(\frac{1}{\beta_{SR}} + \frac{1}{\beta_{RD}}\right)\right\}. \tag{18}$$

Proof Please see Appendix 1. □

At D, SC technique is employed to finally recover the symbol s. Therefore, the closed-form expression of the outage probability for the symbol s at D is given as

$$\hat{P}^{out}_{sc} = \hat{P}^{out}_{SD,\,sc} \times \hat{P}^{out}_{viaR,\,sc}. \tag{19}$$

Substituting (17) and (18) into (19), we get the outage probability for s at D as

$$\hat{P}^{out}_{sc} = \left[1 - \exp\left(-\frac{\gamma_{th}}{\delta\beta_{SD}}\right)\right] \times \left[1 - \exp\left\{-\frac{\gamma_{th}}{\delta}\left(\frac{1}{\beta_{SR}} + \frac{1}{\beta_{RD}}\right)\right\}\right]. \tag{20}$$

3.2 Outage Probability with DF Relay and MRC Technique at D

Theorem 3 *The outage probability for the symbol s for the end-to-end S − D link is given by the following expression*

$$\hat{P}_{mrc}^{out} = 1 - \left\{ \exp\left(-\frac{\gamma_{th}}{\delta\beta_{SR}} \right) \right\} \times \left\{ \frac{1}{\beta_{SD} - \beta_{RD}} \right\} \times \left\{ \beta_{SD} \exp\left(-\frac{\gamma_{th}}{\delta\beta_{SD}} \right) - \beta_{RD} \exp\left(-\frac{\gamma_{th}}{\delta\beta_{RD}} \right) \right\}. \tag{21}$$

Proof Please see Appendix 2. □

3.3 Outage Probability with AF Relay and SC Technique at D

Similarly in this case, the outage probability for s at D is represented as

$$\tilde{P}_{sc}^{out} = \tilde{P}_{SD,\,sc}^{out} \times \tilde{P}_{viaR,\,sc}^{out}, \tag{22}$$

where $\tilde{P}_{SD,\,sc}^{out} = \hat{P}_{SD,\,sc}^{out}$ is the outage probability of s for the direct $S - D$ link. Therefore, from (17), we have the following expression for $\tilde{P}_{SD,\,sc}^{out}$

$$\tilde{P}_{SD,\,sc}^{out} = 1 - \exp\left(-\frac{\gamma_{th}}{\delta\beta_{SD}} \right). \tag{23}$$

Whereas, $\tilde{P}_{viaR,\,sc}^{out}$ is the outage probability for s through R and is derived in the following theorem.

Theorem 4 *The outage probability for the symbol s for the end-to-end link through R is given by*

$$\tilde{P}_{viaR,\,sc}^{out} = 1 - A \exp\left\{ -\frac{\gamma_{th}}{\delta}\left(\frac{1}{\beta_{SR}} + \frac{1}{\beta_{RD}} \right) \right\} \times K_1(A), \tag{24}$$

where $A = \frac{2\gamma_{th}}{\delta}\sqrt{\frac{1}{\beta_{SR}\beta_{RD}}(1 + \frac{1}{\gamma_{th}})}$ and $K_1(A)$ denotes first order modified bessel function of the second kind.

Proof Please see Appendix 3. □

Substituting (23) and (24) into (22), we get the outage probability for s at D as

$$\tilde{P}_{sc}^{out} = \left[1 - \exp\left(-\frac{\gamma_{th}}{\delta\beta_{SD}} \right) \right] \times \left[1 - A \exp\left\{ -\frac{\gamma_{th}}{\delta}\left(\frac{1}{\beta_{SR}} + \frac{1}{\beta_{RD}} \right) \right\} \times K_1(A) \right]. \tag{25}$$

3.4 Outage Probability with AF Relay and MRC Technique at D

Theorem 5 *Asymptotic outage probability for the symbol s in high SNR region for the end-to-end $S - D$ link is given by the expression*

$$\tilde{P}_{mrc}^{out}(\infty) \approx \left(\frac{\beta_{SR} + \beta_{RD}}{\delta^2 \beta_{SD} \beta_{SR} \beta_{RD}}\right)\left(\frac{\gamma_{th}^2}{2}\right). \tag{26}$$

Proof Please see Appendix 4. □

Lemma 1 *At higher values of δ, the outage probabilities for the symbol s at D are approximated as*

$$\hat{P}_{sc}^{out}(\infty) \approx \left(\frac{\beta_{SR} + \beta_{RD}}{\delta^2 \beta_{SD} \beta_{SR} \beta_{RD}}\right)\left(\gamma_{th}^2\right), \tag{27}$$

$$\hat{P}_{mrc}^{out}(\infty) \approx \left(\frac{\gamma_{th}}{\delta \beta_{SR}}\right), \tag{28}$$

$$\tilde{P}_{sc}^{out}(\infty) \approx \left(\frac{\beta_{SR} + \beta_{RD}}{\delta^2 \beta_{SD} \beta_{SR} \beta_{RD}}\right)\left(\gamma_{th}^2\right), \textit{and} \tag{29}$$

$$\tilde{P}_{mrc}^{out}(\infty) \approx \left(\frac{\beta_{SR} + \beta_{RD}}{\delta^2 \beta_{SD} \beta_{SR} \beta_{RD}}\right)\left(\frac{\gamma_{th}^2}{2}\right). \tag{30}$$

Proof According to [29, Eq. (9.7.2)], we can approximate $K_1(x)$ with $\frac{1}{x}$, when $x \to 0$. Also, using the approximation $\exp(-x) = 1 - x$, when $x \to 0$, we can derive Lemma 1. □

Remark 1 Lemma 1 provides some useful insights on employing SC/MRC technique in the CR system working in DF/AF mode. From (27) and (29), we can observe that when the SC technique is employed to decode the symbols, then the outage performances of the DF/AF CR system are the same at the high transmit SNR regime. This is because, at high SNR, the impact of noise amplification in the AF relay and encoding/decoding errors in the DF relay on the outage probability is almost the same. Further, from (28), it may be noted that the outage performance of the DF CR system only depends on β_{SR}, when the MRC technique is employed to decode the symbol at the high transmit SNR regime. This is because, at high SNR, if the DF relay is unable to recover the symbol successfully then the whole system will be in outage. Moreover, by comparing the outage probabilities of the CR system with various conditions, we can conclude the DF CR system with the MRC technique shows the worst outage performance, whereas, the AF CR system with the MRC technique achieves the least outage probability, which is half the value when SC technique is employed.

4 Numerical Results

In this section, the Monte Carlo simulations are performed to validate the analytical findings in Section 3. To represent the distances between all the nodes, (i.e., $S-R$, $S-D$, and $R-D$ channels), the average power is taken into account. We consider the distance parameter of the $S-D$ link as $\beta_{SD}=0.5$. The distance parameters of $S-R$ and $R-D$ links are taken four times of β_{SD}, i.e., $\beta_{SR}=\beta_{RD}=2$ and $\gamma_{th}=1$. The results are averaged over 10^8 realizations of the Rayleigh fading channels.

Figure 2 depicts the outage probability results of s at D for the CR system employing DF relay versus the transmit SNR. It may be noted that the analytical results of the outage probability for the symbol in the presented system closely match with simulation results. In Fig. 2, it is shown that the outage probability of the symbol s at D through R is much better than the direct $S-D$ link. This is due to the fact that R has better channel condition with both S and D in comparison with the channel condition of the direct $S-D$ link. Further, it is evident that in a CR system with DF relay is SC technique performs well in comparison with the MRC technique, which verifies the analytical results in (27) and (28). This is because, R re-transmits the

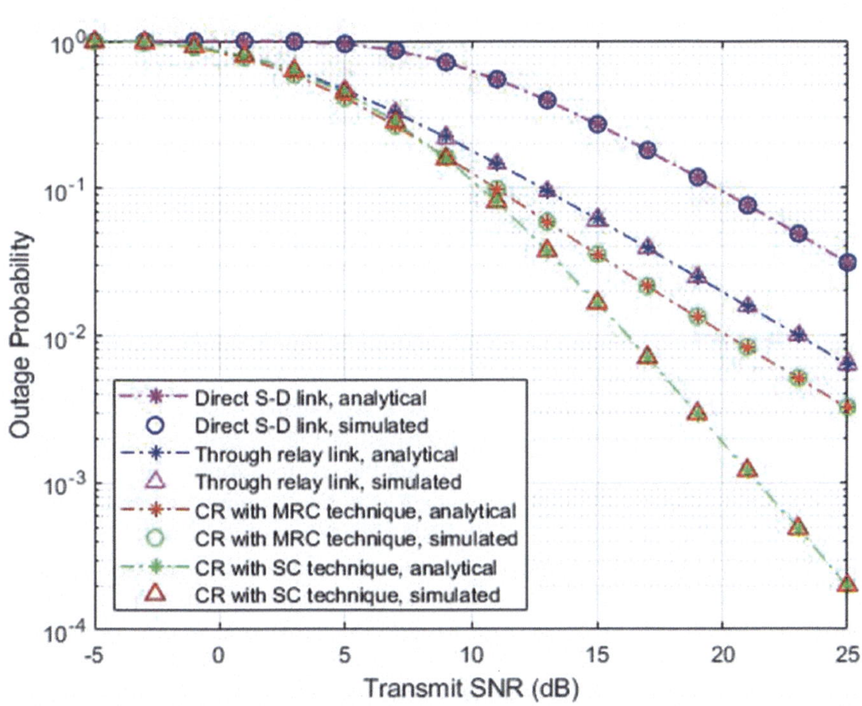

Fig. 2 Outage probability versus transmit SNR of DF CR system

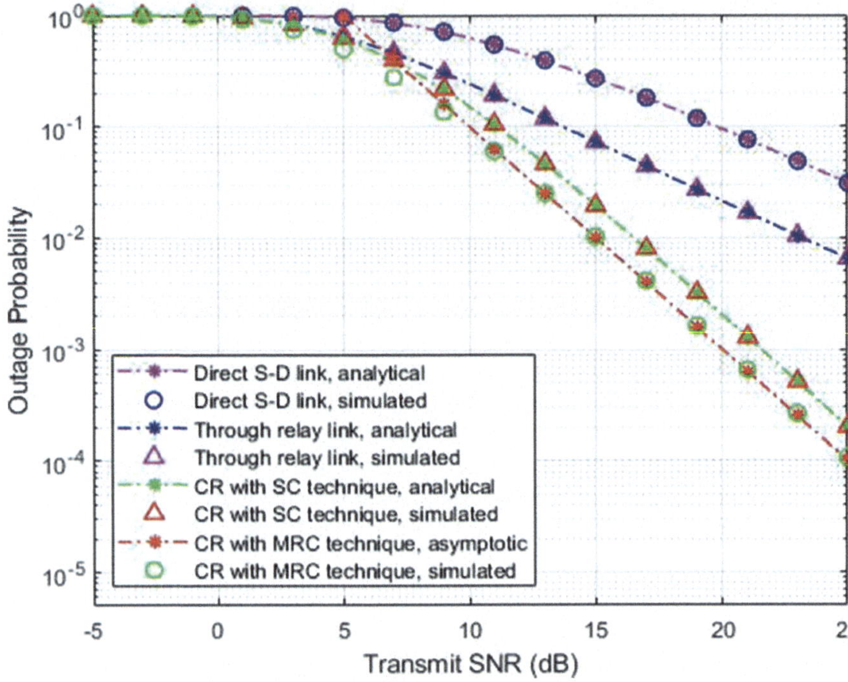

Fig. 3 Outage probability versus transmit SNR of AF CR system

symbol after decoding it in the first time-slot. Therefore, if decoding of the symbol s at R is not perfect then it will impact the outage probability of the symbol at D when the MRC technique is employed.

The outage probability of the symbol s at D for the CR system employing AF relay as a function of the transmit SNR is shown in Fig. 3. Again, the simulation results show a perfect match with the analytical results. In Fig. 3, it is shown that the outage probability of the symbol s through R is better than the direct $S - D$ link. The reason for this is the same as discussed in Fig. 2. However, from Fig. 2 and Fig. 3, the outage performance of s through the AF R is almost the same as the DF R when SC technique is employed, which is in line with our analytical results in (27) and (29). This is because, at high SNR, the impact of noise amplification in the AF relay and encoding/decoding errors in the DF relay on the outage probability become the same. Further, it is evident that the CR system with AF relay, the MRC technique performs well in comparison with the SC technique and the achievable outage probability with the MRC technique is half that of the SC technique. This validates our analytical results in Lemma 1. This is due to the fact that AF R simply forwards the received signal after amplifying it in the second time-slot. Therefore, D receives the same signal from two different channels in two consecutive time slots, which provides a diversity order of 2 when the MRC technique is used.

5 Conclusion

In this chapter, we have analyzed and derived the closed-form expressions for the outage probability of the CR system with DF/AF relay employing the SC/MRC technique. The outage performance of the CR system is compared when R works under DF/AF mode and SC or MRC diversity technique is adopted at D to recover the symbol. For the DF R, the outage performance of the symbol s with the SC technique is better than that with the MRC technique at D. However, for the AF R, the outage performance of the symbol s with the MRC technique is better than that with the SC technique at D. The analysis performed in this chapter demonstrates that when R works under AF mode and the MRC diversity technique is employed at D to recover the symbol gives optimum result in terms of outage performance. Simulations have validated the accuracy of the derived analytical expressions.

Appendix 1

Proof From (9), the outage probability for the symbol s is given by

$$\hat{P}^{out}_{viaR,\,sc} = 1 - Prob\{\min(\gamma_{SR}, \hat{\gamma}_{RD,\,sc}) > \gamma_{th}\}. \tag{31}$$

Since γ_{SR} and $\hat{\gamma}_{RD,\,sc}$ are independent, (31) can be re-written as

$$\hat{P}^{out}_{viaR,\,sc} = 1 - Prob(\gamma_{SR} > \gamma_{th}) \times Prob(\hat{\gamma}_{RD,\,sc} > \gamma_{th}), \tag{32}$$

where from (5), we have

$$\begin{aligned} Prob(\gamma_{SR} > \gamma_{th}) &= Prob\left(\delta|h_{SR}|^2 > \gamma_{th}\right) \\ &= Prob\left(|h_{SR}|^2 > \frac{\gamma_{th}}{\delta}\right) \\ &= \int_{\frac{\gamma_{th}}{\delta}}^{\infty} \frac{1}{\beta_{SR}} \exp\left(-\frac{y}{\beta_{SR}}\right) dy \end{aligned}$$

$$Prob(\gamma_{SR} > \gamma_{th}) = \exp\left(-\frac{\gamma_{th}}{\delta\beta_{SR}}\right), \quad \text{and} \tag{33}$$

$$\begin{aligned}
Prob(\hat{\gamma}_{RD,\,sc} > \gamma_{th}) &= Prob\Big(\delta|h_{RD}|^2 > \gamma_{th}\Big) \\
&= Prob\Big(|h_{RD}|^2 > \frac{\gamma_{th}}{\delta}\Big) \\
&= \int_{\frac{\gamma_{th}}{\delta}}^{\infty} \frac{1}{\beta_{RD}} \exp\big(-\frac{y}{\beta_{RD}}\big)dy
\end{aligned} \tag{34}$$

$$Prob(\hat{\gamma}_{RD,\,sc} > \gamma_{th}) = \exp\Big(-\frac{\gamma_{th}}{\delta\beta_{RD}}\Big).$$

After substituting (33) and (34) into (32), we finally get (18) and the proof is completed. □

Appendix 2

Proof From (12), the outage probability for the symbol s is given by

$$\hat{P}^{out}_{mrc} = 1 - Prob\{\min(\gamma_{SR},\ \hat{\gamma}_{mrc}) > \gamma_{th}\}. \tag{35}$$

Since, γ_{SR} and $\hat{\gamma}_{mrc}$ are independent, (35) can be re-written as

$$\hat{P}^{out}_{mrc} = 1 - Prob(\gamma_{SR} > \gamma_{th}) \times Prob(\hat{\gamma}_{mrc} > \gamma_{th}), \tag{36}$$

From (33), we have

$$Prob(\gamma_{SR} > \gamma_{th}) = \exp\Big(-\frac{\gamma_{th}}{\delta\beta_{SR}}\Big). \tag{37}$$

Substituting $\hat{\gamma}_{mrc}$ from (11), we can write

$$\begin{aligned}
Prob(\hat{\gamma}_{mrc} > \gamma_{th}) &= Prob\left((\delta|h_{SD}|^2 + \delta|h_{RD}|^2) > \gamma_{th}\right) \\
&= Prob\Big((|h_{SD}|^2 + |h_{RD}|^2) > \frac{\gamma_{th}}{\delta}\Big) \\
&= Prob\Big((X+Y) > \frac{\gamma_{th}}{\delta}\Big) \\
&= Prob\Big(Z > \frac{\gamma_{th}}{\delta}\Big)
\end{aligned} \tag{38}$$

$$Prob(\hat{\gamma}_{mrc} > \gamma_{th}) = \int_{\frac{\gamma_{th}}{\delta}}^{\infty} f_Z(z)dz,$$

where, $X = |h_{SD}|^2$, $Y = |h_{RD}|^2$, and $Z = X + Y$ are random variables. The PDF of X and Y are given, respectively, as

$$f_X(x) = \frac{1}{\beta_{SD}} \exp\left(-\frac{x}{\beta_{SD}}\right), \text{ and} \tag{39}$$

$$f_Y(y) = \frac{1}{\beta_{RD}} \exp\left(-\frac{y}{\beta_{RD}}\right). \tag{40}$$

The PDF of Z is given by

$$f_Z(z) = f_X(x) * f_Y(y), \tag{41}$$

where (*) denotes the convolution of two functions. After substituting (39) and (40) into (41) we get

$$f_Z(z) = \left\{\frac{1}{\beta_{SD}} \exp\left(-\frac{x}{\beta_{SD}}\right)\right\} * \left\{\frac{1}{\beta_{RD}} \exp\left(-\frac{y}{\beta_{RD}}\right)\right\}. \tag{42}$$

Evaluating the convolution and after some alzebric manipulation, the PDF of Z is given by

$$f_Z(z) = \left(\frac{1}{\beta_{SD} - \beta_{RD}}\right)\left\{\exp\left(-\frac{z}{\beta_{SD}}\right) - \exp\left(-\frac{z}{\beta_{RD}}\right)\right\}. \tag{43}$$

Substituting (43) into (38), we get

$$Prob(\hat{\gamma}_{mrc} > \gamma_{th}) = \int_{\frac{\gamma_{th}}{\delta}}^{\infty} \left(\frac{1}{\beta_{SD} - \beta_{RD}}\right) \times \left\{\exp\left(-\frac{z}{\beta_{SD}}\right) - \exp\left(-\frac{z}{\beta_{RD}}\right)\right\} dz. \tag{44}$$

After solving the integral, we get

$$Prob(\hat{\gamma}_{mrc} > \gamma_{th}) = \left(\frac{1}{\beta_{SD} - \beta_{RD}}\right)\left\{\beta_{SD} \exp\left(-\frac{\gamma_{th}}{\delta\beta_{SD}}\right) - \beta_{RD} \exp\left(-\frac{\gamma_{th}}{\delta\beta_{RD}}\right)\right\}. \tag{45}$$

After substituting (37) and (45) into (36), we finally get (21), and the proof is completed. □

Appendix 3

Proof The outage probability for the symbol s for the end-to-end link through R is represented as

$$\tilde{P}^{out}_{viaR,\,sc} = Prob(\tilde{\gamma}_{viaR,\,sc} \leq \gamma_{th}). \tag{46}$$

From (14), we have

$$\begin{aligned}\tilde{P}^{out}_{viaR,\,sc} &= Prob\left(\frac{\delta^2|h_{SR}|^2|h_{RD}|^2}{\delta|h_{SR}|^2+\delta|h_{RD}|^2+1} \leq \gamma_{th}\right)\\ &= Prob\left(\frac{\delta^2 UV}{\delta U+\delta V+1} \leq \gamma_{th}\right)\\ &= Prob\Big(U(\delta^2 V-\delta\gamma_{th}) \leq \gamma_{th}(\delta V+1)\Big).\end{aligned} \tag{47}$$

where $U = |h_{SR}|^2$ and $V = |h_{RD}|^2$ are random variables. The PDF of U and V are given, respectively, as

$$f_U(u) = \frac{1}{\beta_{SR}}\exp\left(-\frac{u}{\beta_{SR}}\right), \text{ and} \tag{48}$$

$$f_V(v) = \frac{1}{\beta_{RD}}\exp\left(-\frac{v}{\beta_{RD}}\right). \tag{49}$$

As we are dividing both sides of an inequality by a negative number in (47), the inequality direction are changed. Therefore, if $(\delta^2 V - \delta\gamma_{th}) < 0$, we have

$$\begin{aligned}\tilde{P}^{out}_{viaR,\,sc} &= Prob\Big(U(\delta^2 V-\delta\gamma_{th}) > \gamma_{th}(\delta V+1)\Big)\\ &= Prob\left(U > \gamma_{th}\frac{(\delta V+1)}{(\delta^2 V-\delta\gamma_{th})}\right)\\ &= 1.\end{aligned} \tag{50}$$

If $(\delta^2 V - \delta\gamma_{th}) > 0$, we can write

$$\begin{aligned}\tilde{P}^{out}_{viaR,\,sc} &= Prob\Big(U(\delta^2 V-\delta\gamma_{th}) \leq \gamma_{th}(\delta V+1)\Big)\\ &= Prob\left(U \leq \gamma_{th}\frac{(\delta V+1)}{(\delta^2 V-\delta\gamma_{th})}\right).\end{aligned} \tag{51}$$

Now, the law of total probability leads to

$$
\begin{aligned}
\tilde{P}^{out}_{via R,\,sc} &= \int_0^{\frac{\gamma_{th}}{\delta}} f_V(v)dv + \int_{\frac{\gamma_{th}}{\delta}}^{\infty} \left(1 - \exp\left(-\frac{\gamma_{th}(\delta y + 1)}{\delta(\delta y - \gamma_{th})\beta_{SR}}\right)\right) f_V(v)dv \\
&= 1 - \frac{\exp(-\frac{\gamma_{th}}{\delta}(\frac{1}{\beta_{SR}} + \frac{1}{\beta_{RD}}))}{\delta^2 \beta_{RD}} \times \\
&\int_0^{\infty} \exp\left(-\frac{\gamma_{th}(\gamma_{th}+1)}{t\beta_{SR}}\right) \exp\left(-\frac{t}{\delta^2\beta_{RD}}\right) dt.
\end{aligned} \tag{52}
$$

Finally, using [30, Eq. (3.324–1)], we derive the outage probability of the symbol s for the end-to-end link through the R as (24), and the proof is completed. □

Appendix 4

Proof The outage probability for the symbol S for the end-to-end $S - D$ link is expressed as

$$\tilde{P}^{out}_{mrc} = Prob(\tilde{\gamma}_{e2e,\,mrc} \le \gamma_{th}). \tag{53}$$

From (16), expression (53) can be represented as

$$\tilde{P}^{out}_{mrc} = Prob\left(\delta|h_{SD}|^2 + \frac{\delta^2|h_{SR}|^2|h_{RD}|^2}{\delta|h_{SR}|^2 + \delta|h_{RD}|^2 + 1} \le \gamma_{th}\right). \tag{54}$$

Assuming $M = \delta|h_{SD}|^2$ and $N = \frac{\delta^2|h_{SR}|^2|h_{RD}|^2}{\delta|h_{SR}|^2+\delta|h_{RD}|^2+1}$, the CDF of M is given by

$$F_M(m) = Prob\left(\delta|h_{SD}|^2 \le m\right) = 1 - \exp\left(-\frac{m}{\delta\beta_{SD}}\right). \tag{55}$$

In high SNR regime, using $\exp^{-x} = 1 - x$ when $x \to 0$ [29, Eq. (1.211.1)], (55) is simplified to $F^{\infty}_M(m) \approx \frac{m}{\delta\beta_{SD}}$ and, the PDF of M is given by

$$f^{\infty}_M(m) = \left(F^{\infty}_M(m)\right)' \approx \frac{1}{\delta\beta_{SD}}, \tag{56}$$

where $(.)'$ denotes the differentiation of the function. The CDF of N is determined in Appendix 3. Therefore, from (24), the CDF of N is represented as

$$F_N(n) = Prob\,(N \le n) = 1 - A\exp\left\{-\frac{n}{\delta}\left(\frac{1}{\beta_{SR}} + \frac{1}{\beta_{RD}}\right)\right\} \times K_1(A). \tag{57}$$

where $A = \frac{2n}{\delta}\sqrt{\frac{1}{\beta_{SR}\beta_{RD}}(1+\frac{1}{n})}$ and $K_1(A)$ denotes the first order modified bessel function of the second kind. To approximate (57), when $\delta \to \infty$. According to [29, Eq. (9.7.2)], we can approximate $K_1(x)$ with $\frac{1}{x}$ when $x \to 0$. Applying this approximation and simplifying (57) yields to:

$$F_N^{\infty}(n) \approx 1 - \exp\left\{ -\frac{n}{\delta}\left(\frac{1}{\beta_{SR}} + \frac{1}{\beta_{RD}}\right)\right\}. \tag{58}$$

Moreover, using the approximation $\exp^{-x} = 1 - x$ when $x \to 0$ [30, Eq. (1.211.1)], we get

$$F_N^{\infty}(n) \approx \frac{n}{\delta}\left(\frac{1}{\beta_{SR}} + \frac{1}{\beta_{RD}}\right). \tag{59}$$

Now, from (54), we can write

$$\begin{aligned} \tilde{P}_{mrc}^{out\,(\infty)} &= Prob\,(M + N \leq \gamma_{th}) = Prob\,(N \leq \gamma_{th} - M) \\ &= \int_0^{\gamma_{th}} F_N^{\infty}(n) f_M^{\infty}(m)dm \\ &\approx \int_0^{\gamma_{th}} \frac{(\gamma_{th} - m)}{\delta}\left(\frac{1}{\beta_{SR}} + \frac{1}{\beta_{RD}}\right)\left(\frac{1}{\delta\beta_{SD}}\right)dm. \end{aligned} \tag{60}$$

After solving the integral and performing some manipulation, we get (26), and the proof is completed. □

References

1. J.N. Laneman, D.N.C. Tse, G.W. Wornell, Cooperative diversity in wireless networks: efficient protocols and outage behavior. IEEE Trans. Inf. Theory **50**(12), 3062–3080 (2004)
2. C. Nie, P. Liu, T. Korakis et al., Cooperative relaying in next-generation mobile WiMAX networks. IEEE Trans. Vehicular Technol. **62**(3), 1399–1405 (2013)
3. A. Gupta, R.K. Jha, A survey of 5G network: architecture and emerging technologies. IEEE Access **3**, 1206–1232 (2015)
4. S. Wang, H. Ji, Distributed power allocation scheme for multi-relay shared-bandwidth (MRSB) wireless cooperative communication. IEEE Commun. Lett. **16**(8), 1263–1265 (2012)
5. B. Zhang, X. Jia, Multi-hop collaborative relay networks with consideration of contention overhead of relay nodes in IEEE 802.11 DCF. IEEE Trans. Commun. **61**(2), 532–540 (2013)
6. K.H. Truong, R.W.H. Jr, Cooperative algorithms for MIMO amplify-and-forward relay networks. IEEE Trans. Signal Process. **61**(5), 1272–1287 (2013)
7. Z. Zhang, Z. Ma, M. Xiao et al., Two-time slot two-way full-duplex relaying for wireless communication networks. IEEE Trans. Commun. **64**(7), 2873–2887 (2016)
8. S. Wang, R. Buby, V.C.M. Leung, Z. Yao, A low-complexity power allocation strategy to minimize sum-source-power for multi-user single-AF-relay networks. IEEE Trans. Commun. **64**(8), 3275–3283 (2016)

9. H. Al-Tous, I. Barhumi, Resource allocation for multiuser improved AF cooperative communication scheme. IEEE Trans. Wireless Commun. **14**(7), 3655–3672 (2015)
10. M. Ju, I. Kim, D.I. Kim, Joint relay selection and relay ordering for DF-based cooperative relay networks. IEEE Trans. Commun. **60**(4), 908–915 (2012)
11. H.A. Suraweera, D.S. Michalopoulos, G.K. Karagiannidis, Performance of distributed diversity systems with a single amplify-and-forward relay. IEEE Trans. Vehicular Technol. **58**(5), 2604–2608 (2009)
12. Z. Gao, D. Chen, K. Zhang, et al., Outage performance of cognitive AF relay networks with direct link and heterogeneous non-identical constraints, in *Wireless Communication and Mobile Computing*. Article first published online: 27 Nov. 2014. https://doi.org/10.1002/wcm.2560
13. L. Chen, R. Carrasco, I. Wassell, Distributed amplify-and-forward cooperation through message partitioning. IEEE Trans. Vehicular Technol. **60**(7), 3054–3065 (2011)
14. S. Yang, J.C. Belfiore, Towards the optimal amplify-and-forward cooperative diversity scheme. IEEE Trans. Inf. Theory **53**(9), 3114–3126 (2007)
15. Y. Xiao, L.A. DaSilva, X. Zhang, On the diversity-multiplexing trade-off of an improved amplify-and-forward relaying strategy. IEEE Commun. Lett. **16**(4), 482–484 (2012)
16. A. Behnad, A.M. Rabiei, N.C. Beaulieu, Performance analysis of opportunistic relaying in a poisson field of amplify-and-forward relays. IEEE Trans. Commun. **61**(1), 97–107 (2013)
17. H. Amiriara, M.R. Zahabi, V. Meghdadi, Power-location optimization for cooperative nomadic relay systems using machine learning approach. IEEE Access **9**, 74246–74257 (2021)
18. I.H. Lee, H. Lee, H.H. Choi, Exact outage probability of relay selection in decode-and-forward based cooperative multicast systems. IEEE Commun. Lett. **17**(3), 456–483 (2013)
19. Q. Shi, Y. Karasawa, Error probability of opportunistic decode-and-forward relaying in Nakagami-m fading channels with arbitrary m. IEEE Wireless Commun. Lett. **2**(1), 86–88 (2013)
20. Y. Ma, R. Tafazolli, Y. Zhang et al., Adaptive modulation for opportunistic decode-and-forward relaying. IEEE Trans. Wireless Commun. **10**(7), 1022–2017 (2011)
21. T.Q. Duong, L. Shu, M. Chen et al., Performance analysis of cooperative spatial multiplexing networks with AF/DF relaying and linear receiver over Rayleigh fading channels. Wireless Commun. Mobile Comput. **15**(3), 500–509 (2013)
22. H. Inaltekin, S. Atapattu, J.S. Evans, Optimum location-based relay selection in wireless networks. IEEE Trans. Inf. Theory **67**(9), 6223–6242 (2021)
23. J. Zhang, L. Dai, R. Jiao, X. Li, Y. Liu, Performance analysis of relay assisted cooperative non-orthogonal multiple access systems, submitted to IEEE Wireless Communication Letters (2017)
24. M. Xu, F. Ji, M. Wen, W. Duan, Novel receiver design for the cooperative relaying system with non-orthogonal multiple access. IEEE Commun. Lett. **20**(8), 1679–1682 (2016)
25. Y. Zhang, Z. Yang, Y. Feng, S. Yan, Performance analysis of cooperative relaying systems with power-domain non-orthogonal multiple access. IEEE Access **6**, 39839–39848 (2018)
26. O. Abbasi, A. Ebrahimi, N. Mokari, NOMA inspired cooperative relaying system using an AF relay. IEEE Wireless Commun. Lett. **8**(1), 261–264 (2019)
27. T.T. Kim, H.V. Poor, On the diversity gain of AF and DF relaying with noisy CSI at the source transmitter. IEEE Trans. Inf. Theory **55**(11), 5064–5073 (2009)
28. D. Hu, S. Mao, Cooperative Relay in Cognitive Radio Networks: Decode-and-Forward or Amplify-and-Forward? GLOBECOM 2010, Miami, Florida, USA, 2010
29. M. Abramowitz, I.A. Stegun, *Handbook of Mathematical Functions: With Formulas, Graphs, and Mathematical Tables*, vol. 55 (Courier Corporat, Chelmsford, MA, USA, 1965)
30. I.S. Gradshteyn, I.M. Ryzhik, *Table of Integrals, Series, and Products* (Academic, San Diego, CA, USA, 2014)

Shailendra Singh has received his B.E. degree in Electronics Engineering from Rajiv Gandhi Technical University, Bhopal (M.P.), India, in 2002 and M.E. in Communication, Control & Networking from Madhav Institute of Technology and Science, Gwalior, India in 2005. He is currently pursuing his Ph.D. degree in Electronics and Communication Engineering from PDPM-Indian Institute of Information Technology, Design and Manufacturing, Jabalpur, India. He has qualified national level Graduate Aptitude Test in Engineering (GATE). His current research includes Non-orthogonal multiple access, cooperative relaying diversity, and overlay cognitive radio systems.

Matadeen Bansal has received his B.E. degree from the Department of Electronics and Communication Engineering, Madhav Institute of Science and Technology, Gwalior, India, in 2001, and his M.E. degree in Communication, Control & Networking from Madhav Institute of Technology and Science, Gwalior, India in 2006. He has received his Ph.D. degree from the Department of Electronics and Communication Engineering, ABV Indian Institute of Information Technology and Management Gwalior, India, in 2013. Presently, he has been working as an Assistant Professor at the PDPM-Indian Institute of Information Technology, Design and Manufacturing, Jabalpur, India, since 2013. He has various publications in reputated journals and conferences. He is the reviewer of various SCI indexed journals. His area of interest is wireless communications, cooperative communications, cognitive radios, multiple-input multiple-output (MIMO) systems, non-orthogonal multiple access (NOMA), optimization of communication systems.

Intelligent Decision Making Issues

Programmable Computing in 6G

Zubair Md Fadlullah and Mostafa M. Fouda

Abstract There is a sharp rise in communication and computing needs. The iterative generations (xG) of communication networks typically improve capacity and quality of service (QoS) parameters by an order of magnitude. In the emerging 5G+ and 6G networks, there is a call for embedded intelligence which can bring a revolutionary change in the network fronthaul as well as backhaul. Programmable computing is aligned with the computational concepts from the original inspiration of the Software-Defined Networks (SDNs). However, the essence of the SDN concept has been lost throughout the years due to rapid industry-absorption of the architecture-specific details whereas the programmable computing part has been widely overlooked. In this chapter, we describe a programmable computing architecture in the 6G network system. Then we demonstrate the key enablers that can support this programmable computing architecture. We provide a simple case study to illustrate how programmable computing can be leveraged in the emerging 6G use-cases.

1 Introduction

Communication networks are all around us. As the name suggests, communication networks are typically used for communication, whether this is cellphones for making

The original version of this chapter was revised: The affiliation of the second author has been corrected. The correction to this chapter is available at
https://doi.org/10.1007/978-3-031-04321-5_15

Z. Md Fadlullah (✉)
Department of Computer Science, Lakehead University, Thunder Bay, ON, Canada
e-mail: Zubair.Fadlullah@lakehead.ca

Thunder Bay Regional Health Research Institute (TBRHRI), Thunder Bay, ON, Canada

M. M. Fouda
Department of Electrical and Computer Engineering, Idaho State University, 1776 Science Center, Dr. Stop 8150, Idaho Falls, ID 83402, USA
e-mail: mfouda@ieee.org

A. K. Pathan (ed.), *Towards a Wireless Connected World: Achievements and New Technologies*, https://doi.org/10.1007/978-3-031-04321-5_13

a voice call, video streaming, controlling your IoT (Internet of Things) devices for smart home applications, or drone-based surveillance and monitoring systems [1]. Even the smart, wireless TVs in your living room are part of the massive, wireless communication ecosystem as they serve on-demand high-definition contents which typically require high-speed Internet connection through the home access point or router. These wide examples of communication networks are rolled out every few years in terms of a new generation (xG). The evolution of wireless communication networks from 2 to 5G has already been astounding and associated with a sharp increase in the number of supported users along with much-improved capacity and service performance guarantees. The better the service performance, the more satisfied the users are. As the 5G networks are being rolled out, both academia and industry are drafting the next generation of wireless and mobile networks, which are referred to as the 6G networks [2, 3]. Is 6G just hype? Or does it bring a new flavor of communication? Is it just a business model to sell new networking equipment to your favorite mobile broadband provider and cellular network operators so that they can impose a heftier monthly subscription charge? Or does it really provide something additional to what we already have in 4G (LTE, WiMAX) and 5G networks [4]?

To answer this question, we need to understand what 4G and 5G were designed to deliver to mobile users. A key difference between 4 and 5G is the adoption of millimeter-wave (mmWave) bands (for higher capacity and supporting more users) and much lower delay requirements. There are many small and densified cells in 5G networks that use beamforming and other technologies to deliver 1–30 Gbps (gigabits per second) of speed compared to the speed in the order of hundreds of MBps (megabits per second) in the predecessor 4G networks. On the other hand, the 6G networks are different from 5G networks in terms of providing 100 Gbps to 1 Tbps (1 terabit per second) speed. This is an ambitious requirement drafted, inspired by some of the laboratory experiments conducted recently with Terahertz (THz) frequency and VLC (visual light communication) technology. However, in 6G networks, to accommodate this very high-speed communication, the nodes need to be within even smaller areas than those in 5G networks, referred to as tiny cells. Imagine so many tiny cells in your city, which are dynamically formed on a need basis, then dissolved, and then formed again to cater to a different service need. This dynamism will be much more dominant in 6G networks because of the integrated aerial-terrestrial-satellite networks. In 4G and 5G networks, typically terrestrial network base stations are used to serve the users. However, in 6G networks, drone cells, terrestrial base stations, nanosatellites, etc. will all be connected [5, 6]; and it is very difficult to satisfy the capacity and delay requirements to combat the ultra-high level of dynamism in such a complex mesh of networks. As a consequence, flexible architecture is required to compute the resource allocation, security provisioning, service quality assurance, and so forth. In this chapter, we talk about the softwarized or programmable computing for 6G networks, which can be considered as the key manager of tiny cells of 6G networks.

The softwarized network programming is inspired by the fusion of AI (artificial intelligence)-based computing [7] and the software-defined networks, commonly referred to as SDNs [8, 9]. The reason for this conception is two-fold. 6G networks, in contrast to earlier generations of communication systems, are being designed for embedded intelligence, particularly in the network edge [10]. Therefore, the coordination of 6G should be done in such a way that considers AI natively. You must have read about AI, or watched science fiction movies that talk a lot about intelligent robots! Although we haven't reached that level of intelligence yet, communication networks are a leading area where intelligence may have great success. By intelligence, we refer to the basic pattern recognition algorithms and mainly data-driven models. In popular lingo, we refer to this as machine learning or deep learning models which are built on observing various trends and patterns in the communication networks. These are also known as predictive techniques. If your network operator can predict that there will be 100 cellphone users in your shopping mall area in the suburbs during the next half an hour, it can switch on the tiny cell base stations to serve those users and the other time it doesn't have to keep those base stations active, thereby saving energy. Energy-efficient technology, although appearing to have a negligible impact on the carbon footprint at a tiny cell level, can lead to much reduced emissions on a collective level across a town or city. This is just but one example of the awesome features predictive technologies can provide to communication network users. They can proactively assign network resources (called channels) [11, 12], they can forecast what type of content (movie, music, etc.) the users may want to watch [13], how the users will interact with one another, etc. If such intelligence can be embedded into smartphones and other devices in the edge of the network, they can transform those small user-devices into powerful edge computing devices which can take part in distributed learning to figure out interesting computing problems, for example, contact tracing, pandemic modeling in a distributed environment in real-time, and so forth. We hope that you could get the big picture of the usefulness of embedded AI in 6G systems.

On the other hand, the SDN architecture needs to be fused with the embedded AI in 6G networks also. Typically SDN architecture was coined from a computer science perspective, to make the best use of object inheritance and code reusability [14]. What do we mean by this? If you are familiar with object-oriented programming such as C++, C#, Java, etc., you will definitely be familiar with abstractions such as classes, objects, encapsulation, inheritance, polymorphism. Clever computer scientists thought about using this concept to extend to network entities. On the top level, a network router may have some basic class of operations, and the next level of the router may have further abstraction with some added features. All we have to do is: define the objects, and extend the classes, and map them to the hardware implementation of network routers and other network equipment. This is an abstract but powerful concept, which unfortunately became lost in "translation" from the computer science perspective to network engineering practice. Practitioners of 4G and 5G networks overlooked and forgot the inherent significance of the original concept of SDN, and simply implemented the central coordinator/controller-based management of network nodes and routers. By referring to such implementations as SDN, the programmability feature was entirely missed. If we can bring the programmable computing from the original SDN concept and infuse AI with it, it can do wonders for the embedded AI as a key enabler for emerging 6G network systems.

We hope that we understand the big picture of programmable computing in 6G by now. Let us embark on the journey into the chapter where we will be able to tell you about the machine/deep learning-based AI models that can help the 6G systems to do on-demand model building and deployment. We will first provide some related work regarding programmable computing, inspired by SDNs. Then we will provide preliminary on machine learning and deep learning models. Next we will present the 6G programmable computing architecture with AI as well optimization model incorporation. A use-case will be then provided so that readers can have a high-level understanding of how such programmable computing can provide high service performance in 6G systems. Future challenges and directions for programmable computing in 6G networks will then be briefly discussed.

2 6G Network Requirements

Readers should take note of the fact that at the time of writing this chapter, 6G is still at the conceptualization phase. Even though 6G networks do not exist yet, theoretical and experimental work are encouraging researchers and telecommunication engineers to learn from the limitations of the current generations of networks (4G, 5G, etc.) to help make a completely different generation of networks with several key requirements. The connectivity between devices (not just cell phones and smart devices but also billions of IoT devices) will be complex yet fluid in an integrated aerial-terrestrial-satellite network. Each component of such a complex network system has its own physical specifications and requirements. For example, aerial networks can be formed much faster in remote/rural areas with ease compared to the conventional terrestrial base stations. However, their achilles heel is the battery (energy constraint)! Readers are familiar with drones that can take aerial photography but can be in flight for not more than half an hour (take for example an off-the-shelf DJI Phantom drone). When you use such drones for communication to form a mesh network, they can do wonders for remote sensing to detect fire in the forest bed, or connect rural villages or Indigenous communities that cannot be accessed with hard-to-deploy and much more expensive terrestrial links using fiber optics. Drones can use 2.4 GHz (gigahertz) links which are similar to what you typically have at a home wireless access point, and other high-frequency links for collecting and forwarding packets to the Internet. But their mobility, energy constraint, capacity, wireless channel quality, blocking, and path loss models are quite different when compared to terrestrial base stations and users. Now imagine if you use these two radio access technologies with various inherent requirements together! Now imagine you throw the nanosatellites and low earth orbit satellites into the mix. The 6G network will be so difficult to manage in such a complex co-presence of the different radio access technologies. What is the 6G network management for? There are several requirements of 6G systems considered by researchers

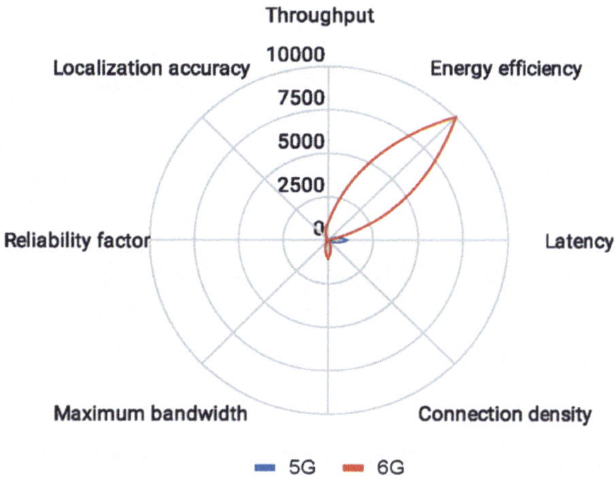

Fig. 1 6G network requirements compared with those considered in 5G settings

which include: up to 1 Tbps capacity, 1 ms or below delay, embedded intelligence for native support for edge computing, and so forth. Now why these requirements? These requirements of 6G systems are needed to support the killer applications of the future that include: tactile internet, remote surgery, haptic connections, immersive infotainment (e.g., using virtual/augmented reality), autonomous cars, etc. These applications typically require a huge amount of bandwidth which justifies the high capacity of 6G networks. Such applications also need a very low communication delay in the order of a few milliseconds, and typically require local computer vision and natural language processing tasks that in turn heavily rely on AI models. Readers can refer to a simple comparative illustration of 6G network requirements compared to the predecessor 5G network settings in Fig. 1. Next we will derive inspiration from the existing work to be able to sketch a programmable computing infrastructure for 6G network systems that can adequately support the aforementioned requirements and applications.

3 Background and Motivation

Software-Defined Networks, commonly referred to as SDNs, have been widely studied and considered in various network technologies from 4G/5G core networks to IoT systems. From SDNs we derive the motivation because it is the focal point of computer science theory for reusable coding for programming/reprogramming network elements and current telecommunication engineering practice. So let us take a closer look at SDNs first and try to understand what is actually missing that makes it not readily scalable to 6G networks with the required embedded intelligence.

Refer to Fig. 2 which describes the operation of the SDN controller. It typically separates (decouples) the control and data plane when packets are

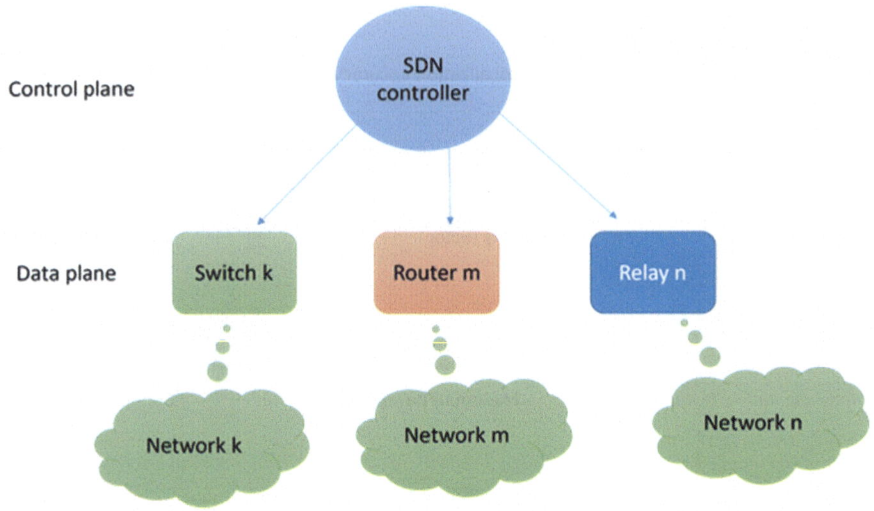

Fig. 2 SDN controller decoupling the control plane from the data plane

forwarded over networks. Traditional routers perform a dual role, they process both control (computing) and data (packet forwarding) planes. But this puts a lot of processing/computing burden on these network nodes, such as bridges, routers, access points, etc. Why do we want to separate these planes? Let us consider a simple example. You are a distribution manager and a distributor at the same time at the pickup corner of a large shopping outlet. As the distribution manager, you receive various calls, emails, etc. for customer orders, and then also need to distribute them to the customers for curbside pickup, home delivery, etc. This is a very complicated job. If you could separate this between you and your colleague, life could be much easier, right? In this simple example, the distribution manager is the example of a control plane while the distribution tasks are analogous to the packet forwarding operations on the data plane of the SDN.

While the control functionality is placed on the SDN controller (or several SDN controllers) to coordinate the entire network which could be your university campus network or the power grid cyber-physical system network, today's SDNs miss the object-oriented design concept. This could strengthen the SDNs to be systematically built and scaled. However, the network engineering practitioners, probably out of the excitement, absorbed the architectural essence of SDN and forgot the programmability part. This is where the motivation for a programmable network starts for us. What does the programmability imply? Think about a network node; which has simple or basic functionality such as packet forwarding. The packet forwarding could be a class and all the simple packet forwarding elements could then be instances of that class. An extended class could be a router with firewall functionality that can forward packets and also have the ability to act as firewalls. The key takeaway of this idea is code reusability. First, we have a basic abstract network element class, and all the instances of this class will be simple network forwarding entities. By simple code reusability and extension, we can have a firewall class to make many other instances of routers with firewall capability. There could be many more extensions with such programmability.

Such a programmable computing-powered SDN controller can simply make new classes and instances and deliver those instances to the off-the-shelf, "empty" routers. In other words, the network controller could decide, for switch k, to add more new rules when the traffic load variation significantly changes in Fig. 1. An even more dramatic example of the programmable computing-aided scenario could be a drone which can act as a "transformer" by changing its role from a flying access point, energy harvesting static access point, a home router in a rural area, or a relay node, or even a surveillance node. How can the node change its modus operandi on the go? To build such a versatile drone (or a versatile network functionality box) will mean there should be lots of hardware resources and multiple software packages on a self-contained system. This raises the cost and practicality of such systems in a significant way.

This brings the next question: how can the controllers have the capability to make the best models and decisions fast enough so that these models could be deployed down onto the network switches, routers, relay nodes, drones, and so on. The answer is: by either a rule-based or data-driven approach. Rule-based approaches include optimization techniques, heuristics, etc.; while the data-driven approaches typically consist of supervised learning, unsupervised learning, and reinforcement learning. Next, we provide the preliminaries of these rule-based data-driven approaches that are the building blocks of the softwarized, programmable computing in 6G networks for performing a diverse set of network and service functions.

4 Preliminaries of Rule-Based and Data-Driven Approaches

As mentioned earlier, there are rule-based techniques which are the traditional techniques. On the other hand, there are data-driven approaches to train a model that can be readily deployed to the network nodes.

4.1 Rule-Based Techniques

There are various approaches to optimize network performance based on the network traffic load dynamics and other variables in the 6G radio access side or the fronthaul. For formulating resource allocation and scheduling problems in various 6G tiny cells, optimization models based on linear programming, convex optimization, Lyapunov optimization, stochastic optimization, and matching algorithms are widely used.

While such techniques are well suited for obtaining closed-form solutions to the optimization problem, collecting all the information typically is a time-intensive process particularly when the search space for finding the solution is large. On the other hand, when optimization techniques do not provide an optimal solution, an acceptable solution is still required. This is when heuristics, such as greedy approaches, are designed. Let us consider a simple example. Suppose there is an array of intelligent reflective surfaces in a 6G network. How to find the optimal angle of the reflecting surface elements with respect to the different frequency bands used in the network? Since there is no unified channel model for multi-band frequencies, it is difficult to obtain a closed-form solution for this problem. So the next best solution is to approximate a solution using trial-and-error-based approaches or heuristics, or greedy algorithms that can provide at least some reasonable or acceptable throughput and delay performances. Typically these algorithms need to be designed manually. Human operators need to observe the various network variables, formulate an objective function subject to various constraints, and then figure out whether the conventional optimization algorithms are computationally hard or not. If this is the case, the problem is typically broken down into simplified problem(s) or subproblems that can be easily solved; and heuristics are developed to opportunistically solve the simplified problem or the subproblems. While for known network configurations, the optimization techniques typically provide optimal or near-optimal solutions, their execution time and the single-shot solution (once at a time) warrant a different method, namely the data-driven approach.

4.2 Data-Driven Approach: Machine Learning and Deep Learning

The data-driven approaches typically build an experiential learning model by discovering various patterns in the network activities. While they do not provide closed-form solutions, they are known to perform very well given large data sizes. The experience culminated by big data originating from the IoT system or a cellular network can allow the network coordinator to predict how much resources need to be allocated in the several next minutes, which frequency bands and channels are likely to be occupied, which contents will be in high-demand, and so on. There are statistical approaches that typically provide descriptive analytics and are widely used for constructing anomaly detection in network intrusion detection systems. There are machine learning-based approaches such as support vector machines and random forest (based on decision trees) to mainly train AI models for deciding various regression and classification tasks. These can provide fast, embedded intelligence for the resource-constrained nodes (e.g., IoT devices and drones) in 6G networks. On the other hand, for complex and large data processing which involves non-linearity and cannot be handled with conventional rule-based approaches, the deep learning methods using various neural networks are gaining popularity. For instance, artificial neural networks (ANNs), convolutional neural networks (CNNs), and recurrent neural networks (RNNs) are used for various types of classification tasks involving non-linear pattern identification in complex networks data. When these models are trained over a long period of time, for similar networks the pre-trained models can be quickly deployed by the network coordinator to the network entities to reprogram their functionality and optimally handle the prevailing network dynamics. This is the power of these AI models, in conjunction with the network manager/coordinators, to provide an on-demand reconfiguration of network nodes.

5 6G Programmable Computing with Optimization and AI Models

With the above preliminary, we are now ready to describe the 6G programmable computing architecture with optimization and AI models. We will use Fig. 3 to describe this architecture.

As shown in Fig. 3, there are three layers in the proposed programmable computing architecture with optimization and AI models: the application plane, hardware plane, and data plane. For detailed description of the inspiring architecture, please refer to the authors' inspiring work in this area in [15]. Let us start from the bottom up, with the data plane. The data plane consists of various base stations (BSs) ranging from terrestrial to drone-managed networks. Satellite networks in the integrated 6G networks could also be shown, but for simplicity, we decided not to make the illustration more complex than it already is. The base stations could be static or mobile cellular base stations managing 6G tiny cells or macrocells. There could be user devices acting as device-to-device (D2D) nodes or hotspots acting as relay entities for packet forwarding where the traditional base stations cannot reach. There could be drones or aerial base stations also to cover areas where the conventional base stations access is not available or they are overwhelmed due to network traffic congestion. These various networks generate a lot of traffic competing for 6G base station resources, and under highly dynamic channel conditions, blocking models, and ultra-high mobility; the serving base stations need to be reprogrammed to deliver the best service and minimize the possibility of a service outage.

Based on the traffic demand, user movement pattern, channel conditions, and other network dynamics, the 6G network manager, located at the 6G network home office (HO) where hardware virtualization is performed to facilitate the control plane and to provide service/application plane functionalities including QoS and security provisioning in terms of network slicing, virtual network function (vNF), signal processing, remote radio resource allocation, mobility control, sleep scheduling of base stations, etc.

The base stations placed on the data planes consist of terrestrial and aerial (drone) base stations, mobile user equipment (UEs), wireless local area network access points

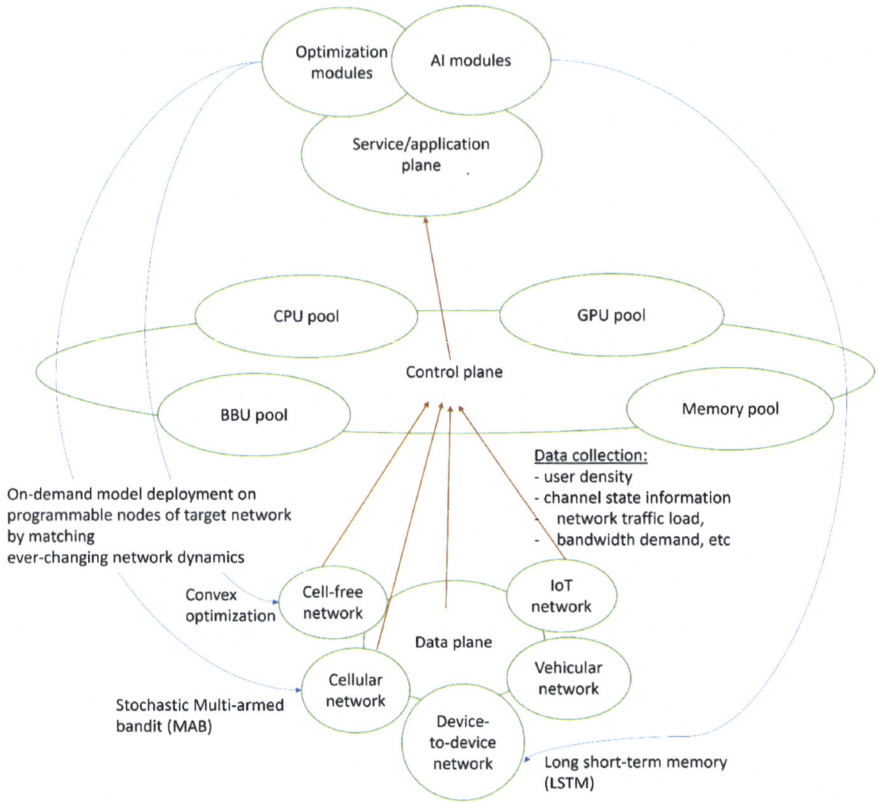

Fig. 3 Programmable computing architecture with optimization and AI models in 6G networks [15]

(WLAN-APs), visible light communication (VLC) base stations, vehicle-to-other (V2x) nodes, IoT devices, etc. The 6G network manager, which could be a centralized or a distributed/virtualized platform, is responsible for deriving and on-demand distribution of optimal and intelligent models and/or policies to these various base stations, access points, and wireless equipment to optimally forward the data packets for satisfying the 6G communication requirement. On the data plane, the wireless nodes, based on the derived optimal and/or intelligent model/policy, observe the current traffic demand, channel conditions, and so forth and then decide which models are ideal for the current situation. Therefore, by matching the network dynamics, they proactively download the optimal/intelligent network policy as shown in Fig. 2. For example, the stochastic multi-armed bandit (MAB) model [16], belonging to the reinforcement learning family, is deployed for stationary D2D nodes while the adversarial MAB model is downloaded by the mobile UEs [17]. Thus, the proposed system model benefits from the reusability of the MAB schemes for new programmable network nodes. Specifically, the application plane has to define a few primary instances of optimization and AI model types. Upon current network topology and dynamics, network devices can simply be reprogrammed to accommodate an appropriate model to combat the prevalent conditions by merely creating an instance or a program based on the base definition that it procures from the repository of optimization/AI modules in the application plane.

As mentioned earlier, the classical optimization techniques are not scalable with the highly varying network dynamics. As a result, it is often challenging to provide closed-form expression on the existence and guarantee of an optimal solution for a well-defined, complex problem. Many of the constraints and conditions are often relaxed upon the utilized algorithm design to find suboptimal solutions. Furthermore, such optimization techniques are typically a one-shot process as they require centralized, oracle-like knowledge to ingest the whole dataset to give the optimal benchmark decision. On the other hand, a supervised learning model is typically trained before decision-making since inference is known to be much faster than the training time. However, such supervised learning models require extensive and versatile training datasets. The lack of an adequate dataset, which is critical to train the existing machine/deep learning models, will be a crucial barrier to maximizing their predictive performance. Moreover, the performances of such supervised learning-based models are typically sub-optimal, and a lack of interpretation as to why they provide such performance still raises a lot of concerns among researchers for mass deployment on networking devices in contrast with the traditional straight-forward, feedback-based decision making. Therefore, ultra-fast online learning techniques are essential to be deployed to the 6G users (e.g., BSs, home APs, mobile UEs, and so forth) for localized, distributed decision making. The type of MAB can also be changed on-demand to cater to the sudden change in the network dynamics experienced by the 6G users. Furthermore, the recent advances in regret analysis for the variants of MAB algorithms can be leveraged to demonstrate their tightly bounded performance guarantee. Thus, MAB emerges as the most viable candidate compared to the classical optimization and supervised learning counterparts.

6 Conclusion: Future Directions and Caveats

Intelligent decision-making is anticipated to be a key embedded feature in the upcoming 6G networks that will realize innovative future applications. Since these services have ultra-reliable requirements easily impacted by varying network dynamics, on-demand ultra-fast learning techniques emerge as a formidable research challenge. In this chapter, we addressed this challenge and proposed a softwarized network consisting of an on-demand policy selector that considers the ongoing network dynamics and accordingly chooses the best intelligence module for deploying to the nodes for that particular network.

Unlike the classical optimization and supervised learning methods, online/sequential learning techniques such as MAB algorithms with different policies can be focused on viable online, sequential learning techniques for 6G node deployment by the proposed on-demand selector.

As a caveat, it is worth noting that for deploying the models on-demand, there could be some connectivity issues causing the AI models not to be timely updated that may cause the target routers/network nodes to be rendered dysfunctional. To combat such a corner case, we may assume a default, basic functionality of programmable routers to cope with such scenarios. How to optimally generalize such a default functionality is left open as future work for 6G softwarized networks and programmable routers.

References

1. J. Hwang, L. Nkenyereye, N. Sung, J. Kim, J. Song, IoT service slicing and task offloading for edge computing. IEEE Internet Things J. **8**(14), Art. no. 14 (2021). https://doi.org/10.1109/JIOT.2021.3052498.
2. M. Alsabah et al., 6G wireless communications networks: a comprehensive survey. IEEE Access **9**, 148191–148243 (2021). https://doi.org/10.1109/ACCESS.2021.3124812
3. S. Elmeadawy, R.M. Shubair, 6G wireless communications: future technologies and research challenges, in *2019 International Conference on Electrical and Computing Technologies and Applications (ICECTA)* (Ras Al Khaimah, United Arab Emirates, Nov. 2019), pp. 1–5. https://doi.org/10.1109/ICECTA48151.2019.8959607
4. M. Agiwal, H. Kwon, S. Park, H. Jin, A survey on 4G–5G dual connectivity: road to 5G implementation. IEEE Access **9**, 16193–16210 (2021). https://doi.org/10.1109/ACCESS.2021.3052462
5. N. Kato et al., Optimizing space-air-ground integrated networks by artificial intelligence. IEEE Wirel. Commun. **26**(4), Art. no. 4 (2019). https://doi.org/10.1109/MWC.2018.1800365
6. J. Liu, Y. Shi, Z.Md. Fadlullah, N. Kato, Space-air-ground integrated network: a survey. IEEE Commun. Surv. Tutor. **20**(4), Art. no. 4 (2018). https://doi.org/10.1109/COMST.2018.2841996
7. Z.Md. Fadlullah et al., State-of-the-art deep learning: evolving machine intelligence toward tomorrow's intelligent network traffic control systems. IEEE Commun. Surv. Tutor. **19**(4), Art. no. 4 (2017). https://doi.org/10.1109/COMST.2017.2707140
8. F. Tang, Z.Md. Fadlullah, B. Mao, N. Kato, An intelligent traffic load prediction-based adaptive channel assignment algorithm in SDN-IoT: a deep learning approach. IEEE Internet Things J. **5**(6), Art. no. 6 (2018). https://doi.org/10.1109/JIOT.2018.2838574
9. F. Tang, B. Mao, Z.Md. Fadlullah, N. Kato, On a novel deep-learning-based intelligent partially overlapping channel assignment in SDN-IoT. IEEE Commun. Mag. **56**(9), Art. no. 9 (2018). https://doi.org/10.1109/MCOM.2018.1701227
10. H. Tataria, M. Shafi, A.F. Molisch, M. Dohler, H. Sjoland, F. Tufvesson, 6G Wireless systems: vision, requirements, challenges, insights, and opportunities. Proc. IEEE **109**(7), Art. no. 7 (2021). https://doi.org/10.1109/JPROC.2021.3061701
11. S. Sakib, T. Tazrin, M.M. Fouda, Z.Md. Fadlullah, N. Nasser, An efficient and lightweight predictive channel assignment scheme for multiband B5G-enabled massive IoT: a deep learning approach. IEEE Internet Things J. **8**(7), Art. no. 7 (2021). https://doi.org/10.1109/JIOT.2020.3032516
12. S. Sakib, T. Tazrin, M.M. Fouda, Z.Md. Fadlullah, N. Nasser, A deep learning method for predictive channel assignment in beyond 5G networks. IEEE Netw. **35**(1), Art. no. 1 (2021). https://doi.org/10.1109/MNET.011.2000301
13. Z.Md. Fadlullah, N. Kato, HCP: heterogeneous computing platform for federated learning based collaborative content caching towards 6G networks. IEEE Trans. Emerg. Top. Comput. 1 (2020). https://doi.org/10.1109/TETC.2020.2986238

14. C. Duan, Q. Zhao, Y. Ma, Object-oriented IPV4/IPV6 distributed network management model, in *2010 3rd IEEE International Conference on Broadband Network and Multimedia Technology (IC-BNMT)* (Beijing, China, Oct. 2010, pp. 238–242). https://doi.org/10.1109/ICBNMT.2010.5705087
15. S. Hashima et al., On softwarization of intelligence in 6G networks for ultra-fast optimal policy selection: challenges and opportunities. IEEE Netw. 1–9 (2022). https://doi.org/10.1109/MNET.103.2100587
16. E.M. Mohamed, S. Hashima, K. Hatano, M.M. Fouda, Z.M. Fadlullah, Sleeping contextual/non-contextual thompson sampling MAB for mmWave D2D two-hop relay probing. IEEE Trans. Veh. Technol. **70**(11), Art. no. 11 (2021). https://doi.org/10.1109/TVT.2021.3116223
17. S. Hashima, K. Hatano, H. Kasban, E.M. Mohamed, Wi-Fi assisted contextual multi-armed bandit for neighbor discovery and selection in millimeter wave device to device communications. Sensors **21**(8), Art. no. 8 (2021). https://doi.org/10.3390/s21082835

Zubair Md Fadlullah (zfadlullah@ieee.org) is currently an associate professor with the Department of Computer Science, Lakehead University, and a Research Chair of the Thunder Bay Regional Health Research Institute, Ontario, Canada. He was an associate professor at the Graduate School of Information Sciences, Tohoku University, Japan, from 2017 to 2019. He received his Ph.D. degree in information sciences from Tohoku University in 2011. His main research interests span the cyber physical system layers in sensing, communication, and computing problems and elegant solutions. He is currently an Editor of IEEE Transactions on Vehicular Technology, IEEE Access, and the IEEE Open Journal of the Communications Society.

Mostafa M. Fouda received his Ph.D. degree in information sciences from Tohoku University in 2011. He is currently an assistant professor with the Department of Electrical and Computer Engineering, Idaho State University. He also holds the position of associate professor at Benha University, Egypt. He served as an assistant professor at Tohoku University. He was a postdoctoral research associate with Tennessee Technological University. He has been engaged in research on cybersecurity, communication networks, wireless mobile communications, smart healthcare, smart grids, AI, blockchain, and IoT. He has published more than 70 papers in prestigious peer-reviewed journals and conferences. He served as the Symposium/Track Chair of IEEE VTC 2021-Fall. He has also served as a Guest Editor of some Special Issues of several top-ranked publications such as IEEE Wireless Communications and IEEE Internet of Things Magazine. He also serves as a referee of some renowned IEEE journals and magazines such as IEEE Communications Standards, IEEE Wireless Communications, TWireless, TPDS, TSG, IEEE Access, TNSM, TETC, and IEEE Network. He is an Editor of IEEE Transactions on Vehicular Technology and an Associate Editor of IEEE Access.

Joint Provisioning of QoS and Privacy with Federated Learning

Zubair Md. Fadlullah and Mostafa M. Fouda

Abstract Combined optimization of Quality of Service (QoS) and security/privacy has always been an overlooked topic among IT professionals. The main reason for this is the difficulty to formulate optimization problems with the various conflicting QoS, security, and privacy attributes. There is no closed-form solution to such an optimization problem and this is why there was a lack of a seamless integration of QoS and privacy in the literature for many decades. In an emerging smart society, data privacy has become a top priority, particularly in the healthcare domain, which deals with massive patient records, health data, and imaging data from a myriad of modalities. Storing and sharing such data across hospital networks requires not only privacy but also presents a considerable challenge in terms of QoS overheads due to the massive bandwidth needed. The closed-source medical record storage and sharing platforms contribute to this challenge even more. We use this as a motivational use case in this chapter to demonstrate the need for seamless QoS and privacy provisioning of medical data exchange among various stakeholders. In this vein, we discuss the applicability of a distributed, decentralized learning framework called federated learning that provides a new paradigm for medical record platforms, biomedical equipment, and even resource-constrained IoT (Internet of Things) devices to participate in collaborative AI (artificial intelligence) model building. The federated learning framework provides two advantages, one from the QoS point of view and the other from the privacy-preserving aspect. As for the QoS assurance, federated learning techniques typically share the wisdom in terms of the locally developed AI model parameters instead of the raw big data, which directly improves the communication delay and bandwidth overheads. On the other hand, it preserves the users' data privacy by eliminating the need to send the original data (which

Z. Md. Fadlullah (✉)
Department of Computer Science, Lakehead University, ATAC 5023, Thunder Bay Campus, 955 Oliver Rd, Thunder Bay, ON P7B 5E1, Canada
e-mail: zfadlullah@ieee.org; Zubair.Fadlullah@lakehead.ca

Thunder Bay Regional Health Research Institute (TBRHRI), Thunder Bay, ON, Canada

M. M. Fouda
Department of Electrical and Computer Engineering, Idaho State University, 1776 Science Center Dr. Stop 8150, Idaho Falls, ID 83402, USA
e-mail: mfouda@ieee.org

A. K. Pathan (ed.), *Towards a Wireless Connected World: Achievements and New Technologies*, https://doi.org/10.1007/978-3-031-04321-5_14

could be sensitive) with remote cloud or third-party stakeholders. In this chapter, we describe an asynchronously weight updating federated learning technique which can further improve the performance of federated learning setups in terms of both QoS and privacy preservation at the same time.

1 Introduction

The Quality of Service, commonly known as QoS, is a widely known and accepted concept by the telecommunication industry to evaluate the performance of the provided services to mobile users. QoS is typically measured by the bandwidth, throughput, packet loss rate, energy efficiency, communication delay, jitter, fairness, and other key performance indicators (KPIs) that may demonstrate the satisfaction of the customer base. Other terms such as Quality of Experience (QoE) has also appeared, which evaluate the service performance from the users' perspective through mean opinion score (MOS) and other ratings. Without their subtle differences, readers may consider QoS and QoE indicators to assess whether you, as a subscriber or customer of a service, are happy with the service quality and express your level of happiness. From the perspective of Internet service providers and mobile operators, it is important to consider certain QoS classes to prioritize network traffic based on the user-subscription type, service type, etc., to guarantee a minimum level of QoS. For example, the home Internet that you subscribed to, typically says a bandwidth rating such as Mbps or Gbps (megabits/gigabits per second). However, if you are an everyday home user, you can go through the contract with the service provider that typically states that the provider does not guarantee that you will get the upper bound of bandwidth specified in the contract. At the same time, there could be some clause stating that they will provide you at least some level of service, above certain 0.5 Mbps, let's say; which is better than not getting any service at all so that you can at least have basic access to the Internet to check your emails and social media messages. This is a very simple, but easy-to-understand QoS provisioning example for a home user. Now, to create various QoS classes and guaranteeing that each user belonging to a class will receive the minimum amount of resources for the desired service with the minimum tolerable amount of delay and energy overheads requires complex computing, i.e., careful optimization. Due to the dynamics of network traffic and user mobility, especially in wireless and mobile broadband networks, QoS provisioning also rapidly changes over time. This makes it quite difficult to consider security and privacy issues to be integrated with the existing notion of QoS parameters.

What are the security and privacy issues in existing networks? Security typically means that you do not want your communication to be intercepted by others. There has been some work to integrate security parameters with QoS within the so-called Quality of Protection (QoP) framework. Beyond security, user privacy needs to be assured in communication since there could be various types of sensitive user data that could be used for analytic purposes for gaining business insight in this era of big data. Optimizing various QoS parameters and security/privacy dimensions at the same time was considered to be a huge challenge for a long time because they typically impact each other. Suppose you want a very high level of security, in this case, you have to invest in more resources and tolerate higher levels of communication delay and other QoS overheads. This is like sacrificing one over the other. This can be computed for a few users for a limited number of QoS, security, and privacy attributes; but when these parameters are many along with a very high number of users, the optimization is computationally hard. Existing solutions to find balanced QoS, security, and privacy levels have also been found to be time-consuming and not practical.

While the researchers left the QoS and privacy (and security) integration as an open research problem, the re-emergence of the machine and deep learning techniques kicked off a completely different race for embedded intelligence in 5G and beyond communication networks. Since many devices are resource-constrained even today, it is difficult to train large AI models individually. Researchers were interested in distributed machine learning techniques, which led to the emergence of collaborating learning frameworks such as federated learning. As a byproduct of federated learning, the framework allows preserving the privacy of the user data while allowing many devices, even resource-constrained ones, to participate in collaborative learning. This is possible by training local AI models individually on distributed devices and sharing the weights of those models with a central aggregator without sharing the actual data. The weights of those models are aggregated by the central aggregator and a robust AI model is generated. This privacy-preservation in federated learning now needs to be somehow tied to service performance assurances. In this chapter, we aim to explain how federated learning can be tweaked in certain ways to assure high QoS performance while preserving the privacy of user data. First, background information on QoS and privacy preservation techniques will be described. Then preliminary on federated learning techniques will be given. The asynchronous weight update technique of federated learning will then be discussed showing implied assurance of QoS and privacy at the same time. Future directions and caveats will be also discussed for readers to understand the unresolved challenges.

2 Background and Motivation

Let us consider an inspiring scenario of hospital networks. Throughout this chapter we will be considering this motivational inspiration with hospitals that store a huge amount of patient data in the electronic medical record (EMR), also called the electronic health record (EHR). There are multi-million dollar platforms including Epic, Meditech Expanse, Cerner, AllScripts that are widely used in the healthcare industry in the USA and Canada. Millions of patient data, their health records, hospital stay records, surgery schedules, rehabilitation, diagnosis/test results, imaging results from magnetic resonance imaging (MRI), computed tomography (CT) scan, X-ray scan, positron emission tomography (PET) scan, etc., are stored in these platforms. Caregivers are given access to the medical imaging data on a need-basis through customizable dashboards. While these platforms have enormous potential, the access levels of doctors, nurses, clinicians, researchers, hospital administrators, insurance companies, etc. typically create a gap in sharing the data due to privacy concerns. Blockchain technology, including smart contracts and hyperledgers, has been highly discussed in recent times to allow secure and privacy-preserving data among these various stakeholders. However, blockchain has a huge impact on the available computing resources and communication bandwidth, and may not scale well. Furthermore, the abovementioned EMR platforms are, all close-sourced. Open-sourced patient data maintenance has been discussed by the cross-disciplinary forums particularly by telecommunication operators and computer scientists who work with the clinicians in the healthcare industry; however, it is yet to be embraced by the hospital community. This is because of the risk-averse nature of the hospital. While we expect that this trend will change in the foreseeable future, there needs to be an agile solution so that the various hospital networks can "talk" to each other, share their findings and expertise, without sharing their raw patient records and imaging data.

So the motivation is: how can these different EMR platforms, the biomedical equipment, and even the resource-constrained IoT and wearable devices in different units of a caregiving facility can exchange patient information without revealing the underlying information. This is a distributed learning and computing challenge which has much more scalability compared to the blockchain and other security/privacy-preservation techniques such as functional encryption. Our study shows that federated learning could be considered as a promising learning framework, distributed across these various EMR platforms and biomedical/IoT devices that has a great potential to transform the smart health industry. From existing proof-of-concept federated learning frameworks, we could see that their promise is huge for circumventing the need for centralizing and sharing sensitive patient data that are often subject to ethical, legal, regulatory, and technical challenges. Recently, some researchers showed the relevance and timeliness of applying federated learning for the ongoing novel coronavirus disease (COVID-19) pandemic data collection while respecting the privacy of the patients. For example, a federated learning approach was adopted to train COVID-19 models in a distributed manner [1], and its performance was compared with the state-of-the-art deep learning models including COVID-Net, MobileNet, and ResNet18. Similar efforts were also made in another recent research work [2], which attempted to detect COVID 19 from radiology images such as X-ray and CT slices. The computing chip-making giant, NVIDIA, also engaged in federated learning-assisted AI model training between hospitals to obtain a generalizable disease prediction model to overcome geographical barriers to detecting diseases [3]. An interesting work, referred to as the EXAM (EMR chest X-ray AI Model) in [4], constructed a federated learning system by utilizing vital signs, lab data, and chest X-ray images to forecast the future oxygen demand of COVID-19 patients. When enough COVID-19 samples were not available, researchers used generative adversarial networks (GANs) to produce synthetic traces and used them to train AI models in a federated learning setup to obtain better prediction accuracies [5, 6].

Academic initiatives, such as the one from Stanford University researchers to generalize AI models via federated learning-assisted training [7], are also worth noting that aimed to establish a trust-based reputation for patients as well as caregivers have also been made recently via federated learning frameworks. Such research work demonstrates how clinics or hospitals having data and computational resources can take part in enriching a centralized repository by only exchanging locally trained AI model parameters. Such work also addresses the problem of data heterogeneity by allowing caregiving facilities in different locations to collaborate with their diverse patient cohorts, various modalities and resolutions of medical imaging, the number of cases, type of dominant strains, etc. However, these existing federated learning methods, even though they consider preserving the data privacy, do not consider the QoS issue, such as reducing bandwidth overhead and other QoS parameters that can significantly restrict the performance of federated learning.

So the motivation of this chapter is to figure out how the existing applications of federated learning, with the use case of pandemic monitoring, can be further improved to assure both QoS and privacy in tandem.

3 Considered Scenario

Figure 1 summarizes our considered scenario showing a caregiving network that has urban, suburban, and rural health care centers and hospitals. The X-ray devices, PET

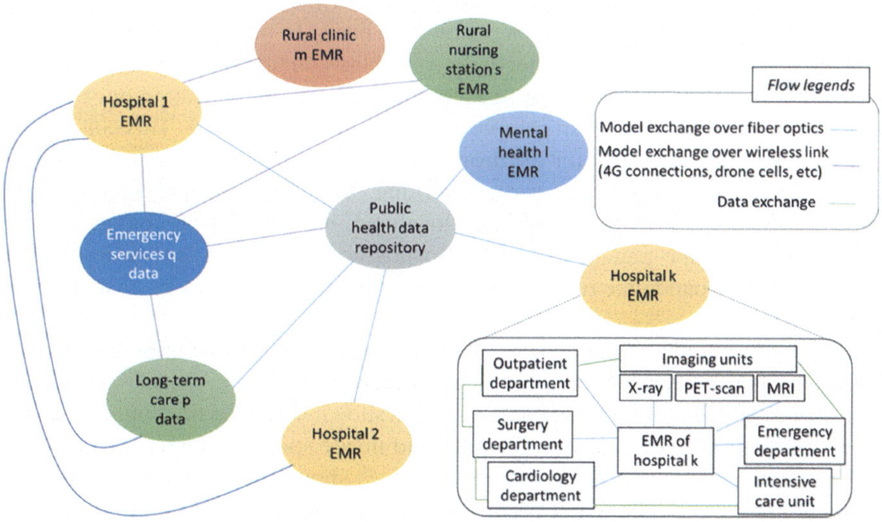

Fig. 1 Considered scenario with various caregiving facilities collaborating in a distributed level to exchange the essence of their EMRs. Note that for hospital k, the EMR building process is showing. The flow between the participating units inside

scanners, MRI scanners, and other imaging devices, as well as bedside monitoring units, are assumed to be part of an IoT data collection system for medical image acquisition and health data monitoring, respectively. In each hospital, various equipment collect the data and transfer them to the data repository of its own unit/centralized framework. It is important to point out that to collect data from remote/rural healthcare facilities such as nursing stations, rural clinics, and so forth, wireless links including 4G/5G connectivity and drone-cell-based dynamic mesh networks for extended coverage are also considered. The end-users are patients, they have mobile devices such as smartphones, tablets, smartwatches, oximeters, and other wearable devices, for monitoring health conditions. For generalization, let us use the term user-device or user equipment (UE) to denote a patient as a user.

Thus, from a communication link point of view, this scenario consists of an integrated aerial and terrestrial communication system over which AI models and data can flow between various caregiving nodes or even the various units within a specific caregiving node. The patient devices for IoT data collection and biomedical devices with IoT capability act as UEs and they are assumed to be able to exchange AI model parameters using the uplink with the wireless (terrestrial/drone-assisted) base stations. This whole system is possible by using a software-defined network to manage the decoupled data, control, and service planes as described in our earlier work [8]. The reason behind considering such a software-defined network controller is to efficiently coordinate the deployment, wireless link quality, and topology control to ensure the required connectivity and capacity for communication between suburban and rural caregiving centers. This is a particularly important assumption because these wireless links are not as high capacity as the fiber optics ones among healthcare centers in urban areas. So, they may not be reliable for transferring big healthcare data on a continuous basis. In other words, the wireless links are used to alleviate the connectivity issue between urban and rural caregiving centers; but they still present a significant bottleneck in communication or exchange of the big medical data and even compressed AI model information. This considered scenario also presents a way for the drones to store AI models received from many UEs and if needed, perform their own learning, to collaborate with other drones in learning pandemic features including localized, high-resolution contact tracing without exposing the patient identities. So this considered system model design provides an agile pandemic recommendation and response network as an additional option.

Now we describe the technical part of our considered communication link model. This is based on our earlier work [8, 9]. The wireless transmission model between UE and a drone (or 4G/5G terrestrial base station) is shown here. The fiber optic communication models between caregiving nodes and inside specific caregiving nodes are omitted since their existing deployment practices should not suffer from capacity and reliability issues. The wireless communication is split into several time rounds, during each of which the local AI model parameters from UEs are collected by a drone. The drone is supposed to forward this collected information through a mesh network (i.e., a combination of other drones, terrestrial base stations, etc.) to another caregiving node as shown in Fig. 1. Since this wireless mesh network link is dynamically formed over naturally untrusty physical terrain, raw data transmission from UEs is highly discouraged, and only the learned AI models from the UEs are collected for transmission to the destination caregiving node. The various caregiving nodes can also use fiber optics or similar types of wireless links (direct 4G/5G connections or drone-based mesh networks) to forward the models to a centralized aggregator in the public health authority node. The unreliability of the wireless transmission in both uplink and downlink directions is modeled using AWGN (additive white Gaussian noise) [10, 11]. The number of communication rounds is estimated depending on the time required to converge the AI model at the central aggregator. For structured AI models, the number of time rounds may be relatively smaller compared with the unstructured AI models dealing with patients' imaging data. To model the communication link mentioned above, we consider the UE and drones to be local nodes (LNs) and consider UE_i's transmission time to be d_i. This is measured using $d_i = Bln(1 + Pt_i g_i/\tau)$, where B represents the available bandwidth for communication at the serving LN. Pt_i denotes the transmission power of UE_i while g_i indicates its channel gain. The background noise power due to AWGN is modeled with the parameter τ. The structured/unstructured data size, θ, is considered for the UE. The transmission time required for a local model update $T_{i_{Tx}}$, given θ, is estimated as: $T_{i_{Tx}} = \theta/d_i$. So, the time needed for UE_i to complete a global iteration becomes $T_{i_G} = (\log\ 1/\varepsilon_i)T_{i_L} + T_{i_{Tx}}$. Here, T_{i_L} refers to the local iteration required by the UE with accuracy ε_i. By this way, the UE can train a localized model using its local data with structured data of patient activity with location as well nonstructured imaging data. If the size of the local dataset in the UE_i is s_i and if c_i CPU cycles are needed each with the frequency f_i, then $T_{i_L} = (c_i/f_i)s_i$.

Let us further elaborate on how the UE data are collected with a pandemic learning use case as follows. What is the important information regarding the user to be collected in a remote nursing station? It would consist of the symptoms (lab result, test-kit result for antigen test and molecular test), vitals (temperature, heartbeat, blood oxygen saturation, fatigue level), history of travel using geolocation data, radiographic imaging with portable PET scanners, etc. As mentioned earlier, the UEs do not share this raw data to the wireless base stations or cloud. So privacy concern is eliminated directly at the source of the data. In this case, consider drones connecting the rural nursing station with a suburban healthcare center or hospital. Only the drones with enough residual battery power for maintaining their flight are considered to form an agile mesh network so that the privacy preservation and learning accuracy for building a distributed pandemic model can be realized. In this vein, the base stations collect the local training model from the UEs, and share this information with other base stations and then the caregiving entities to further boost the decentralized training among drones. The public health repository, upon receiving the locally trained models at the UE layer as well as the healthcare giving nodes layer, can then train a global model to arrive at a reasonable decision with significantly high accuracy. In other words, this is a two-layer federated learning problem. The objective is to minimize a loss function minimize $\sum_{i=1}^{N} \frac{s_i}{s} f_i(w, x_k, y_k)$, where the first term is the loss function to be optimized (minimized) at UE_i given the weight vector w to minimize the gap between the input and output labels x_k and y_k for k features. As mentioned earlier, s_i indicates the training samples used by UE_i. n is the number of participating users (patient UEs, biomedical devices, healthcare nodes, etc.). On the other hand, s is the training samples at the public health repository for building the global AI model. Here, the constraint $w_1 = w_2 = \ldots = w_k$ is needed to guarantee the convergence of global learning such that all the participating users and the public health repository can eventually derive the same AI model without explicitly exchanging the raw health data of patients across the entire system.

With the technical communication and computing model parts presented so far, we are now ready to describe the formal problem setting. The problem, in a high-level language, can be described as: how can the local health data models collected by terrestrial and drone base stations along with other communication networks in Fig. 1 be used to allow convergence of the global AI model being trained at the public health repository, which does not have actual, raw data from the collaborating users and caregiving nodes? Thus, the problem becomes as follows: (1) each UE is capable of training a shared global model with its local model update based on its data, and (2) each UE shares its updated local model with the public health repository through the terrestrial/aerial base stations to revise/update the global model. Similarly, the caregiving nodes confront their own local model training problem to predict UEs' health patterns and participate in second-level federated learning with the global model. Until the loss function is minimized and/or the global model accuracy is deemed reasonable, this distributed training process needs to continue. Therefore, how to design such a decentralized technique with reduced network overhead and privacy-preservation property can be seen as the core research problem in this chapter.

4 Proposed Asynchronous Updating Federated Learning Framework

There exist a plethora of federated learning setups including the basic weighted aggregation, duel-weighted gradient update, gradient compression [12], semi-asynchronous node participation, clustering of nodes according to their response delay, model splitting, and so forth [13]. In this section, we consider the concept of model splitting by enabling the participating patient and caregiver nodes represented by UEs and hospitals/clinics, respectively. The reason for doing as such is to consider the limited computing and energy resources of the mobile user terminals so that they can contribute toward the overall AI model training partially and without revealing the raw data to the other contributors and even the facilitator (i.e., any caregiving node). In this vein, the user nodes (UEs) constitute the initial layer of a hierarchical federated learning setup. By considering the pandemic feature learning through health monitoring of users through their smartphones and other monitoring/screening devices at home or local nursing stations (obviously with constrained resources compared with bigger healthcare facilities), each UE continues to build a local AI model according to the data its terminals collect. The UE splits the model parameters into shallow and deep segments and shares them with the caregiving clinic or hospital. Such clinics and hospitals are considered to be at the second layer of the considered hierarchical federated learning setup. Each caregiving facility at this second layer receives AI models from contributing UEs from the first layer. In addition, it also receives input data from its various units such as radiology department, emergency room, pediatrics, and so forth that helps it to construct its own AI model for pandemic feature learning. The contributed AI models from the UEs from the first layer and the caregiving facility's own AI model are augmented. Each caregiving facility or node in the second layer then divides this augmented AI model into shallow and deep parameters, and schedules them to be exchanged with other hospitals and/or the public health repository cloud, i.e., the third tier of our considered federated learning framework. Here, we can consider using a customized convolutional neural network (CNN) to develop the AI models at UEs and the caregiving nodes as well as the public health cloud repository to identify hidden features from complex pandemic/health data [14–17]. The overall process is shown in Fig. 2.

In the first hierarchy or tier 1, personalized AI model construction of the UEs at residence, vehicular network, campus and shopping district areas are shown so that the users can easily relate to how these user terminals (UEs) capture various information regarding health, for example cold, fever, blood pressure, etc., using smart watch, wearables and other health monitoring IoT devices for capturing oxygen level,

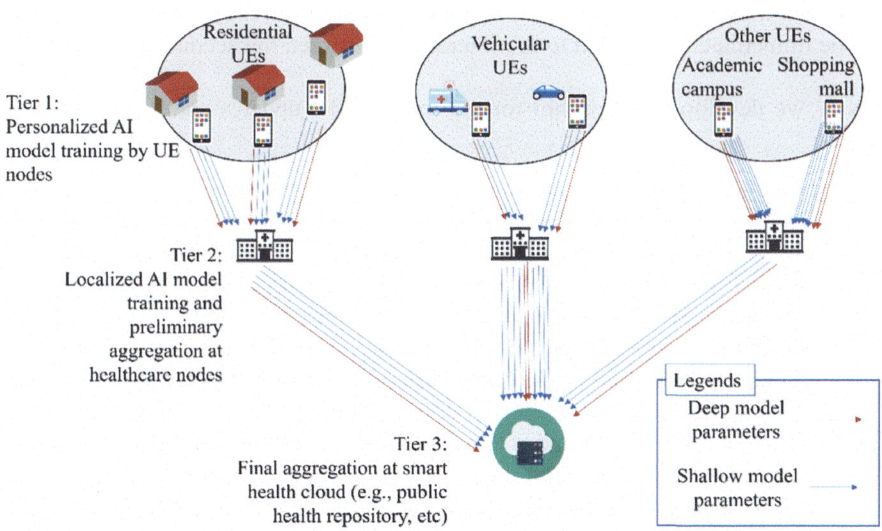

Fig. 2 Proposed asynchronous weight updating federated learning in multiple layers of participating UEs and healthcare entities to train a global model at the public health repository via distributed, privacy-preserving manner while solving QoS issues in terms of communication capacity overhead

temperature, etc. The cameras in shopping districts and school campuses can determine the use of masks in public areas, and could be used to geo-tag users to construct digital contact tracing if required. However, such data should not be revealed to other users/third parties, and this is why the federated learning at individual UE nodes can mask the actual or raw data and provide only the essence by building personalized, local models. By splitting these local models, each UE then needs to schedule when to send the deep and local parameters and the ratio at which these parameters need to be sent so that the network overhead can be minimized. This is depicted in Fig. 2 in terms of the blue and red arrows which indicate the exchanges of shallow and deep parameters, respectively. By doing so, our adopted approach differs from the traditionally utilized federated learning algorithms where the models are periodically obtained and averaged by the model aggregator [18] by considering the parameter update in a single step. Based on [19, 20], we want to avoid the large communication overhead incurred by this traditional approach since the deep model parameters, which are typically large in size, do not change fast compared to the shallow model parameters. The shallow parameters can be considered to embody the generic attributes of the monitored data (e.g., pandemic features in this example). On the other hand, the deep model parameters signify symptom and context-specific information of the users (i.e., patients). This idea inspires the model splitting in such a manner that the shallow model weights are updated and communicated with the aggregator more frequently while the relatively large-sized, deep model parameters are infrequently updated and exchanged with other nodes for model aggregation. The readers should easily understand this from the interaction between the UE nodes in the first tier and the healthcare nodes in the second tier from Fig. 2. This concept can then be immediately extended to the interaction between the second and third tiers also.

Now we develop an algorithm for the local model updates and aggregation as follows:

(1) The first step of the algorithm takes place at the central aggregator in tier 3, i.e., in a bottom up manner. The smart health cloud or the public health repository, acting as the central aggregator, provides an initial AI model and distributes to the local hospitals and clinics. The central aggregator also provides model splitting instructions and other federated learning-specific information (e.g., which models to be used, the optimal hyperparameters including which gradient descent technique to be used, the exchange interval). Readers can consider this as the bootstrapping or initialization phase.

(2) When each caregiving node receives the model training specific information from the central aggregator, it can optionally send an acknowledgment to the central aggregator to indicate its availability for participation or contribution toward the global model averaging. In addition, the caregiving node is considered to be a tier 2 aggregator so that it can aggregate the AI models received from its UE nodes, biomedical equipment, and so forth.

(3) Each caregiving node then transmits model training specific information to each subscribing UEs at tier 1. While the AI model should be the same, the

model exchange parameters in terms of the splitting ratio between shallow and deep model parameters, and model updating/exchanging interval can be different compared to the ones mentioned earlier in step 1.

(4) Each UE node in tier 1 (e.g., residential patient users, biomedical equipment at various units of a hospital or a clinic, etc.) receives the model training related information from its caregiving node. Accordingly, each UE uses its locally monitored/collected health related data to train a personalized AI model based on the partial model received from the tier 2 node.

(5) Each UE maintains a control flag to choose whether all the layers or the shallow layers will be updated. If the UE was instructed by the tier 2 caregiving node to use a particular technique for gradient update, such as the Stochastic Gradient Descent (SGD), then the UE uses it to update its model $w = w - \eta * \partial f(w;b)$. The learning rate is indicated by η while the minibatch is denoted by b. ∂ denotes the vector of partial derivatives of the function f of the weight and minibatch.

(6) Each caregiving node then initializes a clock to synchronize with its subscribing UEs. Then the caregiving node commences polling the UEs one by one.

(7) Next, each caregiving node initiates a temporary buffer and sets a number of rounds equal to the number of subscribing UEs. During each round, the caregiving node at tier 2 receives the relevant parameters from each UE its polls and records into the buffer the timestamp, the control flag for model splitting, and the parameters received from the UE.

(8) When all UEs are polled, each caregiving node now has information of shallow-only or complete weight vector parameters of the UEs. The shallow-only/deep model parameters are exchanged during a flag switch interval. This is regarded to be much smaller than the updated global model receiving interval of each UE.

(9) The caregiving node then aggregates the information receives from all UEs and sends the aggregated model to the cloud. The caregiving node can also locally execute its AI models for predicting pandemic trends in collaboration with the neighboring caregiving nodes. Then, the caregiving node continues to poll the UEs again in the same manner until the model accuracy becomes acceptable.

(10) The central aggregator, i.e., the public health repository-end in our considered example, the temporally weighted aggregation takes place during several communication rounds [17]. The specific and general parameters are fetched from each caregiving node by polling them over these communication rounds. Then, the aggregation is performed by the central aggregator to update the shallow and deep parameters (w_g and w_s, respectively) and construct the overall AI model.

Now we attempt to briefly evaluate the algorithmic performance of this adopted federated learning approach. Does this algorithm converge to a reasonable accuracy within a reasonable number of time rounds or not while solving the original optimization problem of weight optimization of the entire AI model at the aggregator by taking into consideration distributed, piece-wise, locally constructed models through iteration of shallow and deep parameters update. The centralized aggregator transmits the model parameters to the caregiving nodes and in turn to the UEs to initialize and train their respective localized AI models. The UEs and caregiving nodes selectively exchange shallow-only and deep parameters in a manner that the update of each UE's local model parameter depends on the global model. On the other hand, the update of the global model depends on all UEs' local federated learning models. The update of the local model depends on the learning algorithm such as Gradient Descent, Stochastic Gradient Descent (SGD), or Randomized Coordinate Descent (RCD). The update of local model w_i at UEs at time t is expressed as $w_{i,t+1} = g_t - \eta/s_i \sum_{k}^{s_i} = l\ \partial f(g_t,\ x_{ik},\ y_{ik})$, where η denotes the learning rate and $\partial f(g_t, x_{ik}, y_{ik})$ represents the gradient of $f(g_t, x_{ik}, y_{ik})$ with respect to g_t. Then based on the work [21], the federated learning algorithm can converge to an optimal global model $_{g*}$ after the learning steps. Interested readers may view the proof of the expected convergence rate in [18].

While currently we do not have a large-scale dataset to demonstrate the adopted model's performance, we may demonstrate a proof-of-concept performance from our recent work in [9]. This work considered radiology images based on X-ray of normal patients and patients with pneumonia and COVID-19. While contemporary researchers considered scattered and relatively small-sized datasets, a robust dataset preparation was carried out in [9] by combining the PA (posteroanterior) X-ray images from four different datasets. For the asynchronously weight updating AI model construction, both ANN and CNN were considered to compare with a centralized ANN/CNN-based baseline model. For performance comparison, we considered the number of time-rounds, each of which is broken down into several iterations. For example, if the number of iterations is 15 and the number of time rounds is 3, then during the 3th, 9th, 12th, and 15th iteration, the deep model update is carried out in our adopted approach. Thus, only four iterations are dedicated for updating the deep layer parameters. On the other hand, the shallow-layer parameters are updated over the remaining iterations. For detailed simulation parameters including the ANN and CNN structure specifications, epoch size, batch size, activation function, and hyperparameter optimization, readers can refer to our work in [9].

Our adopted asynchronously weight updating federated learning method was shown in [9] to achieve reasonably high accuracy of 0.938 using the CNN structure and 0.91 using the ANN structure when compared to a centralized training (baseline) accuracy of 0.946. What is interesting is how our proposed CNN-based federated learning method reduces the network overhead over just twenty time-rounds while preserving the privacy of the patient's radiology imaging data. This is demonstrated in Table 1. With the larger deep parameter rates, a more overhead reduction is possible by performing more generic parameter exchanges from the local model from the UEs to the cloud and only transferring the specific parameters after a relatively high number of time rounds.

Table 1 Bandwidth overhead reduction ratio for deep parameter rates in the considered federated learning for maintaining user's data privacy [9]

Number of time rounds	Deep parameter rate				
	0.1	0.3	0.5	0.7	0.9
	Overhead reduction				
5	0.0774	0.2394	0.4014	0.5598	0.7218
10	0.0868	0.2699	0.4495	0.6326	0.8121
15	0.0927	0.2793	0.4659	0.6561	0.8427
20	0.0915	0.2816	0.4753	0.6654	0.8556

5 Conclusion, Future Directions, and Caveats

Service quality and privacy properties are natural expectations from the user-side in emerging communication networks to be jointly satisfied at the same time. However, combinatorial optimization of QoS and privacy-preservation along with other security parameters leads to a computationally hard problem. In this chapter, we demonstrated how federated learning can address the privacy-preservation problem in an implicit yet elegant manner. We provided a health analytics scenario to explain the problem and our federated learning-based solution for ease of understanding. To improve the QoS overhead, we made an adjustment to the commonly known federated learning framework by asynchronously updating the shallow and deep model parameters. Experimental results demonstrate how this leads to communication overhead reduction while approximating the centralized deep learning-based AI model training performance. This result is encouraging; however, in the future, more variations of the federated learning methods and their theoretical convergence guarantee need to be considered. Additional security parameters as an added dimension to the QoS and privacy-preservation problem also need to be considered in tandem. The hyperparameter tuning (i.e., how to determine the deep parameter transmission interval, the global AI model update interval) also is critical. In the future, a federated learning framework as an API (e.g., in the existing Python framework) can certainly encourage interdisciplinary researchers to apply and study the full potential of federated learning and its variants to jointly optimize QoS and privacy as well as security parameters under various application scenarios.

References

1. W. Zhang et al., Dynamic-fusion-based federated learning for COVID-19 detection. IEEE Internet Things J. **8**(21), 15884–15891 (2021). https://doi.org/10.1109/JIOT.2021.3056185
2. B. Yan et al., Experiments of federated learning for COVID-19 chest X-ray images, in *Advances in Artificial Intelligence and Security*, vol. 1423, ed. by X. Sun, X. Zhang, Z. Xia, E. Bertino (Springer International Publishing, Cham, 2021), pp. 41–53. https://doi.org/10.1007/978-3-030-78618-2_4

3. COVID-19 Research Hub (NVIDIA), Building a global defense system against coronavirus (SARS-COV-2). https://developer.nvidia.com/research/covid-19
4. I. Dayan et al., Federated learning for predicting clinical outcomes in patients with COVID-19. Nat. Med. **27**(10), 1735–1743 (2021). https://doi.org/10.1038/s41591-021-01506-3
5. L. Zhang, B. Shen, A. Barnawi, S. Xi, N. Kumar, Y. Wu, FedDPGAN: federated differentially private generative adversarial networks framework for the detection of COVID-19 pneumonia. Inf. Syst. Front. **23**(6), 1403–1415 (2021). https://doi.org/10.1007/s10796-021-10144-6
6. D.C. Nguyen, M. Ding, P.N. Pathirana, A. Seneviratne, A.Y. Zomaya, Federated Learning for COVID-19 Detection with Generative Adversarial Networks in Edge Cloud Computing. *ArXiv211007136 Cs Eess* (2021), http://arxiv.org/abs/2110.07136. Accessed 25 Dec 2021
7. I. Burcin, A Pandemic AI Engine Without Borders, *Healthcare, Machine Learning, Scientific Discovery*, (2021), https://hai.stanford.edu/news/pandemic-ai-engine-without-borders. Accessed 25 Dec 2021
8. Z. Md. Fadlullah, N. Kato, HCP: Heterogeneous Computing Platform for Federated Learning Based Collaborative Content Caching Towards 6G Networks, IEEE Trans. Emerg. Top. Comput. 1–1 (2020). https://doi.org/10.1109/TETC.2020.2986238
9. N. Nasser, Z.Md. Fadlullah, M.M. Fouda, A. Ali, M. Imran, A lightweight federated learning based privacy preserving B5G pandemic response network using unmanned aerial vehicles: a proof-of-concept. Comput. Netw. 108672 (2021). https://doi.org/10.1016/j.comnet.2021.108672
10. B. Mughal, Z.Md. Fadlullah, S. Ikki, Centralized versus heuristic-based distributed channel allocation to minimize packet transmission delay for multiband relay networks. IEEE Netw. Lett. **2**(4), 180–184 (2020). https://doi.org/10.1109/LNET.2020.3030870
11. B. Mughal, Z. Fadlullah, M.M. Fouda, S. Ikki, Allocation schemes for relay communications: a multi-band multi-channel approach using game theory. IEEE Sens. Lett. 1–1 (2021). https://doi.org/10.1109/LSENS.2021.3137152
12. X. Lu, Y. Liao, P. Lio, P. Hui, Privacy-preserving asynchronous federated learning mechanism for edge network computing. IEEE Access **8**, 48970–48981 (2020). https://doi.org/10.1109/ACCESS.2020.2978082
13. C. Xu, Y. Qu, Y. Xiang, L. Gao, Asynchronous federated learning on heterogeneous devices: a survey, *ArXiv210904269 Cs*, (2022). http://arxiv.org/abs/2109.04269. Accessed 20 Feb 2022
14. S. Sakib, T. Tazrin, M.M. Fouda, Z.Md. Fadlullah, M. Guizani, DL-CRC: deep learning-based chest radiograph classification for COVID-19 detection: a novel approach. IEEE Access **8**, 171575–171589 (2020). https://doi.org/10.1109/ACCESS.2020.3025010
15. S. Sakib, M.M. Fouda, Z. Md. Fadlullah, N. Nasser, On COVID-19 prediction using asynchronous federated learning-based agile radiograph screening booths, in *ICC 2021—IEEE International Conference on Communications*, Montreal, QC, Canada (2021), pp. 1–6. https://doi.org/10.1109/ICC42927.2021.9500351
16. S. Sakib, M.M. Fouda, Z. Md. Fadlullah, K. Abualsaud, E. Yaacoub, M. Guizani, Asynchronous federated learning-based ECG analysis for arrhythmia detection, in *2021 IEEE International Mediterranean Conference on Communications and Networking (MeditCom)*, Athens, Greece (2021), pp. 277–282. https://doi.org/10.1109/MeditCom49071.2021.9647636
17. S. Sakib, M.M. Fouda, Z.Md. Fadlullah, N. Nasser, W. Alasmary, A Proof-of-concept of ultra-edge smart IoT sensor: a continuous and lightweight arrhythmia monitoring approach. IEEE Access **9**, 26093–26106 (2021). https://doi.org/10.1109/ACCESS.2021.3056509
18. K. Bonawitz et al., Towards federated learning at scale: system design, *ArXiv190201046 Cs Stat*, (2019). http://arxiv.org/abs/1902.01046. Accessed 27 Dec 2021
19. W.Y.B. Lim et al., Federated learning in mobile edge networks: a comprehensive survey. IEEE Commun. Surv. Tutor. **22**(3), 2031–2063 (2020). https://doi.org/10.1109/COMST.2020.2986024
20. Q. Yang, Y. Liu, Y. Cheng, Y. Kang, T. Chen, H. Yu, Federated learning. Synth. Lect. Artif. Intell. Mach. Learn. **13**(3), 1–207 (2019). https://doi.org/10.2200/S00960ED2V01Y201910AIM043

21. X. Yin, Y. Zhu, J. Hu, A comprehensive survey of privacy-preserving federated learning: a taxonomy, review, and future directions. ACM Comput. Surv. **54**(6), 1–36 (2021). https://doi.org/10.1145/3460427

Zubair Md. Fadlullah is currently an associate professor with the Department of Computer Science, Lakehead University, and a Research Chair of the Thunder Bay Regional Health Research Institute, Ontario, Canada. He was an associate professor at the Graduate School of Information Sciences, Tohoku University, Japan, from 2017 to 2019. He received his Ph.D. degree in information sciences from Tohoku University in 2011. His main research interests span the cyber physical system layers in sensing, communication, and computing problems and elegant solutions. He is currently an Editor of IEEE Transactions on Vehicular Technology, IEEE Access, and the IEEE Open Journal of the Communications Society.

Mostafa M. Fouda received his Ph.D. degree in information sciences from Tohoku University in 2011. He is currently an assistant professor with the Department of Electrical and Computer Engineering, Idaho State University. He also holds the position of associate professor at Benha University, Egypt. He served as an assistant professor at Tohoku University. He was a postdoctoral research associate with Tennessee Technological University. He has been engaged in research on cybersecurity, communication networks, wireless mobile communications, smart healthcare, smart grids, AI, blockchain, and IoT. He has published more than 70 papers in prestigious peer-reviewed journals and conferences. He served as the Symposium/Track Chair of IEEE VTC 2021-Fall. He has also served as a Guest Editor of some Special Issues of several top-ranked publications such as IEEE Wireless Communications and IEEE Internet of Things Magazine. He also serves as a referee of some renowned IEEE journals and magazines such as IEEE Communications Standards, IEEE Wireless Communications, TWireless, TPDS, TSG, IEEE Access, TNSM, TETC, and IEEE Network. He is an Editor of IEEE Transactions on Vehicular Technology and an Associate Editor of IEEE Access.

Correction to: Programmable Computing in 6G

Zubair Md Fadlullah and Mostafa M. Fouda

Correction to:
Chapter "Programmable Computing in 6G" in: A. K. Pathan (ed.), *Towards a Wireless Connected World: Achievementsand New Technologies*, https://doi.org/10.1007/978-3-031-04321-5_13

In the original version of the book, the following belated corrections have been incorporated: In the original version of this chapter "Programmable Computing in 6G", the affiliation of the second author "Mostafa M. Fouda" was included wrongly. This has now been corrected as "Department of Electrical and Computer Engineering, Idaho State University, 1776 Science Center Dr. Stop 8150, Idaho Falls, ID, 83402, USA".

The correction chapter and book has been updated with the changes.

The updated original version of this chapter can be found at
https://doi.org/10.1007/978-3-031-04321-5_13

A. K. Pathan (ed.), *Towards a Wireless Connected World: Achievements and New Technologies*, https://doi.org/10.1007/978-3-031-04321-5_15

The manufacturer's authorised representative in the EU is Springer Nature Customer Service Centre GmbH, Europaplatz 3, 69115 Heidelberg, Germany. If you have any concerns regarding our products, please contact ProductSafety@springernature.com

Printed and bound by CPI Group (UK) Ltd, Croydon, CR0 4YY
30/09/2026
02225493-0012